BACTERIOLOGY

D0143782

the text of this book is printed
on 100% recycled paper

About the Authors

Dr. Arthur H. Bryan received a B.S. degree from Washington State College, a V.M.D. degree from the University of Pennsylvania, and an M.A. degree from the University of Maryland. Dr. Bryan, Associate Professor of Biological Sciences at Jacksonville University, Florida, has also taught at Washington State College, the University of Baltimore, the University of Maryland, Baltimore City College, Franklin Square Hospital, and Chipola Junior College. Dr. Bryan is a Fellow of the American Public Health Association. In 1935 he received the Academy of Science research prize in biology. He is the author of *Sciences for the Blind*. He has written more than ninety articles for professional journals.

Charles A. Bryan received his B.S. degree from the University of North Carolina and his M.A. degree from Florida State University. His varied career has included positions as control bacteriologist with the U. S. Food and Drug Administration, bacteriologist for the Orange Memorial Hospital, Orlando, Florida, Research and Development Officer at Cape Canaveral, Florida, and science teacher. He is the author or co-author of some twenty-five published scientific papers.

Dr. Charles G. Bryan received his M.D. degree from the University of Southern California and his F.R.C.S. degree from the University of Edinburgh, Scotland. He is surgical consultant for the Welsh Regional Medical Board in Glamorganshire, Wales.

BACTERIOLOGY
PRINCIPLES AND PRACTICE

ARTHUR H. BRYAN

CHARLES A. BRYAN

CHARLES G. BRYAN

Sixth Edition

BARNES & NOBLE BOOKS

A DIVISION OF HARPER & ROW, PUBLISHERS

New York, Hagerstown, San Francisco, London

PREFACE

The science of medical bacteriology is fast coming to resemble an exposition of ancient history, for with the advent of antibiotics, sulfa drugs, vaccines, and the virus antagonist interferon, virtually all of the viral, bacterial, and rickettsial diseases herein recorded have been eliminated as public health problems. The common cold, influenza, measles, pneumonia, and tuberculosis are still hazards, but major breakthroughs even for these diseases are imminent. The last stands of plagues, epidemics, and pandemics of communicable diseases are in sight, except for possible sporadic outbreaks due to neglect or carelessness in handling positive prophylactic procedures. Isolation hospitals are no longer necessary, and quarantine stations exist largely for immigrants and foreign seafaring men. Compulsory immunization of children against smallpox, diphtheria, tetanus, typhoid, and whooping cough, and against tuberculosis in several European countries, relegates these diseases to medical-history significance only.

The bacteriologist is enlarging his sphere of activities to comprehend new areas in medicine, agriculture, and industry. He can look forward in the near future to electronic automation in the identification of microorganisms. A new fluorescent microscopic method for cytologic diagnosis of cancer cells is being applied to bacterial-cell identification. This method reduces the time for processing to a few minutes, and combines easy recognition of some of the more common microbes with procedural simplicity. In medical chemistry, blood sugar and bound urea nitrogen (BUN) determinations are being mechanized, with like methods in store for the up-to-date bacterial physiologist.

Bacteria are universal biochemists, whose activities provide endless possibilities in research and development. To no one in the sciences more than to the bacteriologist does the biblical challenge apply: Seek and ye shall find.

THE SIXTH EDITION

Because bacteriology is basically a laboratory subject, many useful technics, devices, and practical procedures have been described

v

in this book. Quick reference tables and charts are furnished to sum-
marize at a glance exhaustive subject matter. A glossary of short
utilitarian definitions, and a very complete index of some thousand-
odd items, make possible the prompt location of any information
desired. Review questions, taken from actual modern civil service
examinations for the various positions available to trained bacteriologists, will be found in the Appendices.

The data for the modernization of the factual material in the
text, tables, charts, and illustrations have been combed and condensed from thousands of pages of books, professional and research
journals, pertinent abstracts, and utilitarian articles.

The sixth edition contains newly systematized material on the
rickettsia, revised laboratory technics, and enlarged tables of antibiotics. The discussion of virus infections has been brought up to
date, as have been all references throughout the text that have been
superseded by recent findings.

The bibliographies terminating each chapter have also been
brought up to date, with the latest additions of standard texts in
general, and special works on bacteriology, pathology, public health,
and medicine, etc., listed in alphabetical sequence of authors. They
are suggested as additional reading and source material for further
intensive study.

ARTHUR H. BRYAN

ACKNOWLEDGMENTS

I wish to express my acknowledgments to my wife; to Mrs. Leona W. Greenhill; to Dr. Donald S. Martin, Duke University, and Dr. Edwin B. Steen, Western Michigan University, for their critical reading of the manuscript; and to Mr. Vernon Taberner, for his compilation of the index.

Special acknowledgments are also due to Marvin Bram of the staff of Barnes & Noble; to David and Joyce Bryan of Florida State University; to David Anthony Bryan, M.B., Ch.B., University of London, resident physician Cardiff Royal Infirmary, Cardiff, Wales; and to Anne Bryan, M.B., Ch.B., gynecologist and obstetrician, Brookwood Hospital, Cardiff, Wales, for their valued assistance in accumulating and proofreading research data used in the sixth edition.

A.H.B.

TABLE OF CONTENTS

PART I: PRINCIPLES OF BACTERIOLOGY

PART II: MEDICAL BACTERIOLOGY

PART III: INFECTION AND IMMUNITY

PART I

PRINCIPLES OF BACTERIOLOGY

THE SCOPE OF BACTERIOLOGY

Bacteriology is a biological science which studies bacteria, viruses, yeasts, molds, and protozoa. The activities of these organisms depend largely upon chemical reactions. Consequently, bacteriology is closely related to various branches of chemistry, particularly to biochemistry, which deals with the chemistry of plants and animals. It is a laboratory science with so many applications to the economic welfare and the health of mankind that separate divisions of the field have been developed—sanitary, medical, agricultural, and industrial bacteriology—each with its own problems, technics, and specialized data.

Trained bacteriologists serve on the staffs of federal, state, county, and city health departments to improve sanitation and prevent the spread of disease. They help to maintain the high standards of public water works and sewage disposal departments. In hospitals, pharmaceutical supply houses, and medical research laboratories, they aid in research, diagnosis, immunology, and therapy. They contribute materially to the quality of the products supplied by dairies, ice-cream plants, various food manufacturing and processing industries, fruit-growing and numerous other agricultural enterprises, and the textile, leather-tanning, tobacco-curing, air-conditioning, and similar industries. Their work facilitates the activities of public agencies concerned with crime detection and investigation. In all these areas of human endeavor the contributions of bacteriologists have been rapidly increasing. The science is now a basic study in the science departments of many liberal arts colleges as well as in medical, dental, veterinary, and agricultural colleges. It is a frontier science with many unsolved problems, and it is a well-established discipline with infinite practical applications based upon a long history of research and experimentation.

As bacteriology has continued its steady progress, previous misconceptions concerning processes such as decay, fermentation, reproduction, growth, health, and disease have been replaced by highly significant new conceptions attained through the use of scientific instruments and experiments. In addition to its practical applications, this new science has thus contributed to the broadening of man's view of life and to his understanding of nature.

CHAPTER I

HISTORY OF BACTERIOLOGY

The history of bacteriology is told best in the biographies of some of the world's greatest benefactors, namely, those death-fighting microbe hunters who, in the past one hundred years, have saved more lives than the world's armies have destroyed. Microbiologists, devoting their lives to the problems of public health and preventive medicine, have sought to find the cause and prevention of the devastating communicable diseases which bring death to thousands annually. Almost every nation in the world has to its credit men who contributed to all phases of bacteriology and immunology in an effort to exterminate and control infectious diseases. Many problems of agricultural and industrial importance have been solved through their findings. At present, thousands of men and women throughout the world are engaged in bacteriological research, and the future is bright for the ultimate control of most of the communicable diseases of both plants and animals.

It is hoped that this presentation of a short history of bacteriology and its makers will serve as a source of inspiration and encouragement to workers and students in bacteriology and allied sciences.

ANCIENT HYGIENE AND BACTERIOLOGY

The story of bacteriology can be traced back through millions of years, as evidenced by the finding of fossil remains of bacteria in rock strata thirty-three million years old. Renault describes fossil bacteria found in coal. The diseased bones of heavily armored dinosaurs in the Permian deposits strongly point towards the destruction of these gigantic monsters by bacteria.

Greek Civilization goes back to 3400 B.C. A survey of the Knossian, or Minoan, period (1850–1400 B.C.) reveals evidences of devices for sanitation, ventilation, drainage, and latrines, including some methods which excelled anything else constructed before the nineteenth century. Temples located near springs and woodland hills were used as sanatoria and health resorts.

Hippocrates (460–377 B.C.). With the entrance of Hippocrates upon the scene of medicine, there began a new era of scientific thought and the introduction of the experimental method into medical science.

HIPPOCRATES

Hippocrates attributed disease to disorders in the vital fluids of the body. He divided diseases into four classes—acute, chronic, endemic, and epidemic. He stressed the importance of boiling water for irrigating wounds, of cleansing the hands and nails of the operator, and of using medicated dressings around wounds. Hippocrates is universally considered the "Father of Medicine" and the founder of medical ethics. Empedocles (450 B.C.) drained swamps and checked malaria in Sicily. Aristophanes (422 B.C.) described malaria.

Egyptian Period. Excavations show arrangements for the collection of rain water and copper pipes for sewage disposal, dating from the old kingdom (3400–2450 B.C.) to the new empire (1580–1200 B.C.). Egyptian medicine was progressive, and much was known regarding measures for sanitation. Circumcision was widely practiced in 4000 B.C. In the work of Herodotus are presented accounts which indicate that the Egyptians knew the antiseptic value of extreme dryness resulting from the use of such chemicals as niter and common salt. A mummy, dated 1000 B.C. and found by Elliott Smith, showed evidence of spinal tuberculosis.

The Ancient Hebrews, probably as far back as 1500 B.C., were the founders of medical jurisprudence and prophylaxis. The book of Leviticus contains the sternest mandates regarding the purifying of women after childbirth, the hygiene of menstruation, the prevention of contagious diseases, the touching of unclean objects, the necessity for segregation and disinfection in the control of diseases such as plague, scabies, anthrax, epilepsy, trachoma, phthisis, and syphilis. Diphtheria, leprosy, gonorrhea, and leucorrhea were recognized.

The Ancient Hindus. The earliest Sanskrit in 1500 B.C. contained accounts of spells against the demons, witches, or wizards of disease. Indian and Ceylonese hospitals existed in 368 A.D. The *Susrata* (fifth century) describes 1120 diseases, both natural and supernatural. Malaria was attributed to mosquitoes, and warnings were given to persons to desert their homes when "rats fall from the roof," presumably from plague. In the chapter on obstetrics of *Susrata* is a section on infant hygiene and nutrition. Evidences of variolation, inoculation against smallpox, have been found in the Sanskrit text, *Sacteya*. In Marco Polo's travels a description is given of a netting used to keep away flies and vermin.

Assyria and Babylonia (768–626 B.C.). From some eight hundred medical tablets, now in the British Museum, comes our knowledge of Assyro-Babylonian medicine. Therapy was by special ritual, and exorcism; and incantation was prophylaxis. Filthy remedies, also chains, or whipping, were designed to disgust the devil inside those afflicted in order to entice him out of the body. Contagion was considered to result from seizure of the individual by demons. The Assyro-Babylonians were the first public hygienists to conceive of the transmissibility of leprosy, and exercised this belief practically by expelling lepers from the community some 3500 years ago.

Graeco-Roman Period (146 B.C.–476 A.D.). After the destruction of Corinth (146 B.C.), Greek medicine migrated to Rome, where 10,000 people had died of plague daily during an epidemic. Claudius Galen (130–201 A.D.) referred to the false membranes found in cases of diphtheria. Ovid (43 B.C.–17 A.D.) discussed the epidemiology of plague. Virgil (70–19 B.C.) wrote on veterinary medicine and described anthrax of sheep and its transmissibility. The hygienic achievements of the Romans were associated with military objectives. Some examples were the great aqueducts, sewers, drains, public baths, pure food, and potable water supply.

Ancient Chinese Medicine was purely legendary until the Han dynasty, when the first books on medicine were written. During the third century B.C., clinical cases were recorded. Rabies was known in China during the age of Confucius. The ancient Chinese knew of the prevention of smallpox, which they probably got from India. References to syphilis and gonorrhea are said to go back to the Ming dynasty (1368 A.D.). For centuries, in China, Japan, and Asia, legumes have been grown together with nonlegumes to enrich the soil to help farmers procure vigorous plant growth.

Medicine in the Dark Ages. During this period medicine suffered a tremendous setback, for the progressive Roman and Greek medicine was replaced, about 400 A.D., by mysticism. Witches were boiled instead of water. Consequently, filth, pestilence, and plague came back again to stay till the eighteenth century.

Medicine in the Mohammedan Period. In 707 A.D. one of the earliest hospitals was founded at Damascus, with dispensaries and infirmaries in all the important Moslem clubs. Avicenna, an Arabian (980-1037 A.D.), known as the "Prince of Physicians," described wine as the best treatment for wounds, thus initiating a primitive preventive measure against infection.   In England, Bernard de Gordon wrote on smallpox, scabies, anthrax, and trachoma, and described leprosy as contagious.   The communicability of smallpox was referred to by Gilbert.

THE FOUNDING OF BACTERIOLOGY

Medieval Period.   In the eleventh and twelfth centuries Salerno was the first independent medical school.   Roger and Roland, in this school, described cancer, used mercurial salves for parasitic infestations, and taught that pus should be generated in wounds.   Advocates of this pus therapy dominated until the time of Lord Lister and von Bergmann.   During the Middle Ages humanity was plagued with epidemic diseases.   Leprosy became endemic about 1300, forcing attention to hospitalization.   The Black Death (bubonic plague), in 1348–1350, caused the death of one-quarter of the world's population, killing sixty million people.   Other pandemics, which included leprosy, St. Anthony's fire (erysipelas), epidemic chorea, sweating sickness (probably influenza), and syphilis, suggested that certain diseases might be contagious.   The Venetian republic, in 1348, installed the first public health officers to exclude infected ships, and later quarantined infected areas.   Vinegar was used as an antiseptic.

The Renaissance (1453–1600) combated spiritual medicine with the Greek system of medicine and a rising knowledge of chemistry. Paracelsus (1493–1541) listed five causes of disease as cosmic, predispositions, psychic, pathologic poisons, and divine intervention. Smallpox, typhus, and measles appeared from 1493–1570.   Typhus fever, or gaol fever, as it was called, resulted in appalling mortality rates that ran into millions, ofttimes destroying entire communities.

ANTON VAN LEEUWENHOEK

Development of the Microscope.   Roger Bacon (1214–1294) is credited with having made the first simple lens.   Zacharias Janssen, in Holland in 1590, may have produced the first compound lens system.   Athanasius Kircher (1601–1680), a Jesuit priest, devised a primitive magnifying instrument and may have seen bacteria with it.   Although claims have been made that bacteria were observed by many investigators including Pierre Borrel (1620–1671) and Kircher, the first notable observations on bacteria were recorded in 1675 by Anton van Leeuwenhoek (1632–1723).

With an optical instrument made of a biconvex lens, Leeuwenhoek discovered bacteria in various body fluids, in water, and in pepper infusions, and yeasts in beer. The discovery of the microscope opened a profound new field for study of the causation of fermentation and disease.

The Doctrine of Spontaneous Generation. For a long time it was believed that living things were spontaneously generated, and it took about a century before the world was convinced that all living things must have living parents. Francesco Redi (1626–1697) spread gauze over meat to prevent flies from laying their eggs. The observation that such covered meat was not decomposed, whereas meat which was not covered had undergone decomposition and was swarming with maggots, showed conclusively that these maggots developed from the eggs of flies. But it took many more investigations of a like nature to shake the theory of spontaneous generation. Lazzaro Spallanzani (1729–1799) sealed infusions of organic material in flasks and then boiled them. The flasks were found to be devoid of living organisms and without evidence of decomposition. His opponents' argument that the boiling had produced chemical changes which made spontaneous generation impossible was answered by showing that organisms would grow if the heated infusions were subsequently exposed to air.

Later experiments by Schulze in 1836, by Schwann in 1839, and by Schroeder and von Dusch in 1854, gave new evidence, but many persisted in their doubts. Schulze boiled infusions and then admitted air passed through strong sulfuric acid, but failed to find any living organisms. Schwann passed in air which had been heated to a temperature high enough to destroy germs, but could find no decomposition. Schroeder and von Dusch placed plugs of cotton into flasks of boiled solutions and observed the same result.

The greatest blow to the advocates of spontaneous generation was given by Louis Pasteur in 1860. Using flasks, the necks of which had been drawn out into tubes bent in a U form, he succeeded in showing that contamination of materials takes place only through access of air laden with living organisms. With this arrangement, the bacteria in the dust from the air could enter the open end of the U tube, but the lack of a current of air prevented them from rising and entering the second arm into the liquid. Thus, the flasks were uncontaminated.

The Germ Theory of Disease. The spread of disease had been attributed by some to vapors given off from infected persons (miasmatic theory) and by others to divine causes. In 1546 Fracastorius connected communicable diseases with a living "contagium vivum" which was transmitted by direct contact, by fomites, and through the air.   In 1792, Marcus Antonius von Plenciz, a physician of Vienna, published, without supporting experimental data, the *germ theory of disease*, in which he professed his belief in the views of Fracastorius and went a step further by stating that every infectious disease has its invisible living cause.

THE GOLDEN AGE OF BACTERIOLOGY

The discovery of the microscope, the settling of the dispute on the spontaneous generation of living things, and the formulation of the germ theory of disease paved the way for the foundation of the new

EDWARD JENNER

science of bacteriology and a rapid succession of new discoveries concerning the etiology, immunology, and prevention of infectious diseases.

Edward Jenner (1749–1823), a British physician, was named the "Father of Immunology" because he was the first to institute systematic vaccination against an infectious disease.   Utilizing his observations that persons in rural districts who had had cowpox were immune to smallpox, he performed the first successful vaccination against smallpox with the material from cowpox lesions. *Theodor Schwann* (1810–1882), a botanist, formulated the concept of "fermentum vivum" from studies of yeast cells, which may have motivated the research of Pasteur on fermentation.   He discovered the yeast plant *Saccharomyces cerevisiae*, which is a cause of alcoholic fermentation.

Lord Joseph Lister (1827–1912), a British surgeon, applied the principles of fermentation to surgical infections and postulated the theory that "infection was due to passage of minute bodies capable of self multiplication from infector to infected."   In 1852 he laid the foundations for aseptic surgery, of which he is considered the father. The microbial diseases, favus and thrush, two skin infections caused by a fungus growth and a yeast, respectively, were discovered by *Schoenlein* in 1839.   Anthrax was thought by *Pollender*, in 1861, to be caused by rod-shaped bodies present in the blood of animals dying or dead from the disease.   *Davaine*, in 1863, proved the anthrax bacillus to be the cause of anthrax by injecting animals with the rods and thus infecting them.   In 1846 *Semmelweis* contributed to our knowledge of childbed fever by proving the transmissibility of the disease from physician, midwife, or nurse to patient.

Louis Pasteur (1822–1895) is noted especially for his brilliant work on fermentation and the prevention of anthrax and rabies. Between 1857 and 1862 he found that yeast cells caused the normal fermentation occurring in wine and beer. He discovered that heating destroyed wild yeasts which caused bad flavors, and from this originated the important process of *pasteurization* which is used in the wine, beer, and dairy industries. Later, he solved the problem of pébrine, a silkworm disease, the first disease to succumb to bacteriological procedures. His work on chicken cholera laid the foundation of the science of immunity, with the discovery of the *Pasteurella avicida* as the cause

LOUIS PASTEUR

of the disease, and the finding that chickens were rendered immune to virulent cultures following injections with attenuated cultures. The organism grown under unfavorable environmental conditions was rendered avirulent, but at the same time it was still capable of conferring a potent active immunity. During the years 1879–1880, Pasteur showed that sheep could be immunized against anthrax. The Pasteur treatment for the prevention of rabies in those bitten by rabid

animals was introduced in 1885. Pasteur served mankind to his full capacity and is generally recognized as the "Father of Bacteriology."

Robert Koch (1843–1910) was trained to enter the medical profession but turned to the new science of bacteriology for his researches. He is conceded to be the "Father of Bacteriological Technic," for he discovered and developed the use of solid culture media, and isolated pure mixed cultures of microorganisms. He set forth certain criteria for establishing the etiology of diseases. These are known

ROBERT KOCH

as *Koch's postulates* and state: (1) A specific organism must always be associated with a disease. (2) It must be isolated in pure culture. (3) When inoculated into a healthy susceptible animal it must always produce the disease. (4) It should be obtained again in pure culture. These were long accepted as the only proof of the etiology of a disease, but with the development of serological technics of high specificity it is now believed that an organism may be the etiological agent even if not all of Koch's requirements have been fulfilled.   Koch developed many bacteriological procedures that are still used today, including staining with aniline dyes.   He discovered the anthrax spore and the mode of transmission of the disease.   In 1882 he established the *Mycobacterium tuberculosis* as the cause of tuberculosis and later discovered tuberculin.   He found the etiologic agent of Asiatic cholera and was the first one to transmit disease by artificially cultivated organisms. It is said of him that "he was a teacher who taught students how to get bacteria to yield their secrets."

Friedrich Löffler (1852–1915) worked with Koch to cultivate the diphtheria organism known as the Klebs-Löffler bacillus, and in 1884 obtained it in pure culture.   He discovered the cause of swine erysipelas in 1882.   *Edwin Klebs* (1834–1913), working with Löffler, discovered the diphtheria bacillus.   He investigated traumatic infections, malaria, and the bacteriology of gunshot wounds.   He was the first to produce tuberculosis lesions in animals.   In 1888 *Emile Roux* and *Alexandre Yersin* discovered the diphtheria toxin, and *Emil von Behring* described diphtheria antitoxin.   *Elie Metchnikoff* (1845–1916), in 1882, studied the nature and habits of microbes and in 1901 published his chief work, "Immunity in Infectious Diseases," which contains the phagocytosis theory of immunity.   *Karl Joseph Eberth* (1835–1926) discovered the cause of typhoid fever, now called *Eberthella typhosa* after him.   *George Gaffky* (1850–1918), in 1884, grew *Eberthella typhosa* in pure culture and confirmed its etiologic role in typhoid fever.   *Fehleisen*, assistant to Robert Koch, proved in 1883 that the specific cause of "St. Anthony's fire," erysipelas, was a streptococcus. *Albert Neisser* (1855–1916) discovered the gonococcus, the cause of gonorrhea, in 1879.   *William Welch* (1850–1934), in 1892, found the organism responsible for gas gangrene, *Clostridium perfringens*.   *Karl Weigert* (1845–1904) first used aniline dyes to stain bacteria.   *Otto Obermeier* (1843–1873), in 1873, discovered the cause of relapsing fever, *Borrelia recurrentis*.   *David Bruce* (1855–1931) found the cause of Malta fever, the trypanosomes of Nagana and sleeping sickness, and proved that the tsetse fly carried the trypanosome of sleeping sickness. *Theobald Smith* (1859–1934) was the first to demonstrate, in 1893, that Texas cattle fever could be transmitted by the bite of the tick. *Laveran* in 1880 discovered the malarial organism.   *Ronald Ross* (1857–1932) worked out the life cycle of *Plasmodium malariae* in 1896. *William Crawford Gorgas* (1854–1920) was chief health officer of the

Isthmian Canal Commission in 1907 and made the Panama Canal Zone a model of sanitary conditions, completely ridding it of malaria.

August von Wassermann (1866–1925), in 1906, achieved international fame by his application of the Bordet-Gengou phenomenon to the serodiagnosis of syphilis—the so-called Wassermann reaction. *Paul Ehrlich* (1854–1915) formulated the humoral theory of immunity. *William Hallock Park* (1863–1939), American bacteriologist and leader in public health work, was noted for his research on *Corynebacterium diphtheriae* and the prevention of diphtheria and scarlet fever. He and Zingher, in 1914, introduced immunization against diphtheria with toxin-antitoxin into the United States.

HIDEYO NOGUCHI

Among the Japanese bacteriologists, *Hideyo Noguchi* (1876–1928) is world-renowned for his work on venoms and the novel culture methods he introduced into bacteriology. In 1911 he secured pure cultures of *Treponema pallidum*, and in 1913 he discovered the spirochete in the cerebral tissue of paresis cases. In 1918 he found the *Leptospira icteroides*. *K. Shiga*, in 1898, discovered the type of dysentery bacillus which was named after him. He also reported the cultivation of *Mycobacterium leprae*. *S. Kitazato* (1852–1931) introduced bacteriology into Japan, when in 1894 together with Yersin he discovered *Pasteurella pestis* (cause of bubonic plague). Before this he had worked on the cause of tsutsugamushi fever. He studied under Koch from 1885–1891 and worked with von Behring on tetanus and diphtheria. *K. Futaki* and *K. Ishikawa*, in 1916, isolated the *Spirochaeta morsus muris* or *Spirillum minus*, the cause of rat-bite fever.

Ultramicroscopic bodies or filterable viruses were first indicted as producers of disease by *Iwanowski* in 1892 working with tobacco mosaic and by *Löffler* and *Frosch* in 1898 in foot-and-mouth disease.

The mode of transmission and the virus nature of yellow fever were demonstrated by the spectacular work of *Reed, Carroll, Agramonte,* and *Lazear* in Cuba, 1900–1902.

Many have devoted their studies to the bacteriology of the soil. *Boussingault,* a French chemist, demonstrated in 1878 that nitrates were derived from organic matter in soil and claimed that soil contained living organisms. *Pasteur* believed that nitrification, in which ammonia is changed to nitrates, was a bacterial process. In 1877

Scholsing and *Muntz* noted the action of sand filters in the purification of sewage and suggested that nitrification was due to organized ferments. *Winogradsky* isolated nitrifying organisms, but found they would grow only in media free from organic matter, and concluded that

decomposition of organic substances was necessary to nitrification and that the conversion of ammonia to nitrates was a necessary prelude to the work of a Nitrobacter. *Beijerinck* (1851–1931) found yeastlike organisms in the soil which utilize atmospheric nitrogen to build up complex compounds within the cytoplasm. *Hellriegel* and *Wilfarth* noted that bacteria live in the nodules on leguminous plants which take nitrogen from the air and restore it to the soil as nitrates.

M. W. BEIJERINCK

CONTEMPORARY BACTERIOLOGISTS AND HYGIENISTS

Jules Bordet (1870–   ), a Belgian physiologist and Nobel prize winner in medicine in 1919, is one of the outstanding living immunologists. His discovery, with Gengou, of the complement fixation reaction and other serological technics inaugurated a general method for diagnosis of infectious diseases, now universally employed for the diagnosis of syphilis, gonorrhea, glanders, and tuberculosis. He discovered the cause of whooping cough (*Hemophilus pertussis*). *Albert Calmette* (1863–1933), with *Guérin* in 1924, prepared B.C.G. vaccine from attenuated tubercle bacilli for the prophylactic vaccination of children susceptible to tuberculosis. *Sir Patrick Laidlaw* (who died in 1940) and *Major Dunkin*, British veterinary bacteriologists, in 1929, isolated the filterable virus of canine distemper and worked out a specific method of vaccination and homologous antiserum for preventing the disease. *Bernhard Bang* (1848–1932) found the cause of abortion in cattle (*Brucella abortus variety bovis*) and presented a plan for controlling the disease. *Frost* (1867–   ), American dairy bacteriologist, is the originator of the cellular test for pasteurized milk and the little plate method for counting milk bacteria. *Robert Stanley Breed* (1877–   ), one of the leading milk hygienists and dairy bacteriologists, introduced the Breed method for estimating direct bacterial counts in milk.

George and *Gladys Dick*, in 1923, proved that a specific hemolytic streptococcus, which they called *Streptococcus scarlatinae*, is the cause of scarlet fever. *Marion Dorset* (1872–1935) devised a method of immunization against hog cholera which he effected by the simultaneous use of homologous serum and virus. This practice has saved millions of dollars' worth of hogs from the ravages of this disease. *Mazyck P. Ravenel* was the first to isolate the bovine tubercle bacillus from children and proved its pathogenicity by inoculation into cattle. *Simon Flexner* is noted for a wide variety of researches in the field of bacteriology, especially for his contributions in poliomyelitis. *Thomas Rivers* is the foremost living authority on viruses and has studied many of the problems of vaccinia, lymphocytic choriomeningitis, and other virus diseases. He is known primarily for his cultivation of vaccinia virus in tissue culture for more effectve vaccination against smallpox, and for the isolation of the virus of lymphocytic choriomeningitis. *Michael Heidelberger* has made brilliant advances in the field of immunochemistry by his work on the pneumococcus. The contributions of Heidelberger and others have started a new trend in the utilization and application of principles of chemistry to the study of certain problems in bacteriology and immunity.

In the last decade, great strides have been made in chemotherapy. The sulfonamides have been used most successfully in infections caused by pyogenic cocci.

Dr. Alexander Fleming, British scientist, discovered in 1929 the antibiotic penicillin produced by the mold *Penicillium notatum.* Later *Chain, Florey,* and others standardized penicillin manufacture and expression of its antibacterial activity. *Dr. René Dubos* in 1939 discovered the antibiotic tyrothricin. *Selman A. Waksman* states that he discovered streptomycin in 1943 and neomycin in 1949. *Dr. B. M. Duggar* in 1949 produced aureomycin, an antibiotic useful in rickettsial and some

SELMAN A. WAKSMAN

viral diseases. *Dr. Paul Burkholder* of Yale University isolated chloromycetin from *Streptomyces venezuelae,* a soil-borne antibiotic useful in brucellosis typhoid and other microbial infections. *Dr. Frank Meleney* in 1945 discovered bacitracin from the Trace strain of *Bacillus subtilis,* which is active against Staphylococci, anaerobic cocci, and the *Streptococcus agalactiae. Dr. Edward Jonas Salk,* director of the virus research laboratory at the University of Pittsburgh School of Medicine, in March 1953 announced his discovery of the trivalent viral vaccine that is now in world-wide use in prophylaxis against paralytic poliomyelitis.

Since in this brief review of the development of bacteriology it has not been possible to include all those who have contributed, the student may obtain additional information from references made throughout the book and from publications listed at the end of each chapter.

BIBLIOGRAPHY

Bulloch, William. *The History of Bacteriology*. London and New York: Oxford University Press, 1938.

De Kruif, Paul Henry. *Men against Death*. New York: Harcourt, Brace and Company, 1932.

———. *Microbe Hunters*. New York: Harcourt, Brace and Company, 1926.

Dobell, Clifford. *Antony van Leeuwenhoek and His Little Animals*. New York: Harcourt, Brace and Company, 1932.

Dubos, R. J. *Bacterial and Mycotic Infections of Man*. Third Edition. Philadelphia: J. B. Lippincott Comnany, 1958.

Eberson, Frederick. *The Microbe's Challenge*. Lancaster, Pa.: Jacques Cattell Press, 1941.

Grainger, T. H. *A Guide to the History of Bacteriology*. New York: Ronald Press Company, 1958.

Medical Research Council. *A System of Bacteriology in Relation to Medicine*. Vol. I. London: His Majesty's Stationery Office, 1930. Pp. 15–103.

Oparin, A. I. *The Origin of Life on Earth*. New York: Academic Press, 1957.

Swingle, D. B. *General Bacteriology*, rev. by G. W. Walter. New York: D. Van Nostrand Company, 1947.

Taylor, F. Sherwood. *The Conquest of Bacteria*. New York: Philosophical Library, 1942.

Vallery-Radot, René. *The Life of Pasteur*. Garden City: Garden City Publishing Company, Inc., 1927.

Winslow, Charles-Edward A. *The Conquest of Epidemic Disease*. Princeton, N. J.: Princeton University Press, 1943.

CHAPTER II

THE GENERAL CHARACTERISTICS AND
CLASSIFICATION OF BACTERIA

CHARACTERISTICS OF BACTERIA

Definition of Bacteria. Bacteria are minute, unicellular, plant-like, microscopic organisms which differ from true plants in that they lack chlorophyll. They reproduce by binary fission. They are widely distributed in soil, air, water, and milk, on the surface of fruits and vegetables, and in various parts of the body such as the alimentary canal, skin, etc.

Bacteria may be classed as:

1. *Autotrophic*—organisms that live on inorganic matter.

2. *Heterotrophic*—organisms that live on organic matter.

(a) *Parasites*—those which require living organic matter for growth. These include the *pathogens*, which have an injurious effect on the animal or plant (host) upon which they live.

(b) *Saprophytes* or *saprogens*—those which live on dead organic matter.

Size of Bacteria. The unit of microbial measurement is the micron (μ or mu),* which is equal to 1/1000 of a millimeter or about 1/25,000 of an inch. In terms of this unit the size of bacteria generally varies from 0.2–5 microns. 1. *Average size*, 1.5μ. 2. *Hemophilus influenzae*, 0.2 to 0.3μ in thickness by 0.5 to 2μ in length. 3. *Bacillus anthracis*, 1 to 2μ in thickness by 5 to 10μ in length. 4. *Cocci*, 0.15 to 2μ in diameter.

General Morphological Grouping of Bacteria. Bacteria may be classified according to shape into three main classes. 1. *Bacilli* (bacillus, singular)—rod-shaped organisms. 2. *Cocci* (coccus, singular)—round or spherical organisms. 3. *Spirilla* (spirillum, singular) —small, comma-shaped or spiralled organisms which are motile.

Structure of Bacterial Cell. The bacterial cell is composed of cytoplasm, a cytoplasmic membrane, and a cell wall, and often con-

---

The unit of microbial measurement will be indicated throughout this Outline by the Greek letter mu (μ).

tains one or more vacuoles and granules. The cell wall is usually surrounded by a slime layer.

Nucleus of Bacteria. The exact nature of the nuclear material of the bacterial cell is not known. Most recent investigations indicate that bacteria contain nuclear material which, depending on conditions yet unknown but probably related to environment and development, may be diffuse in the protoplasm or may be partially or totally differentiated into a nucleus.

Cell Wall. This term is preferred over the older "cell membrane" by Knaysi. He ascribes the following properties to the cell wall:

1. It is very thin.
2. It has a certain rigidity, ductility, and elasticity, which vary with the bacteria and the conditions.
3. It is very resistant to chemical agents.
4. It has a low affinity for dyes.
5. Its chemical composition may vary in different bacteria. All are principally composed of complex carbohydrate-containing molecules. Some may contain carbohydrates only—true cellulose or hemicellulose. Others may also have nitrogen-containing substances related to mucins.
6. It may be the site of the bacterial cell's semi-permeability.

(a) *Permeability.* The bacterial cell is permeable to water and maintains an osmotic pressure between the cytoplasm of the cell and the surrounding medium.

(b) *Plasmolysis.* Bacterial cells, when grown in a saline solution which is more dense than the concentration of the cell protoplasm, shrink because water passes out, forming so-called "shadow forms."

(c) *Plasmoptysis.* When the bacterial cell is placed in distilled water or in water of low salt concentration, water passes in and the cell swells or bursts.

Slime Layer. It is affirmed by Knaysi that the cell wall of all bacteria is surrounded by a slime layer.

1. This may vary in thickness, between species, strains of a species, and cells of a strain.
2. It has the structure of a jelly.
3. The slime layer contains carbohydrate; its chemical composition may vary in different bacteria.
4. It may often result from modification of the substances of the cell wall.
5. When the slime layer is sufficiently thick and firm to have a form, it is called a capsule.

Capsules. Many bacteria possess a capsule, which consists of a broad, colorless, mucoid or gelatinous outer wall.

1. The capsule is exhibited by harmful bacteria, such as virulent pneumococci, *Bacillus anthracis, Mycobacterium tuberculosis,* etc.

2. The capsule is usually difficult to demonstrate when bacteria are grown on artificial media.

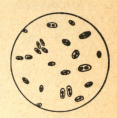

3. It is best shown in preparations of bacteria taken directly from animal tissues and fluids, or from media containing animal serum or milk.

4. Presence of a capsule may denote high virulence.

5. The Neufeld reaction for the typing of pneumococci is based on the swelling of the capsule in the presence of type-specific serum.

Fig. 1. Capsules of pneumococci.

6. Bacteria with capsules generally form slimy growths on culture media.

Motility. Bacteria, like all particles suspended in air or liquid, exhibit *Brownian movement* (swaying or quivering) which may be differentiated from *true motility* due to the presence of flagella. *Flagella* are long filaments which often are waved and undulating and move by a wavy or screwlike motion. Bacteria may be classified as follows on the basis of the number and arrangement of their flagella:

1. Atrichous, possessing no flagella.
2. Monotrichous, having a single flagellum at one pole—e.g., the cholera vibrio.
3. Lophotrichous, having a tuft of flagella at one pole—e.g., *Spirillum undula.*
4. Amphitrichous, with one or a tuft of flagella at each pole—several spirilla.
5. Peritrichous, with flagella surrounding the entire organism—e.g., *Salmonella typhosa (Eberthella typhosa).*

Inclusions or Vacuoles. These consist principally of reserve material in solution or as particles or droplets of one or more pure substances. The reserve material may be a carbohydrate, a fat, or a nitrogen-containing substance called volutin.

1. The *carbohydrate* may be of two kinds, a glycogen or glycogen-like material, or a material related to starch found almost exclusively in some anaerobic spore formers.

2. *Fat* is present in some species.

3. *Volutin* is widely distributed among bacteria, appearing as amorphous granules variable in size or shape.

(a) Volutin probably constitutes most of the so-called *Babes-Ernst bodies* or *metachromatic granules.*
(b) It has a great affinity for basic dyes.
(c) It probably contains zymonucleic acid.

Spore Formation. Certain bacteria, when conditions are favorable, form strongly refractive bodies, called *spores*, in order that they may be able to resist deleterious influences better than in the vegetative state. Optimum conditions for spore formation are the same as those for growth. Spores are formed by well-nourished cells as conditions for vegetative growth become *gradually* unsuitable. Spores are believed by some to be of reproductive significance, for they develop into vegetative forms when the microorganism containing spores is placed in a favorable environment.

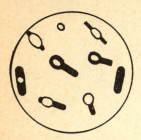

Fig. 2. Position and types of bacterial spores.

1. Spores are highly resistant to heat, drying, and disinfectants.

2. Spores are difficult to stain, and once stained are difficult to decolorize.

3. Spores are formed only in bacilli, and not in cocci or spirilla.

4. Spores may be classified according to shape and position in the bacterial cell as follows:

(a) Terminal oval or elongated spore at one end of the cell—e.g., *Clostridium fallax.*

(b) Terminal spherical spore at one end of the cell—e.g., *Clostridium tetani.*

(c) Central swollen spore in the center of the cell—e.g., *Clostridium multifermentans.*

(d) Central oval, but not swollen, spore in the center of the cell—e.g., *Clostridium botulinum.*

(e) Subterminal oval or elongated spore near the end of the bacillus—e.g., *Clostridium sporogenes.*

Reproduction of Bacteria. True bacteria divide by binary fission, although in some related species other methods of reproduction, such as budding, branching, the formation of conidia, gonidia, arthrospores, and endospores have been described. Spirilla split longitudinally.

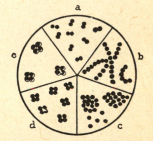

Fig. 3. Arrangement of cocci.

a. Diplococci.
b. Streptococci.
c. Staphylococci.
d. Tetrads.
e. Sarcinae.

1. If chains of bacilli tend to remain together, they are called streptobacilli.

2. Cocci divide in one, two, or three planes to form: (a) Diplococci (pairs), (b) Streptococci (chains), (c) Staphylococci (grapelike clusters), (d) Tetrads (fours), and (e) Sarcinae (bales or packets). (See Fig. 3.)

Growth of Bacteria. Bacteria may reproduce by binary fission at an average maximum rate of every 20 minutes, but this rate is maintained for only a short period. When bacteria are inoculated into a culture medium they pass through the following periods of growth:

1. An initial *lag phase*, during which there is little or no multiplication of the bacteria and some die before they become accustomed to the medium.

2. The *logarithmic period*, or period when the organisms grow at a maximum rate, increasing by geometric progression as a result of reproduction by fission. At this time, if the logarithms of the number of bacteria present are plotted against time, a straight line is obtained.

3. The *stationary period*, or period of the plateau, which may be plotted on a growth curve. During this time the bacteria are dying as fast as they are being formed.

4. The period of *decline and death*, usually at the end of 15, 18, or 24 hours, when the growth rate declines steadily.

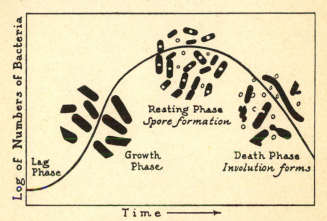

FIG. 4. Growth rate curve.

Growth rate curves (Fig. 4) with each of these periods may be plotted. This curve suggests a *"law of bacterial growth"* which is determined by the proportion between the number of growing bacteria and the amount of available food. Unfavorable environmental factors, such as acid production, may modify or slow up the growth rate.

Involution forms of bacteria appear in cultures during the period of death and decline. These are abnormal, swollen, spindle-shaped, elongated forms; sometimes they are smaller and of different shape from the original culture.

Environmental Factors in Growth of Bacteria.

1. *Nutrition.* Different species of bacteria have different food requirements, which the environment must provide.

Autotrophic bacteria grow in the absence of organic matter, building up their own food supply from simple inorganic substances. In this group are many of the soil bacteria which utilize carbon dioxide, ammonia, or nitrites as sources of carbon and nitrogen.

Heterotrophic bacteria, which comprise the great majority of species, must obtain their carbon and nitrogen from more or less complex organic compounds—carbon from proteins, fats, and carbohydrates; nitrogen usually from amino acids or nitrates.

Moisture is indispensable for bacterial growth. It serves as a solvent for food; it carries waste products from the cell; it enters into most chemical activities; it forms the major portion of the protoplasm; it speeds up all reactions.

Inorganic salts, in very small quantities, are also required. The elements K, Na, Ca, Mg, Fe, Cl, S, and P, in utilizable form, are required by most bacteria, but to a varied extent.

Accessory growth factors in minute quantities have recently been shown to be required. Some that have been identified include thiamin, riboflavin, and biotin.

2. *Oxygen.* Bacteria may be classified according to their oxygen requirements as:

(a) *Obligate aerobes*—organisms which grow only in the presence of free or atmospheric oxygen; e.g., *Bacillus subtilis*.

(b) *Facultative anaerobes*—organisms which are fundamentally aerobes, but can grow in the absence of free oxygen; e.g., *Escherichia coli*.

(c) *Obligate anaerobes*—organisms which grow only in an environment with no free oxygen or minute amounts of free oxygen; e.g., *Clostridium tetani*.

(d) *Facultative aerobes*—a few organisms which are fundamentally anaerobes but can grow in the presence of free oxygen.

3. *Temperature.*

(a) The limit of temperature at which bacterial survival takes place is $-250°$ C. to $160°$ C.

(b) Bacteria carry on activity at temperatures ranging between $0°$ and $90°$ C.

(c) For each organism there is:

(1) *An optimum temperature*, that best suited for its growth.

(2) *A minimum temperature*, the lowest temperature at which it will grow.

(3) *A maximum temperature*, the highest temperature at which it will grow.

(4) *A thermal death point*, that temperature which in a given time destroys all the bacteria present.* The *thermal death time* is the length of time required to kill all the organisms in a given substance at a given temperature. Both of these figures will vary with the moisture, medium, age of culture, etc., and these factors must be specified to make the data significant.

(d) Bacteria may be classified, according to their growth at different temperatures, into the following groups:

(1) *Psychrophilic*—cold-loving bacteria which live at temperatures below 10° C. in cold water and northern soils, with a minimum temperature of 0° C., an optimum temperature between 15° and 20° C., and a maximum temperature of 30° C. These include staphylococci found in the stratosphere, phosphorescent bacteria from fish in ocean depths which live at 0° C., several soil bacteria that live at temperatures even below 0° C. and may account for certain changes in meat in cold storage.

(2) *Mesophilic*—organisms that grow best at a moderate temperature. Optimum of the saprophytes is 25° C.; of the pathogens, most of which are included in the mesophils, 37° C. Minimum temperature of this group varies from 5° to 25° C. The maximum is generally set at 43° C.

(3) *Thermophilic*—heat-loving bacteria. They have a minimum temperature of 25° to 45° C., an optimum temperature of 50° to 55° C., and a maximum temperature of 60° to 85° C. These organisms ferment manure, soils, ensilage, and intestinal contents of man and animals. Bacteria living in hot springs water at 89° C. usually are spore-bearers.

4. *Reaction.* Most bacteria require a neutral or slightly alkaline medium for optimum growth, each species having a definite limiting range and an optimal hydrogen ion concentration. They vary widely in their tolerance to changes in reaction, saprophytic species generally showing greater adaptability.

5. *Osmotic Pressure.* Bacteria vary in their susceptibility to changes in osmotic pressure, but in general show greater resistance to these changes than do higher forms. (See p. 16.) High osmotic pressures are used in preserving some foodstuffs, such as salted meats, jellies, etc.

6. *Light.* Diffused sunlight hinders the growth of most bacteria and the direct rays of the sun, especially the ultraviolet rays, are highly injurious. This factor is utilized in air and water sanitation.

7. *Carbon Dioxide.* Presence of carbon dioxide is apparently essential for growth of some bacteria, and an increase over the normal

---

* For spore-bearing organisms there are two thermal death points, one for the vegetative form and one for the spore form. Freezing does not destroy bacteria; it merely inhibits growth.

concentration of this gas definitely stimulates growth of some species; e.g., the gonococcus.

8. *Salts.*   Chlorides of alkaline earth metals are relatively harmless, and even desirable and stimulating to most bacteria.   Chlorides of the heavy metals, however, although slightly stimulating in extremely low concentrations, exert definitely harmful effects if present in larger amounts.   Salts may act favorably, too.   They may suppress ionization of deleterious substances or react with them to prevent them from affecting bacteria unfavorably.

COLONY FORMATION

A *colony* is a group of bacteria formed from the reproduction of a single microorganism and generally visible to the naked eye.

Colonies may be described as: 1. Discrete; 2. Colored (as in chromogenic bacteria); 3. Opalescent or shiny; 4. Spreading, in which case there is a rapid growth that covers a large area; 5. Mucoid; 6. Pin-point; and 7. Surface.

As observed under the microscope, colonies may be described as shown in Fig. 5.

Areolate Grumose Moruloid Clouded Gyrose Marmorated Reticulate

COLONY AS A WHOLE.

Repand   Lobate   Erose   Auriculate Lacerate Fimbricate Ciliate

EDGE OF COLONY

FIG. 5.   Microscopic structure of colonies.   (Low power.)

Types of Colonies.   For a discussion of variations in colonies, see page 24.

BACTERIAL VARIATION

The amazing phenomena whereby bacteria may undergo changes in morphological, cultural, and physiological characteristics, in colony formation, and in virulence, etc., come under the heading of variation.

Organisms that deviate from the normal characters are called *variants*. There are several factors influencing or modifying variation:

Mutations or transmissible variations are apt to be permanent. They seem to arise suddenly and spontaneously.

Fluctuating Variations. Temporary changes often occur in morphology or physiology, usually due to environmental conditions. Example: pigment-producing organisms may temporarily lose their power to form pigment.

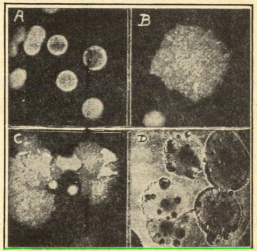

FIG. 6. Types of bacterial colonies. *A*, smooth; *B*, rough; *C*, smooth developing from rough colonies. *A*, *B*, and *C* are from a culture of an acid-fast bacterium. *D*, colonies of the cholera vibrio with secondary daughter colonies. (After Eisenberg.*)

Involution forms, or degenerate cells, often occur when colonies age on artificial media.

Artificial Selection. When bacteria are cultivated under constant conditions of media content, temperature, moisture, surface tension, etc., the organism does not tend to vary. However, by varying the media and other environmental factors, the bacteria may undergo wide ranges of variation in form, structure, reactions, and physiological activities, etc. Example: spore-formers may temporarily become asporogenous.

Adaptations. Bacteria vary due to adaptations to new conditions. Pathogenic bacteria lose their virulence when cultivated on artificial media, but regain it on passage through animals.

---

* From Smith and Martin, *Zinsser's Textbook of Bacteriology*, 1948. Courtesy, Appleton-Century-Crofts Co.

Smooth and rough colony types, together with their respective allied attributes, are perhaps of greatest interest. Both S and R colonies develop in many bacterial species from an individual, apparently pure culture regarded as the "normal." The change from S to R occurs with more ease than that from R to S. The S type is often regarded as the normal, but this may vary with the species. In pneumococcus, the mucoid (M) is the most common type; in typhoid, the S; in anthrax, the R.

The general differences between S and R forms may be tabulated as follows:

S	R
1. Colony margin round and even; surface smooth, convex, glistening.	1. Colony margin irregular; surface flat, uneven, granular.
2. Cells have normal morphology, tend to occur singly.	2. Cells tend toward abnormal form and toward clinging together after fission.
3. Broth culture uniformly turbid.	3. Supernatant fluid clear with granular sediment in broth.
4. Normal suspension in saline.	4. Suspension in saline clumps spontaneously and settles out.
5. Flagellated species usually motile.	5. Motility low or absent.
6. Encapsulated species have capsule.	6. Capsule absent.
7. Good immunizing agents. Somatic and type antigens present, flocculent agglutination.	7. Poor immunizing agents. Only somatic antigens present, granular agglutination.
8. Pathogenic species generally virulent.	8. Virulence greatly reduced or absent.
9. Chemically active.	9. Chemical activity reduced.
10. More common in active disease conditions.	10. More common in convalescents and carriers.
11. Sensitive to bacteriophage.	11. Less sensitive to bacteriophage.
12. Resistant to phagocytosis.	12. Susceptible to phagocytosis.

G colonies have also been observed. These are minute colonies of very small cells varying greatly in morphology. They may revert to the larger colony forms. It has been suggested that these represent growth from minute filterable gonidia (intracellular granules). Others regard them as slow-growing variants at a low level of vigor.

Mucoid colonies, viscous and slimy, are common in many bacteria, both spore-forming and non-spore-forming, rods and cocci. The colonies have a tendency to run together; the cells are often capsulated. It has been reported that this may be the most virulent and toxicogenic form. The mucoid matrix may be secreted by the cells in response to various unfavorable external stimuli; e.g., phenol.

H and O type colonies have also been described, notably in *Proteus vulgaris*. The H type is spreading and consists of motile organisms;

the O is a nonspreading, small, discrete colony of nonflagellated organisms. Considerable significance has been attached to these variants because of the difference in antigens present in flagella and the cell body.

The nature of the H–O variation appears different from the S–R. The former takes place more readily and does not have a semipermanent character.

D (dwarf or diphtheroidal) colonies are very small, measuring about 1 mm. or less in diameter. The cells appear the same, morphologically and physiologically, as in the larger colonies.

No satisfactory explanation has as yet been advanced for the appearance of these various forms. An orderly development toward old age, stages in a complex life-cycle, and reaction to environmental conditions are some of the hypotheses put forth.

CLASSIFICATION OF BACTERIA

In classifying an organism so that we may have a means of referring to it and differentiating it from other organisms, the following points generally employed in taxonomy are determined:

1. Species—a group of organisms possessing certain definite individual characteristics which distinguish it from other groups of organisms.
2. Genus—a group of related species.
3. Family—a group of related genera.
4. Order—a group of related families.
5. Class—a group of related orders.
There may be sub-families, sub-orders, etc.

Bacteria cannot be classified upon the basis of morphology alone, inasmuch as many bacteria which are alike morphologically differ widely in other properties. Bacteria are classified on the basis of biochemical, serological, and pathogenic properties, morphology, physiology, metabolism, chemical composition, and any other distinctive characteristics which may serve as a means of differentiation.

Numerous attempts to classify bacteria have been made, but in this discussion the classification given in *Bergey's Manual of Determinative Bacteriology* of 1948 will be employed. In this system of classification bacteria are delegated to the class *Schizomycetes*, which are typically unicellular, non-chlorophyll-producing plants multiplying by cell fission. There are five orders in this class, the common or true bacteria falling into the order Eubacteriales. The following classification of the Schizomycetes is similar to Bergey's.

ORDER I. *EUBACTERIALES*

Simple and undifferentiated forms, without true branching. Occur as spheres, as short or long rods, or as curved rods. Motility when present by means of flagella. Endospore formation occurs in some species. Some form pigment. Some store materials as volutin, glycogen, or fat. Not acid-fast.

SUB-ORDER I. EUBACTERIINEAE

Do not possess photosynthetic pigments. Cells do not contain free sulfur. Not attached to substrate by a stalk. Do not deposit ferric hydroxide.

Family I. *Nitrobacteriaceae.* Autotrophic. Nonparasitic; commonly soil or water forms. Do not produce spores. Usually Gram-negative rods, occasionally spherical. Nonmotile or single polar flagellum.

Tribe I. *Nitrobacterieae.* Cells oxidize ammonia or nitrites.

Genus I. *Nitrosomonas* (*europaea* *). Oxidize ammonia to nitrites. Cells ellipsoidal.

Genus II. *Nitrosococcus* (*nitrosus*). Oxidize ammonia to nitrites. Cells spherical.

Genus III. *Nitrosospira* (*briensis*). Oxidize ammonia to nitrites. Cells spiral.

Genus IV. *Nitrosocystis* (*javanensis*). Oxidize ammonia to nitrites. Cells ellipsoidal or elongated; unite into compact aggregates surrounded by common membrane to form a cyst.

Genus V. *Nitrosogloea* (*merismoides*). Oxidize ammonia to nitrites. Cells ellipsoidal or rod-shaped. Embedded in slime and united into compact zoogloeal aggregates.

Genus VI. *Nitrobacter* (*winogradskyi*). Oxidize nitrates to nitrites. Rods.

Genus VII. *Nitrocystis* (*sarcinoides*). Oxidize nitrates to nitrites. Rods.

Tribe II. *Hydrogenomonadeae.* These organisms oxidize hydrogen.

Genus I. *Hydrogenomonas* (*pantotropha*). Oxidize hydrogen to water.

Tribe III. *Thiobacilleae.* Oxidize sulfur and compounds of sulfur.

Genus I. *Thiobacillus* (*thioparus*). Oxidize sulfur to sulfate.

Family II. *Pseudomonadaceae.* Straight to spiral rods with polar flagella if present. Gram-negative. No endospores produced. Typically water or soil forms; some are animal or plant parasites.

Tribe I. *Pseudomonadeae.* Mainly soil and water forms; includes numerous plant pathogens. Motile by polar flagella or nonmotile.

Genus I. *Pseudomonas* (*aeruginosa*). Soil and water forms, most often producing water-soluble pigment which diffuses through the medium as bluish-green or yellowish-green pigment.

Genus II. *Xanthomonas* (*hyacinthi*). Cells generally monotrichous with yellow water-insoluble pigment. Most are plant pathogens.

Genus III. *Methanomonas* (*methanica*). Cells monotrichous. Derive energy from oxidation of methane to carbon dioxide and water.

Genus IV. *Acetobacter* (*aceti*). Rod-shaped cells, frequently showing elongated, branched, or swollen forms. No spores formed. Obligate aerobes. Oxidize alcohol to acetic acid.

---

* The type species is given in parentheses after each genus name.

Genus V. *Protaminobacter* (*alboflavum*). Soil or water forms attacking the lower alkylamines. Many pigmented. Can derive energy from phenol and other aromatic compounds. Cells branching. Motile.

Tribe II. *Spirilleae.* More or less spirally curved.

Genus I. *Vibrio* (*comma*). Short, bent rods occurring singly or united into spirals. Generally motile with single polar flagellum. Mostly water forms, few parasitic. (*V. comma* causes cholera.)

Genus II. *Desulfovibrio* (*desulfuricans*). Strict anaerobes. Reduce sulfates to hydrogen sulfide.

Genus III. *Cellvibrio* (*ochraceus*). Long, thin curved rods with rounded ends. Generally motile with single polar flagellum. On the ordinary culture media there is feeble growth. Oxidize cellulose to oxycellulose. Found in soil. Some disintegrate vegetable fibers.

Genus IV. *Cellfalcicula* (*viridis*). Short curved rods having pointed ends. Generally motile with single polar flagellum. There is feeble growth on ordinary culture media. Oxidize cellulose to oxycellulose. Found in soil.

Genus V. *Thiospira* (*winogradskyi*). Colorless rods, bent slightly, pointed at the ends. Granules of sulfur within cells. Motile.

Genus VI. *Spirillum* (*undula*). Cells of varying thickness, and varying length and pitch of spiral, forming either long curves or portions of a turn. Generally motile with several polar flagella. Found in water and putrid infusions.

Family III. *Azotobacteriaceae.* Large rods or oval cells. Can utilize free nitrogen. Depend primarily for growth energy on oxidation of carbohydrates. Soil forms. Produce no spores. Obligate aerobes.

Genus I. *Azotobacter* (*chroococcum*).

Family IV. *Rhizobiaceae.* Heterotrophic, but capable of utilizing inorganic nitrogen. Usually soil or water forms. Usually Gram-negative rods. No endospores are produced. Use dextrose and, in some instances, other sugars, not producing organic acids in any appreciable quantities.

Genus I. *Rhizobium* (*leguminosarum*). Fix free nitrogen when growing symbiotically on roots of leguminous plants.

Genus II. *Agrobacterium* (*tumefaciens*). Cannot fix free nitrogen but can use nitrates or other forms of inorganic nitrogen. Live in soil and on plant roots.

Genus III. *Chromobacterium* (*violaceum*). Produce violet chromogenesis.

Family V. *Micrococcaceae.* Spherical cells growing well on ordinary media. Motility rare. Metabolism heterotrophic. Many pigmented lemon-yellow, red, or orange. Usually Gram-positive.

Genus I. *Micrococcus* (*luteus*). Facultative parasites or saprophytes. Cells in plates or irregular masses, never chains. Found in water, air, milk, and pus.

Genus II. *Gaffkya* (*tetragena*). Facultative parasites and saprophytes. Occur in tetrads in animal body and on special media. May occur in pairs or irregular masses on ordinary media. White to pale yellow.

Genus III. *Sarcina* (*ventriculi*). Facultative parasites and saprophytes. Cells in regular packets. Usually Gram-positive. Formation of yellow or orange pigment on agar.

Family VI.   *Neisseriaceae.*   Gram-negative cocci, occurring in pairs and masses. Strict parasites.   Some species fail to grow, or grow poorly, on ordinary culture media.   Grow best at 37° C.

Genus I.   *Neisseria (gonorrhoeae).*   Occur in pairs, generally with adjacent sides flattened.   Parasites on mammals.

Genus II.   *Veillonella (parvula).*   Cocci of very small size, occurring in masses, rarely in pairs or short chains.   Cells undifferentiated, united by an interstitial substance.   Anaerobes.   Found in saliva, abscesses, appendix.

Family VII.   *Lactobacteriaceae.*   Cocci and rods.   Occur singly, in pairs, and in short chains.   Gram-positive.   Rarely motile.   No spores.   Ferment carbohydrates readily with formation of lactic acid and some volatile acid.   Some produce carbon dioxide and ethyl alcohol from dextrose, and mannitol from levulose.   Rarely liquefy gelatin.   Do not use nitrates.   Pigment, if any, is yellow, orange-red, or rusty brown. Aerobes to anaerobes.   Do not grow well on surface of agar media.

Tribe I.   *Streptococceae.*   Cocci occurring in pairs and chains, never in packets. Parasites and saprophytes.   Fermentative powers high; produce lactic acid from dextrose, lactose, and sucrose.   When gas is formed, it is carbon dioxide without hydrogen.   Gram-positive.

Genus I.   *Diplococcus (pneumoniae).*   Parasites growing poorly or not at all on artificial media.   Cells most often in pairs and lanceolate in shape. Capsules well developed.   High fermentative powers.

Genus II.   *Streptococcus (pyogenes).*   Cells occurring in pairs or short or long chains, never in packets or masses.   Capsules occasionally. Growth slow on artificial media.   Parasites and saprophytes.   Action on blood varies, but some species produce characteristic changes.

Genus III.   *Leuconostoc (mesenteroides).*   Saprophytes.   Form chains of cocci, lengthened to short rods in acid fruit juices.   Grow on ordinary media, but growth is enhanced by addition of yeast cells, yeast extract, or other vegetable tissue.   Generally produce limited amount of acid. Ferment dextrose with production of carbon dioxide, lactic acid, acetic acid, and ethyl alcohol.   Produce mannitol from levulose.   Found in plant material and milk products.

Tribe II.   *Lactobacilleae.*   Rods, often thin and long.   Nonmotile.   Gram-positive.   Ferment carbohydrates, polyalcohols, and lactic acid.   Some grow at relatively high temperatures.

Genus I.   *Lactobacillus (caucasicus).*   Always produce lactic acid from carbohydrates.   When gas is produced it is carbon dioxide without hydrogen.   Microaerophilic.   Widely distributed in milk and milk products, intestines of milk-fed infants.

Genus II.   *Microbacterium (lacticum).*   Form acid (but no gas) from carbohydrates.   Nonmotile.   Gram-positive small rods.   Found in dairy products, decayed beets, fecal material, and soil.

Genus III.   *Propionibacterium (freudenreichii).*   Ferment carbohydrates, polyalcohols, and lactic acid with the formation of propionic and acetic acids and carbon dioxide.   Gram-positive bacteria growing under anaerobic conditions in neutral media, as short diphtheroid rods.   Non-spore-forming.   Nonmotile.   Metachromatic granules.   Under aerobic conditions, grow as long irregular, club-shaped, and branched cells. Strong tendency toward anaerobiosis. Nutritional requirements complex. Optimum temperature 30° C. Found in dairy products.

Genus IV. *Butyribacterium* (*rettgeri*). Nonmotile. Anaerobic to microaerophilic. Gram-positive rods. Ferment carbohydrates and lactic acid to form carbon dioxide and acetic and butyric acids. Intestinal parasites.

Family VIII. *Corynebacteriaceae*. Rod-shaped cells very often banded or beaded with metachromatic granules. Usually Gram-positive. Plant and animal parasites and pathogens. In dairy products, water, and soil.

Genus I. *Corynebacterium* (*diphtheriae*). Rods, frequently with pointed or club-shaped swellings at the ends. Generally aerobic. A powerful exotoxin is produced by pathogenic species.

Genus II. *Listeria* (*monocytogenes*). Pathogenic parasites on warm-blooded animals. Aerobic. Grow freely on ordinary media. Motile.

Genus III. *Erysipelothrix* (*rhusiopathiae*). Rods. Tendency to form long filaments. Nonmotile. Gram-positive. Microaerophilic. Usually parasitic on mammals. Grow freely on ordinary media.

Family IX. *Achromobacteriaceae*. Gram-negative rods. Peritrichous or nonmotile. Feeble attacks on carbohydrates. Growth on agar slants nonchromogenic to yellow to orange. Pigment not diffused through the agar. Generally in salt or fresh water and soil.

Genus I. *Alcaligenes* (*faecalis*). Gram-negative to Gram-variable. When chromogenesis occurs, it is grayish-yellow, brownish-yellow, or yellow. Found in intestinal tract of vertebrates or in dairy products.

Genus II. *Achromobacter* (*liquefaciens*). Nonchromogenic. Gram-negative to Gram-variable rods. Occur in water or soil.

Genus III. *Flavobacterium* (*aquatile*). Rods forming yellow to orange pigment. Usually Gram-negative. Motile or nonmotile. Feeble attacks on carbohydrates. Occur in water or soil.

Family X. *Enterobacteriaceae*. Gram-negative rods widely distributed in nature. Many animal and some plant parasites. Grow well on artificial media. All attack carbohydrates to form acid, or acid and visible gas (hydrogen present). All produce nitrites from nitrates. When motile, flagella are peritrichous. No spores.

Tribe I. *Eschericheae*. Ferment lactose and dextrose with the formation of acid and visible gas. Usually do not liquefy gelatin.

Genus I. *Escherichia* (*coli*). Methyl red positive. Voges-Proskauer negative. Aerobic. May or may not utilize citric acid as the sole source of carbon. Generally cannot use uric acid as sole source of nitrogen. Commonly found in intestinal canal.

Genus II. *Aerobacter* (*aerogenes*). Methyl red negative. Voges-Proskauer positive. Citric acid used as sole source of carbon. Found in the alimentary tract.

Genus III. *Klebsiella* (*pneumoniae*). Short Gram-negative rods, somewhat plump with rounded ends, mostly occurring singly. Encapsulated. Nonmotile. Aerobic. Methyl red usually positive. Voges-Proskauer usually negative. Found in respiratory, intestinal, and genitourinary tracts.

Tribe II. *Erwineae.* Motile rods. For growth they require organic nitrogen compounds. Ferment dextrose and lactose with formation of acid or acid and visible gas. Plant parasites. Usually attack pectin.

Genus I. *Erwinia* (*amylovora*).

Tribe III. *Serrateae.* Chromogens producing red pigment. Small aerobic rods. Rapidly liquefy gelatin. Ferment dextrose and lactose with formation of acid or acid and visible gas.

Genus I. *Serratia* (*marcescens*). Peritrichous. Reduce nitrates. Liquefy blood serum. Coagulate and digest milk. Saprophytic on decaying plants. Found in water.

Tribe IV. *Proteae.* Ferment dextrose but not lactose with formation of acid and visible gas. Usually liquefy gelatin.

Genus I. *Proteus* (*vulgaris*). Actively motile with peritrichous flagella. Characteristically produce ameboid colonies on moist media. Young swarming colonies highly pleomorphic. Decompose proteins.

Tribe V. *Salmonelleae.* Ferment dextrose with formation of acid and usually gas. Do not ferment lactose and sucrose. Gelatin not liquefied. No spreading growth on ordinary agar media.

Genus I. *Salmonella* (*choleraesuis*). Ferment dextrose with formation of acid and usually gas. Usually motile. Do not produce indol or liquefy gelatin.

Genus II. *Shigella* (*dysenteriae*). Ferment dextrose with formation of acid but no gas. Nonmotile.

Family XI. *Parvobacteriaceae.* Small motile or nonmotile rods growing well on media containing blood fluids. Gram-negative. Usually parasitic on warm-blooded animals, infection in some cases taking place by penetration of organisms through mucous membrane or skin.

Tribe I. *Pasteurelleae.* Ovoid to elongated rods. Show bipolar staining.

Genus I. *Pasteurella* (*multocida*). Small ovoid to elongated rods. Facultative aerobes. Slightly carbohydrate fermentation. Parasitic on man, other mammals, and birds. Cause hemorrhagic septicemia, pseudo-tuberculosis, and plague.

Genus II. *Malleomyces* (*mallei*). Short rods with rounded ends, sometimes forming threads and showing tendency toward branching. Tendency toward bipolar staining. Specialized for parasitic life. Cause glanders or glanders-like infections. Grow well on blood serum or on other body fluid media.

Genus III. *Actinobacillus* (*lignieresi*). Aerobic. Show much pleomorphism. Many are coccus-like forms. Tend to bipolar staining. Acid but not gas from carbohydrates. Pathogenic for domestic animals. Tend to form aggregates in culture or tissue.

Tribe II. *Brucelleae.* Small rods or coccoids. Grow well on special media.

Genus I. *Brucella* (*melitensis*). Gelatin not liquefied. No acid or gas from carbohydrates. Parasites in animal tissue. Infect genital, respiratory, or intestinal tracts, mammary gland, or lymphatic tissue.

Tribe III. *Bacteroideae.* Gram-negative rods without endospores.

Genus I. *Bacteroides* (*fragilis*). Rods with rounded ends.

Genus II. *Fusobacterium* (*plauti-vincenti*). Rods with tapering ends. Anaerobic. Stain with more or less distinct granules.

Tribe IV. *Hemophileae.* Minute rod-shaped parasitic forms growing on first isolation only in the presence of hemoglobin, ascitic fluid, or other body fluids, or in the presence of certain growth accessory substances. Commonly found in mucosa of respiratory tract or conjunctiva.

Genus I. *Hemophilus* (*influenzae*). Sometimes thread-forming and pleomorphic. Aerobes to facultative anaerobes. Nonmotile. Strict parasites growing best (or only) in presence of hemoglobin; in general requiring blood serum, ascitic fluid, or certain growth accessory substances. Affect respiratory tract, conjunctiva, or genital region.

Genus II. *Moraxella* (*lacunata*). Short rod-shaped cells. Occur singly or in pairs. Nonmotile. Parasitic. Gram-negative.

Genus III. *Noguchia* (*granulosis*). Aerobes to facultative anaerobes. Motile. Encapsulated. Mucoid growth. Present in conjunctiva of man and animals affected by follicular type of disease.

Genus IV. *Dialister* (*pneumosintes*). Minute rods. Occur singly, in pairs, and in short chains. Nonmotile. Anaerobic. Strict parasites (respiratory tract); media must contain either fresh sterile tissue or ascitic fluid.

Family XII. *Bacteriaceae.* Rod-shaped cells without endospores. Metabolism complex, amino acids being utilized and generally carbohydrates. (A heterogeneous group of genera whose relationships to one another and to other groups are not clear.)

Genus I. *Bacterium* (*troloculare*). Classification not definite.

Family XIII. *Bacillaceae.* Spore-forming rods. Usually Gram-positive. Peritrichous flagella if motile.

Genus I. *Bacillus* (*subtilis*). Mostly saprophytic. Sometimes in chains. Generally oxidize carbohydrates or proteins rather completely.

Genus II. *Clostridium* (*butyricum*). Rods frequently enlarged at sporulation. Anaerobic (occasionally microaerophilic). Often parasitic. Found in soil and in animal and human feces.

SUB-ORDER II. CAULOBACTERIINEAE

Nonfilamentous bacteria growing characteristically upon stalks. Cells asymmetrical: gum, ferric hydroxide, or other materials secreted from one side or one end of cell to form stalk. Cells occur singly or in pairs, never in chains or filaments. Not ensheathed. Typically aquatic. Some parasitic on animals.

Family I. *Nevskiaceae.* Found in water or sugar vats.
Family II. *Gallionellaceae.* Found in iron-bearing waters.
Family III. *Caulobacteriaceae.* Found growing on submerged surfaces.
Family IV. *Siderocapsaceae.* Found growing on surfaces of leaves or on other parts of water plants.

SUB-ORDER III. RHODOBACTERIINEAE

Cells are spherical, rod-, spiral-, or vibrio-shaped. No endospores. Gram-negative. Photosynthetic. May store sulfur globules. Largely found in fresh water. Some found in marine water.

Family I. *Thiohodaceae.* The sulfur purple bacteria. Spheres, vibrios, ovoids, short rods, spirals, long rods, chains. Anaerobic or microaerophilic.

Photosynthetic.   Store sulfur globules when hydrogen sulfide is changed to sulfur.

Family II.   *Athiorhodaceae.*   Require organic growth factors.   Do not store sulfur globules even in presence of hydrogen sulfide.   Usually microaerophilic.   Pigment brownish.

Family III.   *Chlorobacteriaceae.*   Green sulfur bacteria.   Photosynthetic in the presence of hydrogen sulfide.   Generally do not store sulfur globules.

ORDER II.   *ACTINOMYCETALES*

Cells usually elongated, frequently filamentous, and with a tendency toward branching, in some genera forming definite branched mycelium.   Frequently show swellings, clubbed or irregular shapes.   No pseudoplasmodium.   No deposits of free sulfur, iron, or bacteriopurpurin.   No spores.   Conidia developed in some genera.   Usually Gram-positive.   Nonmotile.   Some parasitic on animals or plants.   Usually strongly aerobic and oxidative.   Complex proteins frequently required.   Growth on culture media often slow.   Some show moldlike colonies.

Family I.   *Mycobacteriaceae.*   Slender filaments or straight or curved rods, which are often irregular in form and sometimes slightly branched.   May stain unevenly.   No conidia.   Aerobic.   Gram-positive.

    Genus I.   *Mycobacterium (tuberculosis).*

Family II.   *Actinomycetaceae.*   Branched rods and filamentous forms, sometimes forming mycelia.   Conidia sometimes present.   Some parasites.   Some soil forms.

    Genus I.   *Nocardia (farcinica).*   Thin rods or filaments, often swollen, sometimes branched, forming mycelium.   No conidia.   Nonmotile.   Gram-positive.   No endospores formed.

    Genus II.   *Actinomyces (bovis).*   True mycelium formed which may break up into segments which function as conidia.   Anaerobic to microaerophilic.   Sometimes parasites, with clubbed ends of radiating threads conspicuous in lesions in the animal body.

Family III.   *Streptomycetaceae.*   Vegetative mycelium which does not fragment.   Conidia on sporophores.   Most are soil forms.

    Genus I.   *Streptomyces (albus).*   Take the form of a mycelium of many branches.   Aerobic.   Conidiospores produced in chains from aerial hyphae.   Generally saprophytic soil forms.

    Genus II.   *Micromonospora (chalcea).*   Not forming a true aerial mycelium.   Conidia produced singly at the ends of special conidiospores.   Conidiospores short.   Usually saprophytes occurring in hot composted manure, soil, lake bottoms, and dust.

ORDER III.   *CHLAMYDOBACTERIALES*

Filamentous bacteria, alga-like, typically water forms.   Some are ensheathed.   May be unbranched or show false branching.   Conidia and motile "swarmers" may be developed, never spores.

Family I.   *Chlamydobacteriaceae.*   Filamentous bacteria which often show typical false branching.   Cells divide transversely.   If swarm cells develop they are motile by means of flagella.

Family II. *Crenothrichaceae.* Filaments unbranched, attached to a firm substrate, differentiating base and tip. Cells of filaments divide in three planes forming spherical nonmotile conidia.

Family III. *Beggiatoaceae.* Filamentous, in chains of cells. Filaments contain sulfur globules in presence of hydrogen sulfide. Special organs of reproduction unknown.

ORDER IV. *MYXOBACTERIALES*

Slime bacteria. Cells develop as a colony (pseudoplasmodium or swarm) consisting of slender, relatively flexible elongate rods. Cells move together as an advancing mass by excretion of slime. No flagella. Most species found on dung or isolated from soil. They grow rapidly on nutrient agar, and develop microcysts, or fruiting bodies. The myxobacteriales produce enzymes which may hydrolyse, or dissolve complex molecules such as protein, cellulose, and chitin. They tend to be strict saprophytes and aerobes. The genus *Cytophaga* inhabits sea water and soil with the *C. columnaris* being pathogenic for commercial fish.

ORDER V. *SPIROCHAETALES*

Protozoan-like in certain characteristics. Cells usually slender, flexible spirals. Multiply by transverse division. Motility often characteristic but without polarity.

Family I. *Spirochaetaceae.*

Genus I. *Spirochaeta* (*plicatilis*). Nonparasitic. Flexible, undulating body with or without flagelliform tapering ends. Common in sewage and foul waters.

Genus II. *Saprospira* (*grandis*). Free-living. Actively motile. Distinct periplast membrane. Transverse markings. Found in marine ooze.

Genus III. *Cristispira* (*balbianii*). Peculiar flattened ridge runs length of body. Cross-striations. Parasitic in molluscs.

Family II. *Treponemataceae.*

Genus I. *Borrelia* (*anserina*). Small, parasitic, spiral, flexible with terminal filaments. Spirals large, wavy, three to five in number.

Genus II. *Treponema* (*pallidum*). Undulating body, coiled in rigid spirals, without crista. May have flagelliform pointed ends. Parasitic, frequently pathogenic. (*T. pallidum* causes syphilis.)

Genus III. *Leptospira* (*icterohaemorrhagiae*). Sharply twisted cylinders with flagelliform tapering ends, one or both extremities being sharply curved into a hook.

Rickettsiales. These are small, rod-shaped, coccoid and often pleomorphic microorganisms which occur as elementary bodies. They are usually intracellular but may occasionally be facultative or exclusively extracellular. The rickettsiales are nonfilterable and Gram-negative, and are cultivated outside the host only in living tissues, embryonated eggs, or, rarely, in media containing body fluids. Following are the varieties:

Rickettsia prowazekii. A minute, coccobacillary form, sometimes ellipsoidal or long rod-shaped cells which occasionally are filamentous.

Often seen in pairs and sometimes in chains. Causes louse-borne typhus.

Rickettsia typhi. Resembles *R. prowazekii* morphologically and in staining properties. Nonmotile, Gram-negative.

Rickettsia tsutsugamushi (*R. orientalis*). A small, pleomorphic, bacterium-like microorganism. Gram-negative. Stains well with azur III and methylene blue. The etiological factor in tsutsugamushi fever.

Rickettsia rickettsii. A minute, paired organism surrounded by a narrow clear zone or halo; often lanceloate, resembling in appearance a minute pair of nonmotile pneumococci. Average size is 0.6 by 1.2 microns under the electron microscope. Serologically distinguishable from *R. prowazekii* and *R. typhi* by complement fixation or agglutination with specific antigens.

Rickettsia conorii. Diplobacillary forms predominate which are nonmotile, Gram-negative. Causes boutonneuse fever.

Rickettsia australis. A minute, ellipsoidal or coccoidal form resembling *R. prowazekii* morphologically and in staining properties. Nonmotile, Gram-negative.

Rickettsia akari. Minute, bipolar diplobacilli; Gram-negative rods which stain poorly with methylene blue.

Rickettsia quintana. Coccoidal or ellipsoidal organism, often occurring in pairs. In lice, appear as short rods. Cause of five-day fever.

Rickettsia burnetii (*Coxiella burnetii*). Small, bacterium-like pleomorphic organisms varying in size from coccoidal forms to well-marked rods. Cause of Q fever.

BIBLIOGRAPHY

Brachet, J. *Biochemical Cytology.* New York: Academic Press, 1957.

Braun, W. *Bacterial Genetics.* Philadelphia: W. B. Saunders Company, 1953.

Breed, R. S., E. G. D. Murray, and N. R. Smith. *Bergey's Manual of Determinative Bacteriology.* Seventh Edition. Baltimore: Williams & Wilkins Company, 1957.

Catcheside, D. G. *The Genetics of Microorganisms.* New York: Pitman Publishing Corporation, 1951.

De Robertis, E. D. P., W. W. Nowinski, and F. A. Saez. *General Cytology.* Philadelphia: W. B. Saunders Company, 1954.

Dorland, W. A. N. *American Illustrated Medical Dictionary.* Philadelphia: W. B. Saunders Company, 1944. Pp. 190–196.

Dubos, René J. *The Bacterial Cell.* Cambridge, Mass.: Harvard University Press, 1945.

Edwards, P. R., and W. H. Ewing. *Identification of Enterobacteriaceae.* Minneapolis: Burgess Publishing Company, 1955.

Ephrussi, B. *Nucleocytoplasmic Relations in Microorganisms.* New York: Oxford University Press, 1953.

Frobisher, M., Jr. *Fundamentals of Microbiology.* Philadelphia: W. B. Saunders Company, 1958. Pp. 140–149.

Gale, E. F. *Chemical Activities of Bacteria.* New York: Academic Press, 1948. Pp. 10–129.

Gerard, R. W. *Unresting Cells.* New York: Harper and Brothers, 1949.

Giese, A. C. *Cell Physiology.* Philadelphia: W. B. Saunders Company, 1957.

International Committee on Bacteriological Nomenclature (Editorial Board). *International Code of Nomenclature of Bacteria and Viruses.* Ames, Ia.: Iowa State College Press, 1958.

Jawetz, E., J. L. Melnick, and E. A. Adelberg. *Review of Medical Microbiology.* Los Altos: Lange Medical Publications, 1962.

Jordan, Edwin O., and William Burrows. *Textbook of Bacteriology.* Philadelphia, W. B. Saunders Company, 1945.

Kelser, R. A. *Veterinary Bacteriology.* Baltimore: Williams & Wilkins Company, 1948. Pp. 8–34.

Knaysi, Georges. *Elements of Bacterial Cytology.* Second Edition. Ithaca, N. Y.: Comstock Publishing Company, 1951.

Lanjou, J., ed. *Botanical Nomenclature and Taxonomy.* Waltham, Mass.: Chronica Botanica Company, 1950.

Lederberg, J., ed. *Papers in Microbial Genetics, Bacteria, and Bacterial Viruses.* Madison: University of Wisconsin Press, 1951.

Medical Research Council. *A System of Bacteriology in Relation to Medicine.* London: His Majesty's Stationery Office, 1930. Vol. I, pp. 104–178, 292–374.

Porter, J. R. *Bacterial Chemistry and Physiology.* New York: John Wiley and Sons, 1946.

Stedman's Medical Dictionary. Nineteenth Edition. Baltimore: Williams & Wilkins Company, 1957. Pp. 154–158.

Stephenson, Marjory. *Bacterial Metabolism.* Third Edition. New York: Longmans, Green and Company, 1949.

Swingle, D. B. *General Bacteriology,* rev. by G. W. Walter. New York: D. Van Nostrand Company, 1947. Pp. 41–70.

Taber, C. W. *Cyclopedic Medical Dictionary.* Eighth Edition. Philadelphia: F. A. Davis Company, 1958.

Thimann, K. W. *The Life of Bacteria: Their Growth, Metabolism, and Relationships.* New York: Macmillan Company, 1955.

Top, F. H. *Communicable Diseases.* St. Louis: C. V. Mosby Company, 1947. Pp. 68–111.

Wagner, R. P., and H. K. Mitchell. *Genetics and Metabolism.* New York: John Wiley and Sons, 1955.

Waksman, S. *The Actimomycetes.* Waltham, Mass.: Chronica Botanica Company, 1950.

Waksman, S., and H. Lechevalier. *Guide to the Classification and Identification of the Actimomycetes and Their Antibiotics.* Baltimore: Williams & Wilkins Company, 1953.

Weinrich, D. H., I. F. Lewis, and J. R. Raper, eds. *Sex in Microorganisms.* Washington, D. C.: American Association for the Advancement of Science, 1954.

Werkman, C. H., and O. W. Wilson, eds. *Bacterial Physiology.* New York: Academic Press, 1951.

Wilson, E. B. *The Cell in Development and Inheritance.* New York: Macmillan Company, 1928.

CHAPTER III

PHYSIOLOGICAL AND CHEMICAL CHARACTERISTICS AS DETERMINATIVE PROCEDURES

By inoculating various kinds of culture media and studying the resultant physiological and chemical reactions, the bacteriologist may, by elimination, make specific diagnoses and determinations of unknown organisms. For example: Some organisms decompose carbohydrate media to produce lactic acid, gas, or alcohol. Other groups of bacteria act specifically on protein media, digesting and liquefying them. Some organisms produce a fat-splitting enzyme; others reduce nitrates to nitrites. According to the products of their chemical activity, bacteria may be identified as: 1. Chromogens—those bacteria which produce soluble or insoluble color pigments. 2. Photogens—sea bacteria, which cause putrefaction and produce slight phosphorescence. 3. Zymogens—bacteria which produce fermentation of carbohydrates, splitting them to alcohol, or lactic and acetic acids. 4. Aerogens—gas producers.

ENZYMES PRODUCED BY BACTERIA

Many changes which bacteria bring about in a medium are the result of the action of enzymes. An enzyme is an organic catalyst produced by a living cell. Bacterial enzymes may be:

1. *Constitutive*—essential enzymes carried by the cell and secreted irrespective of the composition of the medium in which it is grown.

2. *Adaptive*—not essential enzymes; elaborated by the cell after continued transfer on a specific substrate.

3. *Extracellular*—secreted into the surrounding culture media.

4. *Intracellular*—remaining within the bacterial cell.

According to the substrates upon which they act, enzymes are classified as:

Proteolytic Enzymes. These bring about the liquefaction of gelatin, coagulated serum, fibrin, or milk curd.

Saccharolytic Enzymes. These attack sugars producing gas, acid, or alcohol. Example: lactase, maltase, etc.

Amylases. These split starch into simple sugars.

Lipases. These attack fats and fatlike substances and break them down into glycerol and fatty acids.

Zymases. These change sugars to alcohol.

Oxidases. These catalyze oxidations in which oxygen plays a part, such as alcohol to acetic acid.

Dehydrogenases. These catalyze anaerobic oxidations, which involve removal of hydrogen.

Coagulases (lab enzymes). These produce coagulation in liquid protein.

Reductases. These enzymes (catalases) change hydrogen peroxide to water and molecular oxygen.

RESULTS OF MICROBIAL GROWTH

Acid Production. In media containing milk or carbohydrates certain bacteria form lactic, acetic, butyric, formic, or proprionic acids.

Gas Production. This is generally observed in media containing a carbohydrate such as lactose, saccharose, or dextrose. The gases formed are carbon dioxide, hydrogen, nitrogen, hydrogen sulfide, ammonia, and methane.

The gas produced by a microorganism from the fermentation of a particular carbohydrate may be ascertained by:

1. *Smith Fermentation Tube.* (Fig. 7.) Some of these tubes are calibrated so that the amount of gas formed may be read off directly on the closed arm. For the study of fermentation reactions use a sugar-free broth to which 1% of pure dextrose, lactose, saccharose, maltose, or other sugars is added. (a) Fill closed arm

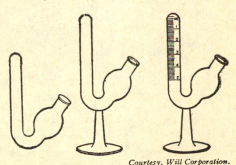

Courtesy, Will Corporation.

FIG. 7. Smith's fermentation tubes, one calibrated for volumetric determinations.

of tube and open arm partially with the medium by placing the thumb over the opening and inverting several times till the closed arm is full. (b) Stopper, sterilize, and inoculate the culture through the open arm, taking sterile precautions by flaming the opening.   (c) Gas, if produced, collects in closed arm after the tubes have been incubated for 24 to 48 hours.

2. *Dunham Tube.*   The Dunham tube, as shown in Fig. 8, consists of a test tube with a smaller inner tube.   The test tube is filled with about 5 ml. of the carbohydrate broth, and the smaller tube, mouth down, is slipped into it.   The tubes are sterilized in the autoclave under reduced temperature and pressure.   During this procedure the broth rises in the inner tube.   Then the tubes are inoculated with the culture.   Gas, if produced, collects in the small inner tube after the tubes have been incubated for 24 to 48 hours.

3. *Gas Analysis.*   A rough estimate of the amount of CO_2 and H_2 produced can be obtained.

(a) Fill the open arm of a Smith fermentation tube completely with N/5 NaOH.

(b) NaOH absorbs CO_2, and gas bubbles leave the closed arm.

(c) The gas lost consists of CO_2.

(d) The gas remaining is H_2.

4. *Shake cultures* in dextrose agar reveal gas formation, indicated by the appearance of bubbles in the agar.

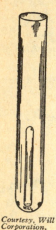

Courtesy, Will
Corporation.

FIG. 8.   Dunham's fermentation tube.   Note small gas tube inside.

Proteolysis.   The action of proteolytic enzymes is observed in the laboratory by studying bacteria that liquefy gelatin, or coagulate serum, or peptonize milk.   The liquefaction of gelatin may be used as a means of differentiation for putrefactive and decay bacteria, e.g. *Proteus vulgaris, Pseudomonas aeruginosa.*

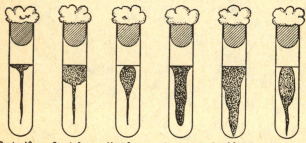

Crateriform Stratiform Napiform Saccate Infundibuliform Fusiform
TYPES OF LIQUEFACTION

FIG. 9.   Tube cultures, showing types of liquefaction of gelatin.

Alcohol Production.   Alcohol is produced by yeasts, molds, and a few bacteria when suitable carbohydrates are present.

Pigment Production.   Certain bacteria produce soluble or insoluble pigments of a characteristic color which aid in the identification of these organisms. The pigments are usually classified as follows:

1. *Carotenoids*. These are red, orange, or yellow pigments, insoluble in water and soluble in alcohol, ether, and chloroform, and produced by *Sarcina lutea, Micrococcus pyogenes* var. *aureus*, etc.

2. *Anthocyanins*. These are red, blue, and intermediate shades. They are soluble in water and alcohol, insoluble in ether, easily decolorized by reduction. An anthocyanin pigment is produced by *Actinomyces waksmanii*.

3. *Melanins*. Certain black, brown, orange, and red pigments, insoluble in almost all solvents, are produced by *Azotobacter chroococcum*, many Actinomyces.

Sulfur Bacteria or Thiobacteria. These bacteria have the power of using hydrogen sulfide which they oxidize to free sulfur, thus producing the energy necessary for their growth.

Denitrifying Bacteria. Certain bacteria as *Escherichia coli* and *Pseudomonas aeruginosa*, as well as many soil bacteria, bring about the reduction of nitrates to nitrites and ammonia (denitrification).

Nitrifying Bacteria. The process of nitrification, which is the conversion of ammonia to nitrites and nitrates, is carried out by certain organisms which help to restore the nitrate content of soils.

Decay. This may be defined as decomposition of organic compounds under aerobic conditions to form amino acids, which are further broken down to carbon dioxide, water, and ammonia. Often hydrogen sulfide and indol are produced.

Putrefaction. This is distinguished from decay in that it takes place under anaerobic conditions in the absence of oxygen and is characterized by the formation of amino acids, mercaptans, indol, skatol, and hydrogen sulfide, with an accompanying unpleasant odor. In addition, certain nitrogenous decomposition products as cadaverine, putrescine, muscarine, etc. are produced. Among the bacteria which play a part in the putrefying of organic substances are members of the Proteus and Clostridium groups.

Hydrogen Sulfide Production. The formation of hydrogen sulfide by bacteria is studied by cultivating the organisms in a peptone solution to which 0.1 ml. of a 1% solution of ferric tartrate or lead acetate has been added. The production of hydrogen sulfide is shown by the turning black of the precipitate which is formed when the peptone solution is mixed with the lead acetate.

Alkali Production. The amount of alkali produced by the growth of a bacterium may be ascertained with the use of indicators and studies of the pH of the medium at various intervals.

Reduction of Nitrates. The ability of certain bacteria to bring about the reduction of nitrates to nitrites may be determined as follows:

1. Make up a solution consisting of

Sulfanilic acid............................	8 gm.
5N Acetic acid...........................	1000 ml.

2. Make up another solution from

Dimethyl-alpha-naphthylamine...............	5 gm.
5N Acetic acid...........................	1000 ml.

3. Add 0.1 ml. of the solution of sulfanilic acid to 5 or 10 ml. of a nitrate broth culture of the bacterium to be studied.

4. Add the solution of dimethyl-alpha-naphthylamine drop by drop until a red color is observed. The formation of the red color indicates that the nitrates have been reduced to nitrites.

Indol Production. 1. Shake up 5 ml. of a 24 or 48 hour culture of the organism in peptone water or sugar-free broth with 1 ml. of ether, which causes the indol produced to go into solution.

2. After the ether rises to the top and forms a clear layer, allow 0.5 ml. of Ehrlich's reagent to flow down the side of the tube so that it forms a layer between the ether and the broth. Ehrlich's reagent is made up of:

Paradimethylaminobenzaldehyde..............	4 gm.
95% Ethyl alcohol..........................	380 ml.
Concentrated hydrochloric acid................	80 ml.

3. The appearance of a rose-red color at the junction of the ether layer and Ehrlich's reagent indicates the presence of indol. The color will spread up into the ether layer.

Cholera Red Reaction. In some instances bacteria produce nitrite from the nitrate present as an impurity in the peptone, and a red color is noted when only acid is added. A few drops of concentrated sulfuric acid are added to a two or three days old culture of *Vibrio comma* in peptone solution. The development of a red color is a positive test, and is also given by any organism, such as the colon bacillus, that both reduces nitrate and produces indol.

Methyl Red Test. This test is of importance in the differentiation of *Escherichia coli* and *Aerobacter aerogenes*. The test is based upon the pH concentration of a culture in 0.5% glucose broth after incubation at 37% C. for 4 days. The addition of methyl red to a culture results in an orange-red color or positive reaction when the pH is 4.5 or less, but in a yellow color or negative reaction when the pH is less acid. A negative reaction is given by *Aerobacter aerogenes*, while *Escherichia coli* is methyl red positive.

Voges-Proskauer Reaction. This reaction is based upon the production of acetylmethylcarbinol from dextrose and is useful to distinguish *Aerobacter aerogenes*, which gives a positive reaction, from *Escherichia coli*, which is Voges-Proskauer negative. To 10 ml. of a glucose-peptone broth culture which has been incubated for 4 days at 37° C. are added 5 ml. of 10% sodium hydroxide. A pink color is a positive test, indicating the reaction of diacetyl (from oxidation of acetylmethylcarbinol in the presence of potassium hydroxide and air) and a constituent of peptone.

Toxin Production. The formation of toxins by bacteria is generally studied by animal tests, although in the case of the toxin of *Corynebacterium diphtheriae* an *in vitro* test (Ramon Flocculation Test) is used. (See footnote on p. 218.)

ANIMAL EXPERIMENTATION

In general, animals are employed in the laboratory when culture media cannot be utilized to study certain properties of an organism, or when it has been found impossible to grow that organism (filterable viruses and a few bacteria). Specific properties, such as pathogenicity, invasiveness, and immunological response, necessitate the use of animals or chick embryos in the preparation of many viral or bacterial vaccines.

Methods of Inoculating Animals. When animals are inoculated, it is necessary to employ some means of identification, such as fastening a numbered aluminum tag to the ear, or staining with various dyes using the tail, hind and forelegs, and back in combinations. In making injections, all aseptic precautions should be carefully observed. Use a glass syringe and a hypodermic needle that have been sterilized by hot air or in boiling water. In the former method, sterilize the needle, barrel, and plunger separately in test tubes, or wrap the barrel and plunger separately in paper. In the latter method, place the needle, barrel, and plunger in a sterilizer and boil for 5 minutes immediately before using. Be careful to allow time for the needle and syringe to cool before filling. In filling the syringe, place the lumen of the needle well below the surface of the material to be injected and draw up into the barrel a little more than is necessary for the injection. Then invert the syringe to a perpendicular position and, covering the lumen of the needle with carbolized cotton if the material is infectious, force out any air bubbles that may have entered. Everything is now ready for the inoculation.   There is a variety of routes by which material **may** be introduced into an animal:

1. *Cutaneous.*   The skin is abraded and the material is placed upon it **or** rubbed into it.

2. *Intracutaneous.*   With the aid of a hypodermic needle and syringe **the** material is inoculated into the skin.   If a rabbit is used, the animal is held by **an** assistant or fastened to a board.   It is held securely, but not so that the skin **is** stretched at the point of injection.   The material is introduced between the layers of the skin, care being taken that the needle does not go through to the subcutane-ous tissue.   During the procedure, the site of injection is watched for the appear-ance of a bleb which forms as the material is being injected, if the technic is correct. (See Fig. 10 for syringes.)   For intracutaneous injections use 26 gauge needles**,** one-quarter to three-eighths of an inch long.

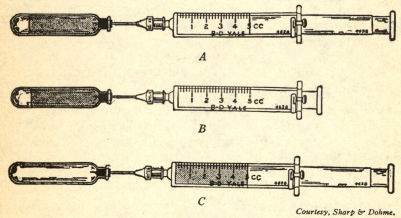

Fig. 10.   *A*. Plunger of syringe set to desired volume and needle inserted through stopper.   *B*. Air in syringe barrel has been forced into the ampule-vial. *C*. Air pressure within ampule-vial has forced solution into syringe barrel.

3. *Subcutaneous.*   The material is inoculated under the skin.   For subcuta-neous injection, use needles of 24–19 gauge, three-quarters to one and one-half inches long.   The animal should be held as for intracutaneous injection.   The material is introduced under the skin, care being taken to avoid puncturing the peritoneum, if injections are made over the abdomen.   To facilitate the procedure, the skin is raised slightly just in front of the point of insertion.

4. *Intravenous.*   The material is injected into a vein, with needles of 26–**19** gauge, three-eighths to one and one-half inches long.   When a rabbit is used, **the** animal is placed in a special box with a sliding top.   The hair is removed from **the** margin of the ear, and the area is rubbed lightly until the veins become prominent. After the area is disinfected, and it has been ascertained that all the air is out **of** the syringe, the ear is rested against the first three fingers of the left hand and held down with the thumb, and the needle, directed toward the head, is introduced into the marginal vein as nearly parallel to the vessel as possible.   A very small amount of the material is injected.   If a bleb forms, it shows that the needle **is**

not in the vein. Therefore, the needle should be withdrawn at once and inserted again nearer the base of the ear. If the needle is in the ear vein, the injection should be continued until all the material is injected. Remove the needle, wipe the ear with 70% alcohol, and if there is any bleeding, place a small piece of cotton over the point of injection.

5. *Intracardial.* The material may be injected directly into the heart in smaller animals.

6. *Intracerebral.* The material is inoculated into the brain with a hypodermic needle and syringe after an opening has been made in the skull.

7. *Intraspinal.* The material is inoculated into the spinal canal.

8. *Intraperitoneal.* The material is injected into the peritoneum.

9. *Intratracheal.* An incision is made in the trachea and the material is inoculated into the mucous membrane, or a catheter is passed through the mouth and larynx into the trachea.

10. *Intranasal.* The material is inhaled by the animal in the form of a spray or powder.

11. *Oral.* The material is fed to the animal through the mouth.

12. *Parenteral.* These injections cover any inoculations made by routes other than oral.

BIBLIOGRAPHY

Barker, H. A. *Bacterial Fermentations.* New York: John Wiley and Sons, 1956.

Bates, R. G. *Electrometric pH Determinations.* New York: John Wiley and Sons, 1954.

Clifton, C. E. *Introduction to Bacterial Physiology.* New York: McGraw-Hill Book Company, 1957.

Conn, H. J. *Manual of Microbiological Methods. Society of American Bacteriologists.* New York: McGraw-Hill Book Company, 1957.

Dubos, R. J. *Bacterial and Mycotic Infections of Man.* Third Edition. Philadelphia: J. B. Lippincott Company, 1958.

Gale, E. F. *The Chemical Activities of Bacteria.* Third Edition. New York: Academic Press, 1951.

Harris, R. J. C., ed. *Biological Applications of Freezing and Drying.* New York: Academic Press, 1954.

Hollaender, A., *et al. Radiation Biology.* New York: McGraw-Hill Book Company, 1956.

Jawetz, E., J. L. Melnick, and E. A. Adelberg. *Review of Medical Microbiology.* Los Altos: Lange Medical Publications, 1962. Pp. 226–231.

Kluyver, A. J., and C. B. van Neil. *The Microbe's Contribution to Biology.* Cambridge, Mass.: Harvard University Press, 1956.

Oginsky, E. L., and W. W. Umbreit. *An Introduction to Bacterial Physiology.* Second Edition. San Francisco: W. H. Freeman and Company, 1959.

Werkman, C. H., and P. W. Wilson. *Bacterial Physiology.* New York: Academic Press, 1951.

CHAPTER IV

MICROSCOPIC EXAMINATION OF BACTERIA

THE OPTICAL MICROSCOPE *

The microscope is the most important instrument used by bacteriologists, and its care and correct usage are vital to successful technical skill. Any difficulties encountered in using it should be instantly drawn to the attention of the instructor.

Theory of the Microscope. It is the function of the microscope to make visible those things too small to be seen by the naked eye. This is accomplished by a system of lenses of sufficient magnification and resolving power so that those small elements which are close together in a specimen examined appear to be distinctly separated, thus giving a clearly defined detailed image, when sufficient illumination is used. The total magnification of a microscope is the product of the magnification due to the objective (found engraved on the side of the objective mount) and the magnification due to the eyepiece (engraved on the eye-lens mount, generally). Thus, the total magnification of a 43 X objective and a 10 X eyepiece will be 430 X. The limit of magnification of the average light microscope is about 2000 diameters, the limiting factor being the size of the wave lengths of visible light rays.

Parts of the Microscope. Biological microscopes may be monocular or binocular. The detailed description of the microscope given here refers to the monocular microscope which has the following parts: (See Figs. 12 and 13.)

1. *Eyepiece.* This may be of different magnifications, the lowest power being used first. The exterior surface of the eye-lens and the field-lens should be carefully cleaned before using. The eyepieces should be carefully placed in the tube. Eyepieces are available which magnify from 5 to 15 times.

2. *Revolving Nosepiece.* This is supplied in most microscopes and facilitates the use of the microscope in that two or more objectives may be mounted.

3. *Objectives.* There are three kinds of objectives generally used:

---

* For details the reader is referred to *The Use and Care of the Microscope*, a booklet published by the Bausch and Lomb Optical Company (Rochester, N. Y.: 1939), from which this information was taken, and to *The Microscope*, 17th edition, by S. H. Gage (Ithaca, N. Y.: Comstock Publishing Co., 1947).

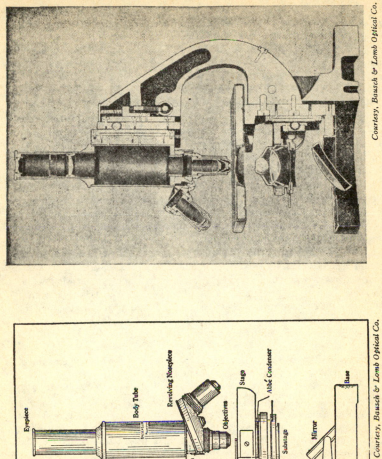

FIG. 13. Bausch & Lomb H type microscope, from photograph of actual cross section.

FIG. 12. Mechanical construction of the microscope.

45

(a) Low-power dry objective (10 X, with an equivalent focus of 16 mm.). With this objective a larger field is obtained in which the object may be more easily found and then moved to the center of the field.

(b) High-power dry objective (43 X or 60 X, with an equivalent focus of 4 mm. and 3 mm. respectively). With this objective greater magnification is obtained so that detailed structure can be studied.

(c) Oil immersion objective (magnification of 91 X, with an equivalent focus of 1.9 mm.). With this objective greater definition is obtained.

Objectives are so adjusted as to be *parfocal*, that is, so fitted to the nose-pieces that as either objective is swung into position, it is so nearly focused as to require the use of only the fine adjustment.

4. *Coarse Adjustment Button.* This button is rotated until the image appears and is used to bring the object into vision.

5. *Fine Adjustment Button.* This button is rotated until the image becomes clear and well defined. It is used to obtain a more sensitive and slower adjustment for focusing through the different planes or depths of the object.

6. *Stage.* This is the place where the slide or object to be examined is put and kept secure with the use of spring clips.

7. *Mirror.* This is usually double, plane on one side and concave on the other. In using sky-light as a source of illumination, it does not matter whether the plane or concave mirror is used. It is easier to eliminate the window bar images when the plane mirror is used. When, however, small artificial light sources are employed, the concave mirror should be used. With daylight illumination, the plane mirror should be used; with artificial illumination, the plane mirror should be used for the low-power objectives, and the concave mirror for the higher magnifications.

8. *Substage Diaphragm.* The diaphragm is used to modify the amount of light so that definition of the object will be obtained.

9. *Substage Condenser.* With objectives of magnification higher than 10 X, the ordinary illuminating mirror is not large enough to provide an illuminating cone of light of sufficient angle to fill the aperture of the objective. This is done by introducing a suitable lens or lens system between the mirror and the slide. Such lenses are called substage condensers, or, frequently, Abbé condensers, and are used only with the plane mirror.

Use of the Microscope. The use of the microscope involves the following steps:

1. Place the eyepiece into the tube carefully, always using the lowest power first.

2. Attach the objectives to the revolving nosepieces. To change one objective to another, grasp two of the objectives between the thumb and forefinger of the right hand and rotate the revolving nosepiece until the desired objective is brought into line with the axis of the body tube. Exact alignment is important, and is indicated by a slight click which manifests the stop for each objective.

3. Adjust the position of the mirror so that the proper illumination is obtained. Avoid direct sunlight. Diffuse daylight is the best source of illumination, although many forms of microscope lamps are available. If daylight is used, place the microscope as near directly in front of a north window as possible. This is the

preliminary adjustment of illumination; the exact adjustment of illumination can be effected only after the microscope has been focused on the object.

4. Place the slide upon the stage under the spring clips and move to a point where the object comes as nearly as possible over the center of the opening of the stage.

5. *Focusing.* To focus a microscope is to adjust the relation between the optical system and the object so that a clear image is obtained. With high-power objectives the distance between the cover-glass and the front lens of the objective is so short that unless the operation of focusing is conducted with care and skill there is danger of damaging the specimen, the objective, or both. In lower-power objectives the danger is less because of the greater working distance (the distance from slide to objective).

It is advisable while watching for the image to appear to move the object slowly in different directions, as the flitting of shadows across the field will give indication that the objective is nearing the focal point. It will be noted upon moving the object to the left that there is an apparent movement to the right of the field. This is because the image in the eyepiece is reversed in position from that of the object.

(a) Lower the head to the level of the stage, to be able to see the bottom of the objective. (b) Lower the tube by the coarse adjustment until the bottom of the objective is within one-quarter inch of the object. (c) Look through the eyepiece and slowly elevate by the coarse adjustment until the image is distinct. (d) Then use the fine adjustment until the greatest visibility is obtained. (e) Having focused with the low-power objective, switch to the high-power objective by revolving the nosepiece.

6. *Oil Immersion Objectives.* Immersion contact between the objective and cover-glass is made with cedar oil. This oil is specially prepared to have the same refractive index as glass. Great care should be used to keep it free from dust.

(a) Apply a small quantity of oil to the outer lens of the objective or to the cover-glass, using the rod in the oil bottle.

(b) Lower the objective very carefully with the coarse adjustment until contact is made. This can be best determined by watching the space between the objective and slide with the eye well down to the level of the stage. At the instant contact is made a flash of light will illuminate the oil. When the flash is seen, the objective will be near enough to focus by means of the fine adjustment.

Care of the Microscope.

The following rules should be followed to keep the microscope at its maximum efficiency and in the best working condition:

1. Keep the microscope free from dust. When the microscope is not in use it should be kept in its case or covered with a bell jar or close-meshed cloth such as cotton flannel. Remove any dust that may settle with a camel's hair brush and then wipe carefully with a chamois skin.

2. When handling the stand, grasp it by the pillar or handle arm.

3. Avoid sudden jars.

4. Remove any Canada balsam or cedar oil which may adhere to the stand with a cloth moistened with xylol. Wipe dry with chamois.

5. Avoid excessive use of xylol on lenses because this loosens their attachments. Xylol should be used on the oil immersion lens only when it is thickly

covered with cedar oil so as to obstruct vision. As a rule, wiping off the oil with lens paper each time the oil immersion lens is used is sufficient.

6. To use the draw tube, impart the spiral motion.

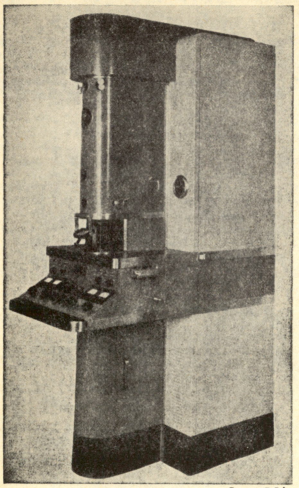

Courtesy, R.C.A.

Fig. 14.   Universal Model electron microscope.

7. Keep the coarse adjustment free from dust. The tube should occasionally be withdrawn from the arm and the slides carefully wiped with a cloth moistened with xylol. Lubricate by applying a small quantity of paraffin oil to a cloth and wiping well over the surfaces, removing the superfluous amount with a dry cloth. The teeth of rack or pinion should never be lubricated.

8. At regular periods unscrew the eye-lens and field-lens and clean the inner surfaces. Keep the objectives clean and free from dust.

THE ELECTRON MICROSCOPE

The recently developed electron microscope (Fig. 14), which utilizes a beam of electrons rather than a beam of visible light, makes possible a magnification up to 100,000 times. The resolving power is so considerably greater than that of the ordinary optical microscope because of the fact that fast-moving electrons possess a wave length much shorter than that of visible light.

In the electron microscope, a heated tungsten filament is used as the source of electrons, and shaped electric and magnetic fields replace the glass lenses of the optical microscope. The stream of electrons enters the magnetic condenser and is concentrated on the object, which intercepts the stream to cast a "shadow." This is recorded on a photographic plate.

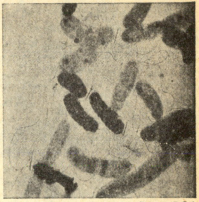

Courtesy, R.C.A.

FIG. 15. Electromicrograph of *Vibrio schuylkilliense*, found in streams. ($\times$ 23,000.)

Courtesy, R.C.A.

FIG. 16. Electromicrograph of *Mycobacterium tuberculosis var. hominis.* ($\times$ 42,000.)

The electron microscope opens up wide possibilities for observing those bacteria and viruses whose size is beyond the limits of the optical microscope and for viewing bacterial structure in greater detail. Living organisms cannot be studied with this instrument at the present time, however, because objects must be prepared as a dry film on collodion and observed in a high vacuum.

MICROSCOPIC METHODS IN STUDY OF BACTERIA

Material may be examined for microorganisms either in the living and unstained state, or after stains have been used to color the microorganism to bring it out more clearly.

Examination of Living Bacteria in Hanging Drop. Bacteria are studied in hanging drop preparations to observe motility, binary fission, and other characteristics of the living organism which may be destroyed upon staining.

1. Using a toothpick, put a little petrolatum around the concavity on a special hanging drop slide (hollow ground slide).

FIG. 17.   Hanging drop preparation.

2. If the culture medium is solid, a small amount of the colony growth on it is first emulsified with a little water.   Then a drop of this liquid containing some of the culture is placed in the center of the cover-slip with an inoculating loop.
3. Lower the prepared slide over the cover-slip so that the drop is centered in the concavity of the slide.
4. Invert the slide and adjust the cover-slip gently over the depression in the slide so that the drop hangs suspended in the concavity, sealed with petrolatum. This prevents evaporation.
5. With the diaphragm closed about two-thirds, examine with the low-power lens to find the hanging drop.
6. Center it and then lower the high-power lens gradually, focusing until the organisms come into view.
7. To remove the cover-slip when the examination is completed, the hanging drop slide should be placed in a solution of bichloride of mercury and the cover-slip floated off.

Examination of Living Bacteria in Hanging Block (Hill Method). The hanging block method makes it possible to study cell division of bacteria.

1. Pour nutrient agar into a Petri dish and allow to solidify.
2. Cut out a piece about one-quarter of an inch square.
3. Place the square on a sterile slide.
4. Spread bacteria over the surface of the block and cover with a sterile dish, allowing the block to dry for a few minutes in the incubator.
5. Place a sterile cover-slip on the surface of the block and seal the edges with agar.
6. Remove the block and cover-slip from the slide and place over a moistened hollow slide and seal with paraffin.
7. Place the preparation upon the stage of a microscope and examine.

Dark Field Illumination. For the study of living bacteria and especially spirochetes which are difficult to stain, the use of dark field illumination is an important microscopic method.   The object is illuminated in such a way that no direct light reaches the objective, but only that which is refracted, diffracted, or scattered by the object itself.   In this way an object is seen as a bright image on a dark background.   Dark field illumination is obtained by blocking out the central portion of the Abbé condenser in the microscope substage by placing a dark ground stop in the carrier underneath the condenser.   The light source should be either an arc lamp with condenser or the convertible substage lamp.   The light should be adjusted to produce a parallel beam of light.   This is done by directing the light beam at a wall at least six feet away and then focusing it until an image of the source is formed on the wall.   With the Abbé condenser and dark ground stop, use only low-power objectives of low numerical aperture.   The 16 mm. and 8 mm. objectives are recommended for this form of dark field illumination.   The oil immersion objective in dark field observations gives higher magnifications than can be secured with dry objectives.

1. Select a slide of 1.3 to 1.4 mm. thickness and mount the material to be examined in water or some other liquid and cover with a cover-slip.   Seal edges of cover-slip with thick oil or shellac.

2. Place a drop of cedar oil on the bottom of the slide just under the preparation.

3. Place another drop on the cover-slip.

4. Place the slide carefully on the microscope, so that the bottom drop of oil spreads evenly without bubbles between the bottom of the slide and the surface of the dark field condenser.   The oil will contact with the condenser and the object slide will then become luminous.

5. Focus on the material under the cover-slip until the smallest possible disc of light is procured.

6. Examine with the oil immersion objective.

Examination of Bacteria in Fixed Preparations.

1. With a sterile needle or loop, mix a small part of a culture in a drop of sterile water and spread over the surface of a slide or cover-slip in a thin film. Cultures in liquid media are applied directly.

2. Dry the film in air.

3. Fix the film by passing the slide through a Bunsen flame three times with the film side up.   This causes the bacteria to adhere to the slide so that the film is not easily displaced by the stain.   Immersion in methyl alcohol, formalin, or other fixatives may be used instead of heat, in which case the preparation should be washed with water before applying the stain.

4. The stain to be used (generally, any of the basic aniline dyes as methylene blue, gentian violet, methyl violet, etc.) is put on the surface and allowed to act for one-half to one and one-half minutes, depending upon the stain employed.   The exact time can be determined only with practice—this is influenced by the concentration of the stain and the organism being stained.

5. Wash off excess of stain with water.

6. Air dry or blot preparation with absorbent paper.

7. Examine under the microscope with the oil immersion lens.

8. If the slide is to be kept permanently, put a drop of Canada balsam on slide, cover with cover-slip, and press down to exclude all air between slide and cover-slip.

Preparation of Blood Smear.

1. The blood is generally obtained from the fingertip or ear lobe.

2. Cleanse the part thoroughly and dry with absorbent cotton.

3. Prick the skin with a sterile needle.

4. Discard the first drop of blood.

5. Touch the second drop of blood with the middle of one of the narrow edges of a microscopic slide, free from grease and dust. (To ensure thorough cleansing, slides are washed with soapsuds solution or placed in a solution of potassium dichromate and sulfuric acid, rinsed with distilled water, and kept in alcohol until required.)

6. Place the narrow edge of a clean slide on the drop of blood and when the blood has spread out in the angle between the slides, push or draw the first slide rapidly along the second.

7. Dry the film as rapidly as possible by waving in the air.

STAINS AND STAINING METHODS

General Considerations.

1. *Staining* is the process of artificially coloring microorganisms with dyes or reagents in order to facilitate their study under the microscope.

2. A *dye* is an organic compound consisting of benzene rings with chromophore and auxochrome groups.

3. *Chromophore*, or color radical, is any chemical group which gives a specific color to a compound uniting with an auxochrome group to form a dye.

4. *Auxochrome group* furnishes salt-forming properties and is responsible for transferring the color of a dye to a substance or material upon which it acts (the process known as dyeing).

5. The type of auxochrome group divides dyes into two classes:

(a) *Basic* dyes have the auxochrome group NH_2 or $N(CH_3)_2$.

(b) *Acid* dyes have the auxochrome group SO_3H or $COOH$ or OH.

6. A *mordant* is any substance which will fix the stain so that the material will retain the stain. Example: ferrous sulfate, tannic acid, iodine, etc.

7. Stains may be:

(a) *Selective stains* for capsules, metachromatic granules, spores, flagella, cytoplasm, and the nucleus in pathogenic protozoa.

(b) *Differential stains*, e.g., Gram stain.

(c) *Direct stains*, e.g., Löffler's methylene blue stain or any of the aniline dyes, which actually color the organism studied.

(d) *Indirect stains*, e.g., nigrosin or India ink, which colors the background and not the organism, which can be seen by contrast.

8. Bacterial stains are usually dilute aqueous solutions prepared from saturated and filtered alcoholic solutions.

9. Dilute carbol fuchsin, methylene blue, crystal violet, and safranin are among the simple staining solutions most commonly used.

Use of Dyes in the Laboratory. 1. To stain bacteria to make them visible under the microscope. 2. To show the structure of the organism studied. 3. For identification of organisms, e.g., acid-fast bacteria, Gram-negative bacteria. 4. To inhibit the growth of certain bacteria so that other microorganisms growing simultaneously may be studied. Crystal violet in certain concentrations inhibits the growth of Gram-positive cocci, as does gentian violet. Brilliant green is added to media to inhibit the growth of *Escherichia coli* which is usually present in large numbers in stools, thus aiding the study of *Salmonella typhosa* which may be present in smaller numbers and is not inhibited by the brilliant green.

Gram Stain. The most useful stain in bacteriology is *Gram's Differential Stain*. Practically all bacteria can be classed as Gram-positive or Gram-negative, depending upon whether the original stain is fixed to the organism. In Gram's method the following solutions are employed:

1. Initial stain (Gram's stain): A solution of a pararosaniline dye —gentian violet, methyl violet, crystal violet, Victoria blue—in distilled water.

2. Gram's iodine, which is prepared by dissolving 1 gm. of iodine and 2 gm. of potassium iodide in 300 ml. of distilled water. This solution should be alkaline, for the presence of acid causes Gram-positive bacteria to become Gram-negative.

3. Decolorizer: 95% alcohol, acetone, or a mixture of equal parts of acetone and alcohol.

4. Counterstain: A 1:20 dilution of carbol fuchsin, a 0.2% aqueous solution of safranin, or a 0.1% solution of Bismarck brown. Usually carbol fuchsin or safranin is used.

Technic.

1. Prepare film, dry, and fix as described on page 51.
2. Flood slide with gentian violet.
3. Add three or four drops of a 5% solution of $NaHCO_3$ and allow to stand for 2 to 3 minutes.

4. Rinse with water quickly until no more stain comes off.

5. Cover with Gram's iodine solution for 1 to 2 minutes.

6. Rinse in water and remove excess by blotting.

7. Decolorize with 95% alcohol until the color ceases to run from preparation, usually from 30 seconds to 1 minute.

8. Counterstain with carbol fuchsin, safranin, or Bismarck brown for 30 seconds.

9. Rinse in water, blot or air dry, and examine.

Bacteria retaining the original stain (violet) are *Gram-positive*, while those which are decolorized by the alcohol and take the counterstain (red or brown) are *Gram-negative*.

Principle of Gram Stain.

1. The initial stain stains practically all bacteria.

2. The iodine is a mordant which fixes the stain in Gram-positive organisms.

3. The organisms which are decolorized by 95% alcohol are Gram-negative.

4. Gram-negative organisms will take the counterstain but Gram-positive bacteria will remain violet.

5. Knaysi believes:

(a) Gram-positive material adsorbs more of the initial stain than does Gram-negative material, and holds it more firmly.

(b) When the mordant is added, a compound is formed that is insoluble in water and sparingly soluble in the decolorizing agent.

(c) The rate at which the cell is decolorized is determined by the quantity of this precipitate formed in the cell and by the permeability of the cell wall to the dye-mordant complex. Gram-positive cells will be decolorized more slowly.

6. According to Churchman, certain Gram-positive bacteria consist of a Gram-negative medulla which is surrounded by a Gram-positive cortex. The Gram-positivity of an organism is attributed to the presence of the cortex, since in its absence the organism is Gram-negative.

7. Other explanations given are that Gram-positivity is due to the presence of unsaturated fatty acids which have a high affinity for iodine, or to differences in permeability of the cell wall.

Stains for Metachromatic Granules (Corynebacterium Diphtheriae).

1. *Löffler's Methylene Blue.* Of a saturated alcoholic solution of methylene blue 30 ml. are mixed with 100 ml. of a 1:10,000 solution of KOH in distilled water. The alkali renders the stain more penetrating and intensifies its staining power.

2. *Neisser Stain.*

(a) Prepare Solution No. 1 as follows:

Methylene blue......................	1 gm.
95% Alcohol........................	20 ml.
Acetic acid (glacial)..................	50 ml.
Distilled water......................	930 ml.

(b) Prepare Solution No. 2 as follows:

Crystal violet.......................	1 gm.
95% Alcohol........................	10 ml.
Distilled water......................	300 ml.

(c) Prepare Solution No. 3 by dissolving 1 or 2 gm. of chrysoidin in 300 ml. of hot water.

(1) Stain films for 10 seconds in a mixture of 2 parts of Solution No. 1 and 1 part of Solution No. 2.

(2) Wash the film.

(3) Stain for 10 seconds in Solution No. 3.

(4) Wash quickly in water and blot.

(5) Examine under oil immersion lens.

The bacilli either appear entirely brown or show a dark blue round body at both ends.

Capsule Stains.

1. *Welch's Method.*

(a) Prepare slides as usual and air dry.

(b) Cover with glacial acetic acid for a few seconds.

(c) Pour off acetic acid.

(d) Cover with aniline water-gentian violet, renewing stain several times until the acid is all gone, usually three or four times, finally leaving it on for 3 minutes.

(e) Wash in 2% salt solution and examine.

The capsule stains a pale violet.

2. *Gin's Method.*

(a) Place a drop of India ink (diluted with an equal amount of sterile distilled water) near one end of a clean slide and carefully mix in a loopful of bacterial suspension.

(b) Spread mixture across slide evenly with the edge of a second slide.

(c) Dry in air; fix with heat or by dipping in methyl alcohol.

(d) Stain with carbol fuchsin or gentian violet for 1 to 2 minutes.

(e) Wash in water, dry, and examine.

The bacteria will be stained; the capsules unstained with their margin delineated by the ink.

3. *Hiss's Copper Sulfate Method.*

(a) Add a drop of animal serum to organisms on slide and make film.

(b) Air dry and fix by heat.

(c) Stain with a saturated alcoholic solution of fuchsin or gentian violet prepared by adding 5 ml. of the dye to 95 ml. of distilled water.

(d) Allow to act for a few seconds and hold over a free flame for a second until it steams.

(e) Wash off with 20% aqueous copper sulfate.

(f) Blot dry and mount.

The organism with a capsule appears as a dark purple body with a faint blue capsule around it.

4. Wadsworth's Method.

(a) Prepare smear.

(b) Fix in 40% formalin for 2 to 5 minutes.

(c) Wash in water for 5 seconds.

(d) Cover with aniline gentian violet for 2 minutes.

(e) Then add Gram's iodine solution for 2 minutes.

(f) Decolorize with 95% alcohol.

(g) Cover with a dilute aqueous solution of fuchsin.

(h) Wash in water for 2 seconds, dry, and examine.

Spore Stains.

1. General Method for Staining Spores.

(a) Make smear, dry, and fix with heat.

(b) Stain with carbol fuchsin, steaming for 5 minutes by heating over the Bunsen flame until vapor rises.

(c) Decolorize with 5% acetic acid until film assumes a light pink color. This takes a few seconds.

(d) Stain for 3 minutes with Löffler's methylene blue.

(e) Wash with water.

(f) Air dry or blot and examine.

The bacteria are blue and the spores are red.

2. Dorner's Method.

(a) Add an equal quantity of freshly filtered carbol fuchsin to a heavy suspension of the organism in distilled water in a test tube.

(b) Place in a boiling water bath for 10 to 12 minutes.

(c) Mix a loopful of the stained preparation and a loopful of a saturated aqueous solution of nigrosin (10 gm. nigrosin in 100 ml. distilled water boiled for 30 minutes and filtered) on a slide.

(d) Smear and dry quickly.

The bacteria are almost colorless against a dark gray background, and the spores are red.

Flagella Stains. For the examination of bacteria for flagella it is necessary that slides and cover-slips be free from grease and scrupulously clean. A young agar culture is prepared by transferring a loopful of growth from an 18-hour agar slant culture of the organism to 2 ml. of sterile tap water, and incubating at 37° C. for 10 to 15 minutes. A loopful is transferred to a clean slide, and the film dried in air.

1. *Casares-Gil's Method.*

(a) Prepare a suspension of a young agar culture of the organism to be stained.

(b) Make smear of the bacteria with caution not to break up organisms.

(c) Dry in air.

(d) Pour on mordant which is prepared just before use as follows:

Tannic acid	10 gm.
Aluminum chloride	18 gm.
Zinc chloride	10 gm.
Rosaniline hydrochloride	1.5 gm.
60% alcohol	40 ml.

For use, dilute with two parts distilled water, filter, and collect filtrate on smear.

(e) Allow to act for 1 to 2 minutes. Note precipitate and metallic sheen that forms.

(f) Wash with distilled water.

(g) Cover with carbol fuchsin for 1 to 2 minutes.

(h) Wash in distilled water, dry without blotting, and examine.

2. *Löffler's Method.*

(a) Prepare film, dry in air, and fix by heat.

(b) Cover with mordant solution prepared as follows:

20% aqueous tannic acid	10 parts
Saturated aqueous solution of ferrous sulfate	5 parts
Saturated alcoholic fuchsin solution	1 part

(c) Allow to act for 30 seconds to 1 minute, gently heating, without boiling.

(d) Wash in water.

(e) Pour on a 5% alkaline aniline water solution of fuchsin or gentian violet.

(f) Warm gently and allow to act for 1 to 2 minutes.

(g) Wash in water and dry in air.

The stains should be filtered directly on to the slide when they are added.

3. *Van Ermengen's Method.*

(a) Prepare Solution No. 1 as follows:

20% tannic acid solution	60 ml.
2% osmic acid solution	30 ml.
Glacial acetic acid	4–5 drops

(b) Prepare Solution No. 2 as follows:

Gallic acid	5 gm.
Tannic acid	3 gm.
Fused potassium acetate	10 gm.
Distilled water	350 ml.

(c) Place preparation in Solution No. 1 for 1 hour at room temperature or for 5 minutes at 100° C.

(d) Wash in water and then in absolute alcohol.

(e) Cover with a 0.25–0.5% solution of silver nitrate for 1 to 3 seconds.

(f) Without washing, cover with Solution No. 2 for a few minutes.

(g) Cover with silver nitrate again until preparation turns black.

(h) Wash in water, blot, and examine.

Stains for Acid-Fast Bacteria. Acid-fast bacteria are those organisms which are not decolorized by acid once they have been stained. This ability to resist decolorization by acid has been associated with the presence of a high content of lipoidal substances. When such organisms are treated with fat solvents, such as chloroform to which minute amounts of hydrochloric acid have been added, the acid-fastness is removed.

1. *Ziehl-Neelsen Method.*

(a) Prepare a solution of carbol fuchsin by mixing 10 ml. of a saturated alcoholic solution of basic fuchsin in 90 ml. of a 5% aqueous phenol solution.

(b) Prepare a decolorizer by mixing 3 ml. of concentrated hydrochloric acid with 97 ml. of 95% alcohol.

(c) Cover preparation with carbol fuchsin and steam over Bunsen flame for 5 minutes without boiling. Add stain from time to time to prevent drying.

(d) Wash with water and decolorize with acid alcohol.

(e) Stain with methylene blue for 10 to 30 seconds.

(f) Wash with water, blot dry, and examine.

Mycobacterium tuberculosis in sputum stained by this method appears red against a blue background. Bacteria which are not acid-fast are stained blue.

2. *Pappenheim's Method.* This procedure is used to differentiate *Mycobacterium lacticola* (*Bacillus smegmatis*) from *Mycobacterium tuberculosis* when the former is found in urine, feces, or sputum. Decolorization and counterstaining are accomplished by one solution.

(a) Prepare smears and fix by heat.

(b) Stain with carbol fuchsin and steam for 2 minutes.

(c) Pour off dye without washing and cover with mixture of:

Rosolic acid........................   1 gm.
Absolute alcohol....................   100 ml.
Methylene blue added to saturation
Glycerin...........................   20 ml.

(d) Pour mixture on and off four or five times.

(e) Wash with water, blot, and examine.

Mycobacterium lacticola is decolorized by the mixture of alcohol and rosolic acid. *Mycobacterium tuberculosis* is stained red.

3. *Baumgarten's Method.* This method is used to differentiate *Mycobacterium tuberculosis* from *Mycobacterium leprae.*

(a) Prepare films and fix by heat.

(b) Pour on alcoholic fuchsin for 5 minutes.

(c) Pour off dye and, without washing films, add mixture of 10 parts of 95% alcohol and 1 part of nitric acid.

(d) Wash off films, blot, and examine.

Mycobacterium tuberculosis does not take up the dilute alcoholic fuchsin as readily as does *Mycobacterium leprae,* and therefore is stained blue, while the latter is stained red.

Polychrome Stains. These stains are usually mixtures of eosin and methylene blue and are used for the staining of pus and exudates in which the bacteriologist observes the relation of bacteria to cellular elements. They are also widely used as blood stains.

1. *Wright's Stain.* This stain is prepared by heating in a steam sterilizer 0.9 gm. of methylene blue hydrochloride and 100 ml. of a 0.5% aqueous solution of sodium carbonate. The mixture is cooled and filtered. To filtrate is added:

Eosin Y........................... 1 gm.
Distilled water..................... 500 ml.

After thorough mixing, the mixture is filtered and the precipitate is dissolved in the proportion of 0.1 gm. in 60 ml. of neutral, acetone-free, absolute methyl alcohol. The stain should stand for a day or two and then should be filtered.

(a) Prepare film and cover with stain for 1 minute.

(b) Dilute by adding equal amount of distilled water. A metallic sheen forms on the surface.

(c) Allow to act for 3 to 15 minutes.

(d) Wash in distilled water by flooding slide.

(e) Blot and examine.

2. *Giemsa's Stain.* This stain is prepared for use in staining blood smears as follows:

Azure II eosin...................... 3 gm.
Azure II........................... 0.8 gm.
Glycerin........................... 250 ml.
Neutral, acetone-free, methyl alcohol... 250 ml.

For tissues the same ingredients are used, except that 125 ml. of glycerin and 375 ml. of methyl alcohol are added instead of the above amounts.

(a) Prepare film and cover with stain for 1 minute, applying in same manner as described above for Wright's stain.

(b) If the stain is diluted 1 to 10 with water, the preparations previously fixed in methyl alcohol should be allowed to remain in the diluted stain for 6 to 18 hours.

(c) Wash off stain, dry by blotting, and examine.

Stains for Spirochetes.

1. *Burri's India Ink Method.*

(a) Prepare a 1:10 solution of India ink in water which is sterilized in the autoclave for 15 minutes.

(b) Mix a loopful of the bacterial suspension with an equal amount of India ink solution, and prepare film.

(c) Dry and examine.

The spirochetes or bacteria appear white upon a dark field. This method is generally employed for organisms with poor staining properties.

2. *Fontana-Tribondeau Method.*

(a) Prepare Solution No. 1.

Acetic acid (glacial)...................	1 ml.
40% formalin........................	10 ml.
Distilled water......................	100 ml.

(b) Prepare Solution No. 2.

Tannic acid.........................	5 gm.
1% phenol..........................	100 ml.

(c) Prepare film and dry in air.

(d) Cover with Solution No. 1 for 1 minute.

(e) Wash with distilled water, and cover with Solution No. 2.

(f) Heat until fluid steams.

(g) Wash with distilled water and cover with a 5% ammoniacal silver **nitrate** solution.

(h) Heat until fluid steams and allow to act for 30 seconds.

(i) Wash with water, dry in air or blot.

Spirochetes are stained a dark brown or black.

Stain for Negri Bodies.   Williams' Modification of Van Gieson's Method.

1. Prepare smear of brain tissue by pressing the tissue on a slide with another slide.

2. Fix slide for 10 seconds in neutral methyl alcohol to which 0.1% picric acid has been added.

3. Remove slide and blot.

4. Flood slide with the following solution:

Saturated alcoholic solution basic fuchsin...	0.5 ml.
Saturated alcoholic solution methylene blue..	10.0 ml.
Distilled water.........................	30.0 ml.

5. Heat to steaming, and then wash film in tap water.

6. Blot and examine.

Stain for Protozoa.   Heidenhain's Iron-Hematoxylin Stain.

1. Prepare film and fix for 15 to 20 minutes in the following solution **heated** to 60° C.

Saturated mercuric chloride solution....	2 parts
Absolute alcohol......................	1 part
Glacial acetic acid.... 5 ml. to 100 ml. of mixture	

2. Rinse in 50% alcohol, and then place successively into:

70% alcohol-iodine....................	2 min.
70% alcohol..........................	2 min.
50% alcohol..........................	2 min.

3. Rinse in water, and place for 10 minutes in mordant consisting of 2 to 4% iron alum heated to 30° C.

4. Wash in running water for about 5 minutes.

5. Stain for 30 minutes in 0.5% Heidenhain's hematoxylin at 30° C.

6. Wash in running water for 5 minutes.

7. Decolorize in 2 to 4% iron alum, watching under the microscope until **nuclei** are clearly visible.

8. Wash in running water for 5 minutes and then immerse successively in the following:

50% alcohol	2 min.
70% alcohol	2 min.
85% alcohol	2 min.
95% alcohol	5 min.
Absolute alcohol	5 min.
Absolute alcohol and xylol	5 min.
Xylol	5 min.

9. Apply Canada balsam and mount.

Negative Staining. Used for the studying of spirochetes, this method involves staining the background with nigrosin, which does not penetrate the bacterial cell.

1. *Preparation of Stain.*

(a) Mix 10 gm. nigrosin with 100 ml. water and boil 30 minutes.

(b) Add 0.5 ml. formalin.

(c) Filter through filter paper and store in 2 ml. amounts.

2. *Examination.*

(a) Place a loopful of suspension on a slide.

(b) Add an equal amount of the nigrosin mixture, spread, and make a thin film (avoid a thick film, which may crack).

(c) Air dry with no heat or moisture. Examine as in Burri's India ink method.

BIBLIOGRAPHY

Benford, J. R. *The Theory of the Microscope.* Rochester: Bausch and Lomb Optical Company, 1952.

Bennet, A., *et al. Phase Microscopy.* New York: John Wiley and Sons, 1951.

Conn, H. J. *Biological Stains.* Sixth Edition. Baltimore: Williams & Wilkins Company, 1953.

Hall, C. E. *Introduction to Electron Microscopy.* New York: McGraw-Hill Book Company, 1953.

Jawetz, E., J. L. Melnick, and E. A. Adelberg. *Review of Medical Microbiology.* Los Altos: Lange Medical Publications, 1962.

Schaub, I. G., M. K. Foley, E. G. Scott, and W. R. Bailey. *Diagnostic Bacteriology.* Fifth Edition. St. Louis: C. V. Mosby Company, 1958.

Society for General Microbiology. *Constituents of Bacteriological Culture Media* Cambridge: Cambridge University Press, 1956.

Wadsworth, Augustus B. *Standard Methods of the Division of Laboratories and Research of the New York State Department of Health.* Third Edition. Baltimore: Williams & Wilkins Company, 1947.

CHAPTER V

PREPARATION AND INOCULATION OF CULTURE MEDIA

GENERAL CONSIDERATIONS

Any material in which microorganisms find nourishment and in which they can reproduce is called a *culture medium*. The growth or crop of organisms obtained in such a medium is designated a *culture*. When microorganisms in a culture are all of the same species, the culture is commonly referred to as a *pure culture;* when there are two or more different species of microorganisms growing in a medium the culture is called a *mixed culture*. If a culture accidentally contains more than one species of bacteria it is called a *contaminated culture*.

Cultivation of Bacteria. Bacteria may be grown in any one of the following types of glassware after these have been cleaned and sterilized: (1) test tubes, (2) Petri dishes, (3) Florence and Erlenmeyer flasks, (4) fermentation tubes, and (5) other tubes which may be easily manipulated. Into these appliances are placed culture media in which is incorporated the necessary nutritive material, with moisture and osmotic relations, surface tension, oxygen requirements, etc., best adapted to the specific needs of the bacteria being cultured.

Media Requirements. Many bacteria require only the ordinary culture media. Some need special media for growth (see page 20) or for aiding in identification, where the cultural characteristics and specific reactions may be the determinative factors. Media for autotrophic bacteria are generally prepared from entirely inorganic materials, the selection depending partly on the substance required by the species as a source of energy.

Many culture media contain meat macerated in water usually as commercially prepared extracts (Armour's or Liebig's beef or pork) together with peptone and common salt. If solid media are required for studying colony characteristics, isolating species, etc., substances which cause the liquid to coagulate in the form of a jelly, either agar or gelatin, are added. Agar is a complex mixture of much carbohydrate and a little protein obtained from Ceylon moss; gelatin is a protein substance lacking essential amino acids.

Types of Media. The various media may be identified as:

1. *Solid Media.* These are media which consist essentially of agar, gelatin, or albumin. *Agar* dissolves in water near the boiling point; this hot solution then "jells" or "sets" at about 38° C. and does not liquefy again until reboiled. Agar is generally added in a concentration of 2 to 3%. *Gelatin* is usually added to culture media in

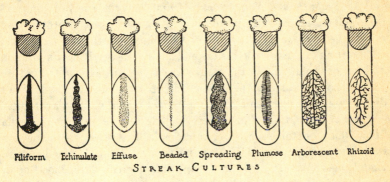

Filiform   Echinulate   Effuse   Beaded   Spreading   Plumose   Arborescent   Rhizoid

STREAK CULTURES

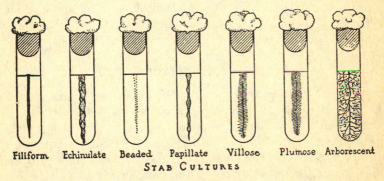

Filiform   Echinulate   Beaded   Papillate   Villose   Plumose   Arborescent

STAB CULTURES

FIG. 18. Tube cultures, showing (a) streak slants and (b) anaerobic stab culture growths.

proportions of 10 to 15%, which solidifies the media only at temperatures below 25° C. (the melting point of gelatin); above this temperature the media become liquid again, which is an advantage in the isolation of some bacteria. Gelatin is a diagnostic medium because some bacteria liquefy gelatin, and others do not. *Albumin*, unlike gelatin and agar which melt upon heating, is coagulated by heat, forming a medium which has a favorable amount of food and a desired degree of firmness for certain bacteria—e.g., Löffler's blood serum for *Corynebacterium diphtheriae.*

2. *Semisolid Media.* These are prepared with agar added in a proportion of 1 to 1.5%, or with a mixture of agar and gelatin. They are especially useful for keeping routine stock cultures alive for long periods of time.

3. *Fluid Media.* These are prepared without substances like gelatin or agar, and therefore are not solidified or jelled.

4. *Synthetic and Nonsynthetic Media.* On the basis of composition, culture media may be divided into two classes, synthetic and nonsynthetic. A *synthetic medium* is one in which the exact chemical composition of the ingredients is known. Example: Ringer's solution and Locke's solution. In the case of a *nonsynthetic medium* the precise composition of some or all of the nutritive substances used is not definitely known. Example: Meat extract broth, vitamin agar, etc.

5. *Differential Media.* These include a wide variety of media upon which organisms grow characteristically so that their isolation is facilitated. Examples of such media are: Löffler's blood serum, Endo's agar, Wilson-Blair medium, brilliant green agar, blood agar, chocolate agar, Krumwiede's triple sugar medium.

6. *Dehydrated Media.* These are media in dried powdered form ready to be prepared into various liquid or solid media by merely restoring the moisture with distilled water, and sterilizing.

TITRATION OF CULTURE MEDIA

Variations in the acidity or alkalinity of culture media affect the growth of bacteria, since each microorganism has a definite pH range within which the maximum growth occurs. Excessive changes inhibit growth and alter cultural characteristics. Therefore, it is necessary to titrate media in order to ascertain the concentration of hydrogen ions present. A pH of 7 signifies a medium that is neutral; below 7, acid; and above 7, alkaline.

Definition of pH. The hydrogen ion concentration (pH) is expressed as the logarithm of the reciprocal of the hydrogen ion concentration [H$^+$]. The pH value of 7, which is the hydrogen ion concentration of pure water, is arrived at as follows: Water dissociates according to the equation:

$$HOH \rightleftharpoons H^+ + OH^-$$

From the law of mass action, it follows that:

$$\frac{[H^+] \times [OH^-]}{[HOH]} = K, \text{ the ionization constant.}$$

Since the concentration of the undissociated water is very great, it can be taken as a constant. The equation can be expressed as follows:

$$[H^+] \times [OH^-] = K_w$$

The dissociation constant for water (Kw) has been measured to be 1.012×10^{-14} at 25° C. Since, however, there are for one molecule of water that dissociates, one hydrogen ion and one hydroxyl ion,

$$[\text{H}^+] = [\text{OH}^-] = \sqrt{\text{Kw}} = 1.006 \times 10^{-7} \text{ or approximately } 10^{-7}.$$

In 1909 Sörensen designated the hydrogen ion concentration in terms of the logarithm (to the base 10) of its reciprocal. Because of the convenience of this method, it has been used universally. A neutral solution has a hydrogen ion concentration of 1×10^{-7}. The pH value of such a solution is therefore $\log \dfrac{1}{1 \times 10^{-7}}$, or $\log 10^7$, or 7. If the hydrogen ion concentration of a solution is given, the corresponding pH may be calculated from the formula $\text{pH} = \log \dfrac{1}{[\text{H}^+]}$.

A normal solution of acid contains sufficient acid to furnish 1 gm. of ionizable hydrogen per liter.

A normal solution of alkali contains sufficient alkali to furnish hydroxyl ion equivalent to 1 gm. of hydrogen.

A buffer is any substance in a solution that inhibits changes in its hydrogen ion concentration upon the addition of relatively large amounts of acid or alkali.

An indicator is a substance, usually a weak organic acid or base, which changes color when the reaction of a solution changes. The following table lists indicators most commonly employed in the laboratory, with their color changes and pH ranges:

Indicator	Acid Color	Alkaline Color	pH Range
Thymol blue	Red	Yellow	1.2– 2.8
Bromphenol blue	Yellow	Blue	3.0– 4.6
Congo red	Blue	Red	3.0– 5.0
Methyl orange	Orange red	Yellow	3.1– 4.4
Bromcresol green	Yellow	Blue	3.8– 5.4
Methyl red	Red	Yellow	4.4– 6.0
Litmus	Red	Blue	4.5– 8.3
Bromcresol purple	Yellow	Purple	5.2– 6.8
Bromphenol red	Yellow	Red	5.4– 7.0
Bromthymol blue	Yellow	Blue	6.0– 7.6
Phenol red	Yellow	Red	6.8– 8.4
Cresol red	Yellow	Red	7.2– 8.8
Cresolphthalein	Colorless	Red	8.2– 9.8
Phenolphthalein	Colorless	Red	8.5–10.5

Colorimetric Method of Titrating Culture Media. With this method the hydrogen ion concentration of culture media is determined by using standard solutions of certain indicators.

1. *Titration Method with Phenolphthalein. Reagents and apparatus required:* (a) 1% solution of phenolphthalein, (b) N/10 sodium hydroxide, (c) N/1 sodium hydroxide, (d) ring stand, (e) burette, (f) casserole.

1. Wash all utensils with distilled water.
2. Place 5 ml. of the medium in the casserole and add 45 ml. of distilled water.
3. Add 1 ml. of phenolphthalein.
4. Heat to boiling.
5. Add N/10 sodium hydroxide from the burette with constant stirring until a faint pink color persists.
6. Read the burette and record the amount of N/10 sodium hydroxide used.

Calculation: Suppose it takes 1 ml. of N/10 sodium hydroxide to neutralize 5 ml. of the medium; then 1000 ml. will require 200 ml. of N/10 sodium hydroxide, or 20 ml. of N/1 sodium hydroxide.

2. *Comparator Block Method with Standard Solutions. Reagents and apparatus required:* (a) indicator of suspected pH of the medium, (b) N/10 sodium hydroxide, (c) N/1 sodium hydroxide, (d) set of standard color tubes of the indicator selected, (e) comparator block.

1. Two ml. of the medium are diluted with 8 ml. of distilled water in each of two test tubes.
2. To one of these tubes 0.5 ml. of the indicator is added, that indicator being chosen whose range includes the pH value desired.

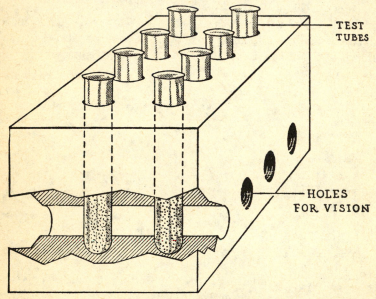

Fig. 19.   Comparator block.

3. The tubes are placed in the comparator block (Fig. 19). A tube of distilled water is placed behind the tube with the medium and indicator, while the standard solution tubes with the indicator are placed behind the tube with the medium but without the indicator. Those standard solution tubes are selected which correspond to the desired pH concentration of the medium.

4. If the shades are not the same and the pH value of the medium is below that of the standard (i.e., on the acid side) then N/10 NaOH is added to the medium from a burette slowly and with constant stirring until the color is identical with that of the standard tube.

5. The amount of N/10 NaOH used is read from the burette and recorded.

Calculation: Suppose 0.2 ml. of N/10 solution of NaOH was used for 2 ml. of medium, then for 1000 ml. of medium, 100 ml. N/10 solution of NaOH, or 10 ml. of N/1 NaOH, would be added for the correct titration.

Electrometric Method of pH Determination. Recently an electrometric method, more precise than the colorimetric one, has come into general use. It is essential for determining the pH of solutions of high turbidity or strong color.

CLEARING OF CULTURE MEDIA

Egg Clarification. This is accomplished by the coagulation of albumin, the coagulum entangling the particles of insoluble matter in the medium, which are then removed by filtration. In media rich in albumin, such as those made from meat infusion, it is unnecessary to add any albuminous material; but in others, fresh egg or a solution of dried egg albumin should be added. When fresh egg is added, proceed as follows: 1. Beat it up with an equal amount of water. 2. Pour the egg mixture into the medium cooled to from 50° to 55° C., and stir thoroughly, one egg being added for each liter. 3. Dried egg albumin when used is dissolved by allowing it to stand overnight in water in the proportion of 10 gm. to 20 ml. of water. 4. Twenty ml. of this solution is added to each liter of cooled medium (50° to 55° C.). 5. The medium is then heated in the autoclave for half an hour, or in an Arnold sterilizer for 45 minutes, to form a firm coagulum.

Modified Filtration. Filtration through paper pulp in a Buchner funnel will give a clear solution. A filter pump is used in this method. Solidifying media may be cleared by sedimentation, by allowing them to cool slowly in a straight-sided vessel. The coarser particles fall to the bottom. The solidified whole mass is then cut away from the upper cleared portion. This method is used in clearing vitamin agar, as vitamins are lost when coming in contact with paper or cotton.

FILTRATION OF CULTURE MEDIA

Filtration of fluid media is accomplished by passing the media through filter paper moistened with cold water to prevent the passage of fats. A piece of absorbent cotton placed in the neck of the funnel adds to the efficiency of the method.

Agar and gelatin are best filtered through absorbent cotton placed in a glass funnel, especially if they are kept warm by a hot water jacket. The cotton should be split horizontally and the two layers arranged so that their fibers are at right angles to each other. The cotton should be moistened with water and supported on a wire mesh to prevent jamming into the neck of the funnel.

Filtration for Sterilization. When heat cannot be applied for the sterilization of culture media, such as sugars, sera, tissue extracts, etc., either diatomaceous or unglazed porcelain filters may be used. Of these, the Berkefeld, the Chamberland, and the asbestos disc-metal Seitz are most frequently employed. The fluid is sucked through the filter candle into a sterile container by the aid of a filter pump. The filter candles are sterilized before they are used. They may be employed to sterilize sera, to separate toxins, to obtain bacterial filtrates, and to obtain filterable viruses from tissues containing them. See Chapter XXXI on Viruses.

PREPARATION OF MEDIA FOR USE

The filtered medium is put in either test tubes, bottles, or flasks. In filling them, it is necessary to avoid getting the solution on the necks of the containers. This is done by using a glass tube, connected with a funnel by a short piece of rubber tubing fitted with a Mohr's pinch-cock. Six-inch test tubes are usually used. Generally 3 to 5 ml. amounts are sufficient. For plates, 10 ml. are necessary, while 6 to 7 ml. are sufficient for slants. The tubes are plugged with cotton plugs which prevent access of microorganisms while allowing passage of air. *Care should be taken, however, to prevent the plugs from becoming damp for then bacteria and molds may pass through them.* After the medium is filtered, it may then be sterilized by any of the methods described on pages 79–82.

FORMULAE AND PREPARATION OF CULTURE MEDIA

MEAT EXTRACT (OR NUTRIENT) BROTH

Beef extract.................	3 gm. or 0.3%
Peptone....................	10 gm. or 1.0%
Sodium chloride.............	5 gm. or 0.5%
Distilled or tap water........	1000 ml.

1. Mix the ingredients in a vessel and dissolve by heating with stirring over a free flame.
2. Make up loss of water by adding water up to the mark which was originally made on vessel to indicate the level of the mixture.

3. Titrate the medium (see pp. 64–67) and adjust reaction.
4. Filter medium through a wet filter paper to clarify, then check the reaction, tube, and sterilize in autoclave.
5. For some purposes, sodium chloride is omitted or the amount of beef extract and peptone varied. (For water analysis, 3 gm. beef extract, 5 gm. peptone, and no NaCl are used.)

MEAT EXTRACT (OR NUTRIENT) AGAR

Agar......	20 gm. or 2.0%
Peptone.....................	10 gm. or 1.0%
Meat extract.................	3 gm. or 0.3%
Sodium chloride..............	5 gm. or 0.5%
Distilled water..............	1000 ml.

1. Mix the ingredients in a pan and dissolve by heating with constant stirring. Boil until all the agar is dissolved.
2. Make up loss of water.
3. Titrate and adjust reaction to pH 7.6–7.8.
4. Filter through absorbent cotton and gauze.
5. Tube medium in tubes or flasks.
6. Sterilize in autoclave at 15 lbs. for 15 minutes.
7. For water analysis, 15 gm. agar, 5 gm. peptone, 3 gm. beef extract, and no NaCl are used.

MEAT INFUSION BROTH

1. Allow chopped meat (veal or beef), 500 gm. to 1000 ml. of water, to infuse in the icebox for 12 to 24 hours, usually overnight.
2. Skim off all the fat, filter through gauze, and boil vigorously for one-half hour.
3. Restore the volume of water lost.
4. Filter through gauze to remove meat particles and then allow sediment to settle.
5. Restore the volume of water lost.
6. To each 1000 ml. of this infusion add:

Peptone......................	10 gm. or 1.0%
Sodium chloride..............	5 gm. or 0.5%

7. Heat to dissolve peptone, titrate, and adjust reaction to pH 7.6–7.8.
8. Boil for one-half hour, allow to cool to about 30° C., filter through paper, tube, and sterilize in autoclave at 15 pounds for 15 minutes.

MEAT INFUSION AGAR

1. To 1000 ml. of meat infusion broth add 20 gm. of agar and boil with constant stirring until the agar is dissolved.
2. Add water to restore volume.
3. Filter through absorbent cotton and tube.
4. Sterilize in autoclave at 15 pounds for 15 minutes.

SEMISOLID MEDIA.

In these media the amount of agar varies according to the use. For slants, 0.75% to 1% of agar is necessary. For stab cultures, 0.5% is sufficient.

1. Warm meat infusion made from 1 pound of meat, 500 ml. of water, 20 gm. of gelatin, and 20 gm. of peptone.
2. Dissolve 10 gm. of agar in 500 ml. of water; cool to below 50° C.

3. Mix the two mixtures, adjust the reaction, and heat to coagulate the albumin.
4. Clear the medium, readjust the reaction, and filter.

SUGAR–FREE BROTH. For fermentation reactions it is important to start with a medium which is known to be sugar-free in order to be certain that the bacterium is acting on only known sugars.

1. Inoculate one liter of meat infusion broth with a young culture of *Escherichia coli* and incubate for 24 to 48 hours. The bacteria will ferment and destroy any sugar which may be present in the broth, and thus a medium free from sugar will be obtained.
2. Place in Arnold sterilizer for 1 hour to kill *Escherichia coli.*
3. Titrate and adjust the medium.
4. Place in Arnold sterilizer for 30 minutes.
5. Filter through paper until clear.
6. To this medium different sugars are added in a concentration of 1%.
7. The medium is sterilized for 3 successive days in the Arnold, for higher temperatures of the autoclave tend to split complex sugars into simpler ones.

MEAT EXTRACT GELATIN

Gelatin......................	150 gm. or 15%
Peptone (Difco)..............	10 gm. or 1%
Meat extract.................	3 gm. or 0.3%
NaCl........................	5 gm. or 0.5%
Distilled water..............	1000 ml.

1. Dissolve all ingredients, except the gelatin, in a pan and bring to boil.
2. Remove from the stove and add the gelatin.
3. Stir until the gelatin is almost entirely dissolved.
4. Heat on stove or in Arnold only long enough to dissolve all the gelatin, stirring frequently.
5. Add water to restore volume lost in heating.
6. Titrate and adjust reaction to pH 7.6–7.8.
7. Filter solution while hot through cotton.
8. Tube the medium and sterilize in the Arnold sterilizer.

MEAT INFUSION GELATIN. This is prepared in the same way as described above, except that meat infusion broth is substituted for the meat extract broth.

MILK MEDIUM

1. Heat fresh milk (raw, not pasteurized) in the Arnold for 15 minutes.
2. Place in icebox for 12 hours to allow separation of cream.
3. Siphon off the milk, leaving the cream behind.
4. Tube with or without indicator and sterilize in Arnold by fractional method. Bromcresol purple is the most satisfactory indicator in milk. Andrade's indicator may be used but it is not desirable since it discolors the milk to some extent.
5. For routine laboratory work, 125 gm. of dehydrated skim milk powder may be rubbed with water in a mortar to make a smooth paste and enough water added to make 1000 ml.

POTATO MEDIUM

1. Cut out semi-cylinders or triangular potato slants from a fresh potato, sized to fit into tubes.
2. Wash the pieces in running water overnight, or immerse in 1% sodium carbonate solution for one-half hour.
3. Push some absorbent cotton into the bottom of the tube and wet with water.
4. Push cut potato down upon the wet cotton.
5. Sterilize by fractional sterilization for 20 to 30 minutes in Arnold sterilizer on 3 successive days.

BLOOD AGAR

1. Melt and cool to 45° C. sterile meat infusion, or meat extract agar of a pH of 7.6.
2. Add 5 to 10% of sterile defibrinated whole blood.
3. Make slants or tubes.
4. Incubate for 24 hours at 37° C. to determine sterility.

PEPTONE WATER, DUNHAM'S SOLUTION

Peptone (Difco).............	10 gm. or 1.0%
Sodium chloride.............	5 gm. or 0.5%
Distilled water.............	1000 ml.

1. Dissolve materials and boil for 5 minutes.
2. Restore volume with distilled water.
3. Titrate and adjust reaction to pH 7.6–7.8.
4. Filter through wet filter paper until clear.
5. Place in tubes or flasks.
6. Sterilize in autoclave at 15 pounds for 15 minutes.

Dunham's solution is used in testing for the formation of indol by bacteria and as a base for other media. Combined with 1 to 2% of various carbohydrates it serves as a fermentation medium for tests with organisms of the coliform group.

MEDIUM FOR FERMENTATION TESTS

Difco peptone...............	5 gm. or 0.5%
Beef extract................	3 gm. or 0.3%
Distilled water.............	1000 ml.

1. Mix ingredients in pan and boil about 2 minutes.
2. Restore volume by adding water.
3. Titrate and adjust reaction to pH 7.4–7.8.
4. Add a 1.6% alcoholic solution of bromcresol purple, 1 ml. per liter of medium.
5. Cool and filter through paper.
6. To make a 1% solution of the carbohydrates, add 10 gm. of the desired carbohydrate to 1 liter of medium.
7. Tube in tubes containing inverted ampules (fermentation tubes).
8. Sterilize in Arnold by fractional sterilization.

It is preferable to prepare a 20% solution of the carbohydrate and to sterilize it by filtration through a Berkefeld filter. The carbohydrate from this solution is then added in a concentration of 1% to the medium which has been sterilized in the autoclave.

NITRATE BROTH

Difco peptone............... 10.0 gm. or 1.0%
Potassium nitrate (nitrite-free)   0.2 gm. or 0.02%
Distilled water.............. 1000 ml.

1. Dissolve ingredients and adjust reaction to pH 7.4–7.6.
2. Filter through paper, tube, and sterilize in autoclave at 15 pounds for 15 minutes.

SERUM–WATER MEDIUM (HISS)

1. Add to 1 part of clear beef serum 2 or 3 parts of distilled water.
2. Heat mixture for 15 minutes in Arnold sterilizer at 100° C. to destroy any diastatic ferments present in the serum.
3. Add 1 ml. of a 1.6% alcoholic solution of bromcresol purple, or 1% of Andrade's indicator may be substituted.
4. To the medium add 1% concentrations of various sugars.

INULIN SERUM WATER

Beef or horse serum.................. 100 ml.
Peptone........................... 2.5 gm.
Inulin............................ 5.0 gm.
Water............................. 400 ml.

1. Dissolve the peptone in 25 ml. of water and heat slightly.
2. Adjust reaction to pH 7.4–7.8.
3. Dissolve inulin in 25 ml. of water and heat slightly.
4. Add 350 ml. of water to 100 ml. of serum.
5. Add the diluted serum to the peptone and inulin solutions.
6. Add bromcresol purple until the medium is colored.
7. Tube and sterilize in Arnold sterilizer for 30 minutes on 3 successive days.

INDICATORS IN MEDIA. Indicators are put in culture media to detect acid production by growing bacteria.

A. **Andrade's Indicator.** This consists of a mixture of 16 ml. of normal NaOH and 100 ml. of 0.5% aqueous acid fuchsin. The mixture is sterilized in the autoclave at 15 pounds for 15 minutes. To 100 ml. of the medium 1 ml. of the indicator is added. When the medium is hot, the mixture is pink, but when the medium is cold, the color disappears. The indicator is colorless at pH 7.2. The color range of the indicator is pale yellow or colorless when the medium is alkaline and pink when it is acid.

B. **Indicator in Endo's Medium.** This is Schiff's reagent which is 0.3 ml. of a 10% solution in 95% alcohol of basic fuchsin decolorized by 5 to 10 ml. of a 2.5% aqueous solution of sodium sulfite. The indicator changes from colorless to deep red in the presence of both aldehydes and acids.

ENRICHING SUBSTANCES IN MEDIA. In order to facilitate the growth of certain organisms which require special substances, glucose, nutrose, glycerin, sodium formate, carbohydrates, serum, blood, or animal proteins may be added to media for enrichment. Generally, the procedure is to add 1 part of the enriching substance to 2 or 3 parts of the medium, taking precautions for sterility. When additions are made to melted agar, it should be cooled to 45° C.

Living Chick Embryo serves as an excellent culture medium for many microorganisms. It has been used both for study purposes and for preparing vaccines for yellow fever, smallpox, influenza, Rocky Mountain spotted fever, etc.

BORDET–GENGOU MEDIUM FOR HEMOPHILUS PERTUSSIS.

This medium is used in the cough-plate method for the diagnosis of whooping cough.

1. The base of this medium is prepared by mixing 500 gm. of peeled potato, 1000 ml. of distilled water, and 40 ml. of glycerin in a flask and heating in the Arnold for 1 hour.
2. The extract is pressed through cheesecloth.
3. To each 500 ml. of the pressed fluid are added 1500 ml. of 0.6% NaCl and 50 gm. of agar.
4. Sterilize in Arnold sterilizer on 3 successive days. At this point the medium may be stored indefinitely.
5. Cool the medium after sterilization, and add from 5 to 10% of fresh citrated horse or rabbit blood.
6. The pH range of the medium before the blood is added is 5.8–6.4; after it is added, the reaction is pH 6.6–7.3.

MEDIA FOR CULTIVATION OF HEMOPHILUS INFLUENZAE

A. **Avery's Sodium Oleate Agar.** This medium is based on the principle that sodium oleate inhibits the growth of many of the Gram-positive cocci present in nasal secretions or in sputum, thus facilitating the isolation of *Hemophilus influenzae*.

1. A 2% solution in water of sodium oleate, neutral (Kahlbaum's), is prepared and sterilized in the autoclave.
2. Human or rabbit blood is defibrinated, centrifuged, the serum removed, and the volume restored by the addition of broth.
3. To 94 ml. of a 2% hormone agar with a pH of 7.4 add 1 ml. of the red blood cell suspension and 5 ml. of the 2% sodium oleate solution.

B. **Chocolate Agar.**

1. To agar add 5 to 10% defibrinated rabbit, beef, horse, or human blood.
2. Heat mixture gradually to about 75° C. until the blood begins to coagulate and turns a dark brown color.

ASCITIC FLUID AGAR FOR CULTIVATION OF NEISSERIA INTRACELLULARIS

1. Heat 150 gm. of minced lean beef with 250 ml. of distilled water, and add 250 ml. of a 0.8% solution of anhydrous sodium carbonate.
2. Cool to 45° C. and add 5 ml. of Cole and Onslow's pancreatic extract and 5 ml. of chloroform.
3. Incubate the mixture at 37.5° C. for 6 hours, with frequent stirring.
4. Add 40 ml. of normal hydrochloric acid, boil mixture for 1 hour. The reaction is adjusted to pH 7.6–7.9. Filter through paper after cooling to about 45° C.
5. Add 20 gm. of agar shreds to 1000 ml. of the broth (Douglas Broth).
6. Heat to boiling to dissolve agar.
7. Filter.

8. To 100 ml. of the agar (Douglas Agar) prepared as described above, cooled to 48° to 40° C. and of a pH 7.4–7.8, add 20 ml. of sterile, bile-free ascitic fluid with sterile precautions.
9. Then add 5 ml. of a sterile 20% aqueous solution of carbohydrate and 1 ml. of sterile Andrade's indicator, with sterile precautions.
10. Pour into test tubes and harden in slanting position.
11. Incubate for 24 hours at 37° C. and test for sterility.

MEDIA FOR CULTIVATION OF CORYNEBACTERIUM DIPHTHERIAE

A. Löffler's Blood Serum.

1. Mix 3 parts of horse, beef, or sheep serum with 1 part of extract broth of pH 6.8–7 to which 1% of dextrose has been added.
2. Tube, and heat medium gradually in inspissator at temperature between 80° and 90° C. in order to coagulate the serum evenly.
3. Sterilize in Arnold for 20 minutes on 3 successive days.

B. Blood Tellurite Medium.

1. To 15 ml. of nutrient agar, pH 7.6, melted and cooled to 50° C., add 1.5 ml. of defibrinated rabbit's blood and 1.5 ml. of a 2% solution of potassium tellurite.
2. Mix and pour mixture into a Petri dish, taking the proper sterile precautions.

Upon this medium the colonies of *Corynebacterium diphtheriae* are black and are easily distinguished from pseudodiphtheria bacilli.

DIEUDONNE'S MEDIUM FOR CULTIVATION OF VIBRIO COMMA

1. Add 30 parts of a sterile mixture of defibrinated beef blood and normal NaOH to 70 parts of 3% agar neutral to litmus.
2. The agar is sterilized by autoclaving at 15 pounds for 15 minutes before the mixture of blood and NaOH is added.
3. The alkaline blood agar is poured into plates and allowed to dry for several days at 37° C. or at 60° C. for 5 minutes.

MEDIA FOR CULTIVATION OF MYCOBACTERIUM TUBERCULOSIS

A. Dorset's Egg Medium.

1. Mix whites and yolks of eggs in wide-mouthed flask.
2. To every 4 eggs add 25 ml. of distilled water, and strain through sterile cloth.
3. Pour 10 ml. of the medium into sterile test tubes.
4. Slant in inspissator, allowing exposure to 73° C. for 4 to 5 hours on 2 days.
5. On third day raise temperature to 76° C.
6. Sterilize by exposure to 100° C. for 15 minutes in Arnold.
7. Add 2 or 3 drops of sterile water to each tube before inoculation.

B. Löwenstein's Medium.

Sodium citrate......................	1 gm.
Monopotassium phosphate............	1 gm.
Asparagin...........................	3 gm.
Magnesium sulfate...................	1 gm.
Glycerin............................	60 ml.
Distilled water.....................	1000 ml.

1. Add 6 gm. of potato flour to 150 ml. of the above ingredients and boil over a double boiler for 15 minutes, shaking frequently.
2. Keep at 56° C. for 1 hour.
3. Add 4 whole eggs and 1 yolk prepared by washing the eggs in tapwater and then placing them in 5% phenol for 15 minutes. Then drain and place in alcohol to neutralize the phenol.
4. Shake the flask to distribute the eggs evenly.
5. Add 5 ml. of a sterile 2% aqueous solution of Congo red.
6. Filter through sterile gauze, tube, and inspissate for 2 hours at 80° to 85° C. on 2 successive days.

MEDIA FOR COLON–TYPHOID–DYSENTERY GROUP DIFFERENTIATION

A. Endo's Medium.

1. Prepare a 3% agar which contains 1% peptone and 0.5% meat extract.
2. To 100 ml. of this agar add 1 gm. of lactose dissolved in about 10 ml. of distilled water and sterilize in autoclave for 15 minutes at 15 pounds. To each 100 ml., then add 1 ml. of a 3% solution of basic fuchsin in 95% alcohol, and then 5 ml. of 2.5% sodium sulfite.
3. Pour into Petri dishes and allow to harden.

The colonies of *Salmonella typhosa* and other non-lactose-fermenting organisms are colorless; those of *Escherichia coli* become red.

B. Eosin Methylene Blue Agar.

Agar	15 gm.
Difco peptone	10 gm.
Dipotassium phosphate (K_2HPO_4)	2 gm.
Distilled water	1000 ml.

1. Boil to dissolve all ingredients.
2. Restore volume by adding distilled water.
3. Filter through cotton.
4. Measure 100 or 200 ml. into flasks and sterilize in autoclave at 15 pounds for 15 minutes.
5. Before using, melt the agar and add to each 100 ml.:

Lactose, sterile 20% solution	5 ml.
Eosin, 2% aqueous solution	2 ml.
Methylene blue, 0.5% aqueous solution	1.3 ml.

6. Mix thoroughly and pour into Petri dishes.

Colonies of *Escherichia coli* have a green metallic sheen with a dark center. Colonies of *Aerobacter aerogenes* are usually larger, not so dark, and with brown centers. Colonies of non-lactose-fermenters appear pink or the same color as the medium.

C. Brilliant Green Agar (Krumwiede).

Beef extract	3 gm.
Sodium chloride	5 gm.
Peptone	10 gm.
Agar	15 gm.
Water	1000 ml.

1. Dissolve ingredients in autoclave, clear, and filter.
2. Adjust reaction to be neutral to Andrade's indicator, or pH 7.2.
3. Bottle agar in 100 ml. amounts.
4. Before use, add to each 100 ml. bottle:
   (a) Andrade's indicator, 1%.
   (b) An amount of acid sufficient to bring the medium to the neutral point. This is evident from the observation that the agar when hot is a deep red, but upon cooling becomes colorless.
   (c) Lactose, 1%.
   (d) Glucose, 0.1%.
   (e) Brilliant green, 0.1% aqueous solution.
5. Mix the bottles and pour the contents into Petri dishes.

On this medium the growth of *Escherichia coli* and of *Shigella dysenteriae* is inhibited, but *Salmonella typhosa*, *Salmonella paratyphi*, and *Salmonella schottmuelleri* grow freely.

D. Wilson and Blair Bismuth Sulfite Medium.

1. Prepare nutrient agar base from 5 gm. Bacto beef extract, 10 gm. Bacto peptone, and 30 gm. Bacto agar dissolved in 1000 ml. of hot distilled water by heating in Arnold or autoclave.
2. Prepare sulfite-bismuth-phosphate-iron mixture:

Anhydrous sodium sulfite (Eastman Kodak), 20% aqueous solution	100 ml.
Ammoniacal solution bismuth citrate	50 ml.
Anhydrous dibasic sodium phosphate (Merck)	10.5 gm.
Ferrous sulfate crystals (Merck), 8% aqueous solution	10 ml.

   Boil for 2 minutes. This forms a stock solution which can be kept for months in refrigerator if it is well corked.
3. To 1000 ml. of hot melted nutrient agar add:

Glucose, 20% aqueous solution	25 ml.
Sulfite-bismuth-phosphate-iron solution	70 ml.

4. Mix well and add 3 to 5 ml. of a 1% aqueous solution of brilliant green.
5. Mix well and distribute in 100 ml. amounts.
6. Sterilize in autoclave at 10 pounds for 10 minutes.

On this medium the growth of *Escherichia coli* is inhibited; the colonies of *Salmonella typhosa*, *Salmonella paratyphi*, and *Salmonella schottmuelleri* are dry, flat, and black, with a metallic sheen due to the reduction of the sulfite to sulfide in the presence of glucose.

E. Koser's Citrate Medium.

Sodium ammonium phosphate (microcosmic salt)	1.5 gm.
Potassium dihydrogen phosphate (KH_2PO_4)	1 gm
Magnesium sulfate	0.2 gm.
Sodium citrate (crystals)	3 gm.
Distilled water	1000 ml.

1. Dissolve all ingredients in the water.
2. Tube, and sterilize at 15 pounds for 20 minutes.

Sodium citrate as the only source of carbon is utilized by nonfecal organisms like *Aerobacter aerogenes*. Fecal *Escherichia coli* fails to grow.

F. Russell's Double Sugar Medium.

Agar..	20 gm.
Difco peptone..................................	20 gm.
Sodium chloride................................	5 gm.
Lactose..	10 gm.
Glucose...	1 gm.
Distilled water.................................	1000 ml.
Andrade's indicator............................	10 ml.

1. Dissolve ingredients in pan and boil until all the agar is dissolved.
2. Restore volume with distilled water.
3. Titrate and adjust reaction to pH 7.6–7.8.
4. Filter through cotton, tube, placing 10 ml. in each tube, and sterilize in autoclave at 15 pounds for 15 minutes.
5. The agar is allowed to harden in a slanting position.

Color changes and presence or absence of gas along the slant and butt are very valuable differential aids.

G. Krumwiede's Triple Sugar Medium.   The Russell double sugar medium has been modified by the addition of 1% saccharose.   This addition gives sharper color changes, and the fermentation of saccharose is a means of excluding the paratyphoid-like intermediate organisms frequently encountered in stools.

The following reactions are noted: *Escherichia coli*—acid and gas in butt and acid in slant; *Salmonella typhosa* and *Shigella dysenteriae*—acid in butt only; *Salmonella paratyphi* and *Salmonella schottmuelleri*—acid and gas in butt only.

MEDIUM FOR CULTIVATION OF DIPLOCOCCUS PNEUMONIAE

1. Add 5 ml. of a sterile 20% solution of glucose to 90 ml. of meat infusion broth, pH 7.4–7.8.
2. Add 5 ml. of rabbit's blood.
3. Mix in flask.
4. Transfer 4 ml. to small test tubes, using a sterile pipette and sterile precautions.
5. Incubate for 24 hours at 37° C. to test for sterility.

MEDIUM FOR CULTIVATION OF PASTEURELLA TULARENSIS:
Blood Glucose Cystine Agar

1. To beef infusion agar of pH 7.3 containing 1% peptone, 1.5% agar, and 0.5% sodium chloride, add 0.1% cystine and 1% glucose.
2. Heat in Arnold sterilizer to melt agar and to sterilize cystine and glucose.
3. Cool to 50° C. and add 5 to 8% defibrinated or whole rabbit blood.

MEDIA FOR CULTIVATION OF ANAEROBES
A. Cooked Meat Medium.

1. To a tube of meat infusion broth containing 0.1% dextrose, pH 7.8, add a few pieces of chopped meat.
2. Sterilize by autoclaving at 15 pounds for 15 minutes.

B. Fluid Thioglycollate Broth (Brewer).

Ground fresh lean meat	500 gm.
Sodium chloride	5 gm.
Dipotassium phosphate (K_2HPO_4)	2 gm.
Peptone	10 gm.
Distilled water	1000 ml.

1. Mix meat and water and allow to stand at 5° C. for 24 hours.
2. Strain through cloth.
3. Heat in Arnold for 1 hour and then in autoclave at 15 pounds for 30 minutes.
4. Filter while hot through moistened filter paper and make up to volume.
5. Add remaining ingredients, stir, and adjust to pH 7.5.
6. Heat in Arnold for 30 minutes.
7. Filter and sterilize in autoclave at 15 pounds for 20 minutes.
8. To 1000 ml. of this stock broth, add:

Dextrose (anhydrous)	10 gm.
Sodium thioglycollate	1 gm.
Agar	0.5 gm.
Methylene blue (1:500 solution)	1 ml.

9. Add agar to cold broth, mix, and heat gradually to boiling point.
10. Cool to about 80° C., add remaining ingredients, stir, and adjust to pH 7.5.
11. Tube, and sterilize in autoclave at 15 pounds for 20 minutes.

Sodium thioglycollate is a reducing substance. With the above medium, anaerobes may be cultivated using the usual aerobic technics.

MEDIA FOR TOXIN PRODUCTION

A. Diphtheria Toxin Broth.

1. Infuse one pound of Bacto veal powder with 7 liters of distilled water at 50° C. for three-quarters of an hour.
2. Boil the infusion for a few minutes and strain through cheesecloth.
3. Add 2% of proteose peptone and 0.5% of NaCl.
4. Adjust reaction with sodium hydroxide to pH 7.5.
5. Boil thoroughly to prevent precipitate from forming and filter through cotton and filter paper.
6. Distribute in 800 ml. amounts in special toxin bottles and autoclave at 15 pounds for 30 minutes.
7. Before inoculating with *Corynebacterium diphtheriae* add 0.15% sterile glucose and 0.3% maltose.

Recently a complex, organic synthetic medium has been developed for the cultivation of *Corynebacterium diphtheriae* and production of its toxin by Pappenheimer, Mueller, and Cohen.

B. Tetanus Toxin Broth.

1. Infuse one pound of lean veal in a liter of water overnight in icebox.
2. Heat to 45° C. for 1 hour and boil for ½ hour.
3. Strain and add 1% Berna peptone, 0.5% sodium chloride, and 1% glucose.
4. Boil until the ingredients are dissolved and adjust reaction to pH 7.2.
5. Sterilize in 2 liter Erlenmeyer flasks in Arnold sterilizer, 1½ hours on first day and 1 hour on second day.

C. Mallein Broth.

1. Macerate one pound of lean veal in one liter of water overnight at room temperature.
2. Heat at 45° to 50° C. for 1 hour.
3. Boil vigorously.
4. Strain through cheesecloth, and add 1% Fairchild's peptone and 0.5% sodium chloride.
5. Boil again to dissolve peptone.
6. Titrate and adjust reaction to pH 6–6.4.
7. Autoclave for 15 minutes at 15 pounds of pressure.
8. Filter through cotton and paper.
9. Measure and add 5% glycerin.
10. Put 250 ml. broth in each quart Blake bottle or liter flask.
11. Sterilize in autoclave at 15 pounds for 30 minutes.

D. Tuberculin Broth.

1. Infuse one pound of chopped lean beef in one liter of water overnight in icebox.
2. Heat to 45° C. and hold at that temperature for 1 hour.
3. Boil thoroughly and strain.
4. To filtrate add 1% peptone and 0.5% sodium chloride.
5. Adjust reaction to pH 7.1.
6. Boil, filter through paper, add 5% glycerine, and distribute in quart Blake bottles, adding 250 ml. to each bottle.
7. Sterilize in autoclave at 15 pounds for 30 minutes.

MEDIA FOR FUNGI

A. Pennsylvania Medium.

American crude dextrose..................	40 gm.
Imported French peptone................	10 gm.
(Fairchild's peptone also may be used.)	
Agar..................................	18 gm.
Distilled water.........................	1000 ml.

1. Boil mixture to dissolve ingredients.
2. Place in tubes and flasks.
3. Sterilize in autoclave at 15 pounds for 15 minutes.

It is not necessary to adjust reaction which is sufficiently acid, pH 5.5–6.

B. Sabouraud's Conservation Medium.

1. Dissolve 30 gm. of peptone, granulée de Chassaing, and 15 gm. of agar in 1000 ml. of water.
2. Tube, and autoclave at 15 pounds for 15 minutes.

STERILIZATION OF CULTURE MEDIA

There are three generally accepted methods for sterilization: *heat, filtration*, and *the use of chemicals*. Filtration as used for sterilization is discussed on pages 68 and 286. The use of chemicals for sterilization will be described in Chapter VI on the Effect of Physical and Chemical Agents on Bacteria.

Steam Under Pressure—Autoclave.

1. *The most rapid and efficient method* of sterilization by heat is the autoclave, which consists of a closed chamber or boiler for utilizing steam under pressure. The steam may be introduced directly, or by boiling water in the bottom of the chamber, first driving out the air, after which the chamber is filled with live steam under pressure. It is cautioned that, after autoclaving, the steam pressure should be

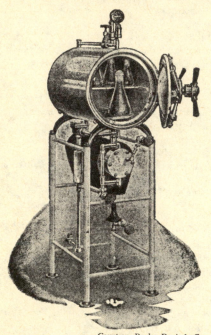

Courtesy, Parke, Davis & Co.

FIG. 20. Horizontal autoclave for sterilization with steam under pressure.

reduced by opening the valve to let out all the steam before opening the chamber door. Otherwise the cotton plugs may be forced out.

2. *Most culture media* are sterilized in the autoclave at 15–20 pounds pressure per square inch for 15 to 20 minutes at a temperature of 120° C. Large flasks of media must be autoclaved for a longer period.

3. *Sugars* decompose in the autoclave. Therefore, intermittent or fractional sterilization in the Arnold steam sterilizer may be used. However, some sugars such as maltose are split even by this method; these are sterilized by filtration through Berkefeld or Seitz filters.

Streaming Steam—Arnold Sterilizer. This method of sterilization utilizes the principle of *fractional* or *intermittent sterilization.*

1. *Moist heat is most efficient.* The material is sterilized by live steam which comes in contact with it. Exposure for 15 to 20 minutes at 100° C. is sufficient to kill all vegetative forms of bacteria, but not the spores.

2. *To kill spores* exposure for 15 to 30 minutes on 3 successive days is necessary. This is called *intermittent* or *fractional sterilization.*

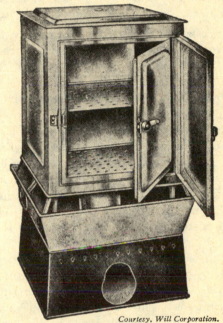

Courtesy, Will Corporation.

Fig. 21. Arnold steam sterilizer. Utilizes streaming live **steam.**

3. Certain media, such as *sugars*, *gelatin*, and *potato*, which are injured by excessive heating, are sterilized by *fractional* sterilization.

Inspissator. This is an apparatus in which the temperature can be slowly raised to between 80° and 90° C. It is used primarily for the uniform heating of blood serum to make Löffler's blood serum for the cultivation of *Corynebacterium diphtheriae*, and for media that cannot withstand a temperature of 100° C.

Sterilization by Boiling. Surgical instruments, needles, hypodermic syringes, etc., may be sterilized by boiling in water.

Sterilization by Dry Heat—Hot Air Sterilizer. A temperature of 150° C. for one-half to one hour is used to sterilize glassware, flasks, tubes, and other apparatus made of glass, as pipettes, etc. This method cannot be used for liquids in glassware or for solid culture media, or for any media containing organic substances. The period of exposure is measured from the time when the temperature reaches 150° C. This is maintained for one hour. Then the sterilizer is allowed to cool off gradually. Finally, when the temperature is down, the door is opened and the material removed.

METHODS OF OBTAINING PURE CULTURES

To study the characteristics and properties of a particular species of bacteria it is necessary to obtain it separated from other organisms, that is, in *pure culture*. Any one of the following methods may be employed to accomplish this.

Pour Plates. The method calls for making a series of dilutions of the bacterial culture in the medium and then pouring into Petri dishes. 1. A loopful of the bacterial culture is inoculated into a tube of the medium, which has been melted and then cooled to 50° C., and thoroughly mixed by tilting and rolling the tube. 2. From this tube three loopfuls are transferred to another tube of medium, which is mixed in the same way. 3. From the second tube five loopfuls are transferred to a third tube and mixed. 4. The mouths of the tubes are flamed and the contents of each poured into a Petri dish, while the lid of the dish is raised slightly. (The lid prevents the entrance of contaminants.)

When there are few organisms present in the material, more of it must be inoculated. Therefore, in such a case a pipette may be employed instead of a loop. 1. The tip of the pipette is touched to the culture and then transferred to the first tube of medium. 2. From this 0.1 ml. is transferred to a second tube. 3. From the second tube 0.1 ml. is carried over to the third, care being taken to mix the contents thoroughly before each transfer.

Streak Plates. Melted agar is poured into Petri dishes and allowed to harden. With a loop or sterile bent-glass rod the culture is streaked over the surface of the first plate. Without sterilizing the loop or the rod, two or more other plates are streaked in the same way. Discrete colonies may thus be obtained. For descriptions of colonies, see p. 22.

INCUBATION OF CULTURES OF BACTERIA

Culture media, after they have been inoculated with bacteria, are exposed to a temperature which is most suitable for the growth of the particular organism. For temperatures above room temperature, an *incubator* is used. In this chamber, constructed of well insulated material, the inoculated culture media are placed and kept for a certain period of time to allow the growth of the organism.

BIBLIOGRAPHY

Burrows, W. *Textbook of Microbiology*. Eighteenth Edition. Philadelphia: W. B. Saunders Company, 1963. Pp. 24–27.

Diagnostic Procedures and Reagents. New York: American Public Health Association, 1941. Pp. 1–39.

Difco Manual of Cutlure Media. Sixteenth Edition. Detroit: Difco Laboratories, 1949. Pp. 20–187.

Gershenfeld, Louis. *Bacteriology and Allied Subjects*. Easton, Pa.: Mack Publishing Company, 1945. Pp. 48–49.

Jawetz, E., J. L. Melnick, and E. A. Adelberg. *Review of Medical Microbiology*. Los Altos: Lange Medical Publications, 1962.

Knaysi, Georges. *Elements of Bacterial Cytology*. Ithaca, N. Y.: Comstock Publishing Company, 1944. Pp. 133–147.

Kolmer, J. A. *Approved Laboratory Technique*. Fifth Edition. New York: Appleton-Century-Crofts, 1951. Part 3, pp. 18–19.

Park, William Hallock, and Anna Wessels Williams. *Pathogenic Microörganisms*. Eleventh Edition. Philadelphia: Lea and Febiger, 1939. Pp. 136–208.

Schaub, I. G., *et al*. *Diagnostic Bacteriology*. Fifth Edition. St. Louis: C. V. Mosby Company, 1958.

Swingle, D. B. *General Bacteriology*, rev. by G. W. Walter. New York: D. Van Nostrand Company, 1947. Pp. 70–88.

Wadsworth, Augustus B. *Standard Methods of the Division of Laboratories and Research of the New York State Department of Health*. Third Edition. Baltimore: Williams & Wilkins Company, 1947.

Chapter VI

EFFECT OF PHYSICAL AND CHEMICAL AGENTS ON BACTERIA

The growth of bacteria is affected by physical agents, such as heat, drying, electric current, light, pressure, or agitation, and by a vast number of chemical agents.

EFFECT OF PHYSICAL AGENTS ON BACTERIA

Temperature. The influence of various degrees of heat upon the growth of bacteria has been described on pages 20, 21, and 80. The effect of heat on bacteria is utilized in sterilization of materials. Resistance of the organism depends in large part upon its growth stage, whether it forms spores, and whether the heat is dry or moist. The length of exposure is also important. Bacteria resist low temperatures better than high temperatures.

Desiccation. The removal of water from a bacterial cell usually results in the death of the organism. Drying in air kills the vegetative forms of most pathogenic bacteria in a few hours. *Mycobacterium tuberculosis* is killed within a few days of exposure to the air. When, however, drying is done together with freezing, as in Flosdorf-Mudd lyophile apparatus, then it is possible to preserve biological products, such as vaccines, antisera, and cultures of bacteria, for long periods of time.

Light. Sunlight has a toxic effect upon bacteria. The ultraviolet, violet, and blue rays of the spectrum are bactericidal; green, red, and yellow rays are less deleterious. X rays inhibit growth only slightly. Operating rooms are often irradiated to reduce sharply the number of organisms in the air. *Photodynamic sensitization* increases the effect of ultraviolet rays of low intensity and of visible light. When certain dyes such as eosin and erythrosin are added to bacterial suspensions exposed to visible light, bacteria are killed by rays which would otherwise be relatively harmless. This phenomenon may be due to fluorescence, selective absorption of light, or peroxide formation.

Electric Current. Bacteria usually pass to the positive pole (anode) when an electric current is passed through a suspension, de-

pending on the pH of the environment. An electric current of sufficiently high voltage passed through a suspension of bacteria can kill
them.

Agitation. Extended and excessive shaking will result in disruption of the bacterial cell, with liberation of its contents.

Pressure. Mechanical pressure, unless very high and prolonged,
has little effect on bacteria. For the effect of osmotic pressure,
see p. 21.

Vibration. Sonic and supersonic vibrations will reduce the bacterial content of material.

EFFECT OF CHEMICAL AGENTS ON BACTERIA

General Considerations. Before discussing the disinfection or
destruction of bacteria by chemicals it will be necessary to distinguish
between certain related terms.

1. *Disinfectant.* A chemical agent which kills bacteria or other
microorganisms. The disinfection process does not take place at once
but is a gradual operation in which the number of organisms killed
per time unit is greater at the beginning and smaller and smaller as the
exposure period is increased. *Germicide* and *bactericide* are synonyms
for disinfectant.

2. *Antiseptic.* A substance which prevents, stops, or inhibits
growth of microorganisms, without necessarily killing them. *Antibiotic* and *bacteriostat* have the same definition as antiseptic, but
different connotations, as will be illustrated later.

3. *Asepsis.* Precautions taken to prevent access of microorganisms to materials. Asepsis is practiced in surgery where utensils,
instruments, clothing, towels, bandages, and the operating room are
freed from infective bacteria.

4. *Sterilization.* The complete destruction or removal of all living forms of microorganisms.

Types of Disinfectants. The disinfectants cover a wide range of
substances including:

1. Salts of heavy metals—$AgNO_3$, $HgCl_2$, $KMnO_4$, etc.

2. Halogens—chlorine, bromine, iodine, and their salts.

3. Aldehydes and acids—formaldehyde, nitric acid, sulfuric acid,
boric acid, salicylic acid, etc.

4. Organic compounds—coal tar derivatives as cresol, tricresol,
camphor, creosote, carbolic acid, etc.

5. Oxidizing agents—hydrogen peroxide, sodium perborate, potassium permanganate, chloride of lime.

6. Gaseous agents—formaldehyde gas, sulfur dioxide, hydrocyanic acid, and the halogens.

7. Alkalies—KOH, NaOH, LiOH, NH_4OH, $Ba(OH)_2$.

8. Soaps.   9. Aerosols.

Ideal Disinfectant.   The ideal disinfectant has the following assets:

1. High coefficient of disinfection.   2. Stability.   3. Solubility. 4. Nontoxicity to animals.   5. Noncorrosive and nonbleaching power. 6. High power of penetration.   7. Cheapness.   8. Deodorizing power. 9. Ease of application.

Phenol Coefficient.   The relative efficiency of disinfectants is determined by testing their ability to kill a test organism and comparing it with that of phenol (carbolic acid) under the same conditions.   The F.D.A. test* for the determination of phenol coefficients is conducted as follows:

1. *Apparatus and Reagents Required.*   (a) Test tubes, (b) pipettes, (c) a 5% stock solution of phenol, (d) the test culture of *Salmonella typhosa*, Hopkins strain, grown in nutrient broth at 37° C., and (e) a 1% stock solution of the disinfectant to be tested.

2. Make up dilutions of phenol of 1:80, 1:90, 1:100, 1:110, 1:120, and 1:130, and place in water bath at 20° C. for 5 minutes.

3. Make up dilutions of the disinfectant to be tested—1:150, 1:175, 1:200, 1:250, 1:275, 1:300, etc., depending upon the expected range, and place in water bath at 20° C. for 5 minutes.

4. To 5 ml. of successive dilutions of the disinfectant and of phenol add 0.5 ml. of the culture of *Salmonella typhosa*.

5. Transfer a loopful of each mixture to a separate subculture tube, properly labeled, at intervals of 5, 10, and 15 minutes.

6. Incubate the subculture tubes and a control seeding tube for 48 hours at 37° C.

7. Record readings of growth (+) or no growth (−) in each tube.

Example:

PHENOL				TEST DISINFECTANT			
	Minutes of Exposure				Minutes of Exposure		
Dil.	5	10	15	Dil.	5	10	15
1:80	−	−	−	1:150	−	−	−
1:90	−	−	−	1:175	−	−	−
1:100	+	−	−	1:200	−	−	−
1:110	+	+	+	1:250	+	−	−
1:120	+	+	+	1:275	+	+	+
1:130	+	+	+	1:300	+	+	+

* There are in general three methods of determining phenol coefficients: 1. The U. S. Hygienic Laboratory (H. L.) method (see *Disinfectant Testing by the Hygienic Laboratory Method*, Reprint No. 675, *U. S. Public Health Reg.*, 36:1559, 1921); 2. the Rideal-Walker (R. W.) method; 3. the Food and Drug Administration (F.D.A.) method described here.

The phenol coefficient is calculated as the ratio of the greatest dilution of the test germicide killing the test organism in 10 minutes but not in 5 minutes to the greatest dilution of phenol showing the same results. In our example this would be $\frac{250}{100}$ or 2.5.

A standard culture of *Micrococcus pyogenes* var. *aureus* may be used instead of *Salmonella typhosa*, in which case the phenol coefficient is expressed in terms of *Micrococcus pyogenes* var. *aureus*.

In connection with determination of phenol coefficients, certain limitations and cautions should be noted:

1. The test was intended and, for most reliable results, should be used only for compounds chemically related to phenol.

2. A standard procedure should be strictly adhered to and the method used reported with the results.

3. The test reveals nothing of a germicide's toxicity to living tissues.

4. The test disinfectant must be completely soluble in or miscible with water.

Testing Insoluble Products. To determine the inhibitory properties of disinfectants or antiseptics which are insoluble in or immiscible with water, the agar plate method has been devised by the United States Public Health Service. Ointments, dusting powders, catgut, suppositories, etc., may be tested in this way.

1. Seed nutrient agar while in the liquid state at a temperature of 42° to 45° C. with a 24-hour broth culture of *Micrococcus pyogenes* var. *aureus*.

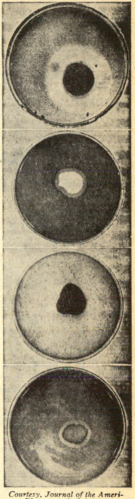

Courtesy, Journal of the American Pharmaceutical Assn.

Fig. 22.   Ointment tests.*

---

* Reading from top to bottom, note inhibition zones against seeded pyogenic cocci by (*a*) tincture of iodine, (*b*) 5 per cent salicylic acid ointment, (*c*) sulfur ointment with which there is no inhibition zone, indicating no antiseptic activity, and (*d*) 5 per cent phenol ointment in petrolatum base. The author has devised a similar test for liquid antiseptics, in which the antiseptic is first soaked in blotting paper and dropped on the usual seeded culture media. Inhibition zones forming around the impregnated blotting paper may be measured and compared, using 5 per cent aqueous phenol as a standard.

2. Pour into a sterile Petri dish and allow to harden.

3. Place test substance in intimate contact with the agar surface.

4. Incubate under unglazed porcelain top at 37° C. for 24 to 48 hours.

5. Observe whether there is a clear zone, indicating inhibition of growth, around the place where the substance has been in contact.

6. The width of the zone indicates the diffusibility of the inhibitory agent.

7. A variation is the agar cup-plate method, used on products liquid at the temperature of the test. A cup is cut in the agar into which a few drops of the liquid to be tested is placed.

Determining Toxicity. A new method for evaluating germicides intended for clinical use has been devised by Salle and coworkers. The materials are tested for effect on the growth of living embryonic chick heart tissue, as well as for ability to kill bacteria. A toxicity index is calculated, which is the ratio of the highest dilution of germicide required to kill the tissue in 10 minutes to the highest dilution required to kill the test organism in the same period of time under identical conditions.

Natural Disinfection. The organisms causing the more communicable diseases tend to die out in an unfavorable environment. Dilution, sunlight, drying, all play a part in the destruction of pathogenic bacteria. Polluted streams become self-purified after a lapse of sufficient time.

Chemotherapy. The prevention and treatment of disease by chemical disinfection or inhibition of the parasitic causes is known as chemotherapy. The action may be bactericidal or bacteriostatic. The chemotherapeutic agents most commonly used include:

1. *Organic Arsenic Compounds.*

(a) Valuable in combating spirillosis, treponematosis (syphilis, yaws), and trypanosomiasis and other protozoan infections.

(b) Include arsphenamine (salvarsan or 606), neoarsphenamine, and compounds derived from them.

2. *Bismuth Compounds.*

(a) Preferred in syphilis therapy.

(b) Include bismuth and sodium tartrate, sodium bismuth thioglycollate, bismuth salicylate, etc.

3. *Antimony Compounds.*

(a) Used successfully in various protozoan and other parasitic diseases, as trypanosomiasis, kala azar, etc.

(b) Tartar emetic (antimony and potassium tartrate) is used most frequently.

4. *Sulfonamides.*

(a) Include sulfanilamide (*p*-aminobenzenesulfonamide or

NH_2⟨ ⟩$SO_2.NH_2$) and derivatives in which one or both of the hydrogen atoms of the sulfonamide group are replaced. The most important compounds and their uses are listed in the table on the next page.

(b) Bacteriostatic action primarily. Woods' theory is most widely accepted as the probable mode of action. It notes that *p*-aminobenzoic acid (paba), probably synthesized by bacteria, is an essential growth substance. Sulfanilamide, if present in excess, is believed to displace paba in the enzyme reaction necessary for its utilization, thereby inhibiting growth.

(c) In testing sulfonamides for sterility and body fluids for identification of specific bacteria during the course of sulfa therapy, paba must be added to the culture medium to inactivate or neutralize the bacteriostatic properties of the sulfas present.

(d) Drug-fast strains of bacteria may be produced, *in vivo* or *in vitro*, by continued use of certain chemotherapeutic agents, notably sulfonamides, arsenicals, bismuth and antimony compounds, and many dyes. The resistance, which is not necessarily permanent, is usually developed to an entire class of drugs.

5. *Quinine.*

(a) A natural alkaloid specific for malarial parasites, obtained from cinchona bark.

(b) Has also been suggested as specific for pneumococcus.

6. *Atabrine* and *plasmochin* are recent synthetics used to combat malarial parasites.

7. *Emetine*, active principle from ipecac, is a specific for amoebic dysentery.

8. *Dyes.*

(a) Used for preoperative skin antisepsis, wound disinfection, in localized mucous membrane infections (e.g., Vincent's angina), in general treatment of surface infections.

(b) Many dyes have selective action for certain organisms, are of low toxicity, are nonirritating, are penetrative, persisting for long periods of time.

(c) Synthetic coal tar dyes belonging to acridine (e.g., proflavine and acriflavine in urinary infections), azo (e.g., scarlet red), diphenylamine, nitroso, pyonine, and triphenylmethane or rosaniline (e.g., brilliant green, gentian violet, acid fuchsin) classes have been used for their bacteriostatic or, less often, for their bactericidal properties.

9. *Penicillin.* See section on antibiotics following.

Compound	Advantages	Disadvantages	Drug of Choice In	Other Uses
Sulfanilamide (*p*-aminobenzenesulfonamide)	When administered orally, readily absorbed, well distributed, promptly excreted (if kidney function is normal).		Meningococcus and hemolytic streptococcus infections, chancroid, gas gangrene, trachoma.	Equally as effective as sulfapyridine in *Streptococcus viridans* infections, brucellosis, gonorrhea, lymphogranuloma venereum.
Sulfapyridine (2-(*p*-aminobenzenesulfonamido)-pyridine; Degenan)	Passes quite readily from blood to body tissues and fluids.	Irregularly absorbed. Lower blood concentration levels than with corresponding doses of sulfanilamide. Nausea, vomiting, dizziness often result.	Pneumococcus pneumonia.	Preferred by some in gonorrhea, lymphogranuloma venereum, Friedländer bacillus infections, staphylococcus meningitis, brucellosis, trachoma, *Hemophilus influenzae* infections.
Sulfathiazole (2-(*p*-aminobenzenesulfonamido)-thiazole)	Readily absorbed. Usually readily excreted. Low toxicity. Tolerated well.	Amount present in specific fluids is inconstant. Does not pass readily into spinal fluid.	Staphylococcus infections, gonorrhea. Used in ointment bases topically for local infections.	Preferred by some in urinary infections. As effective as sulfanilamide and sulfapyridine in pneumococcus pneumonia. Suggested as prophylaxis in gonorrhea and chancroid, and for local use in larynx and nasopharynx.
Sulfadiazine (2-(*p*-aminobenzenesulfonamido)-pyrimidine)	Readily absorbed. Low toxicity. High therapeutic activity.	Excreted slowly.	Widely used in pneumococcus, streptococcus, staphylococcus, and meningococcus infections.	Topical and intraperitoneal implantation in peritonitis.
Sulfamerazine (2-sulfanilamido-4-methylpyrimidine)	Rapidly and highly absorbed from gastrointestinal tract resulting in higher blood levels and longer intervals between doses. More efficient therapeutically than sulfadiazine on dosage basis.	Excreted slowly.	Vibrio cholera.	Calf diphtheria.
Sulfaguanidine (sulfanilyl guanidine monohydrate)	Effective in intestinal tract infection due to poor absorption.	Poorly absorbed—low blood levels. Ineffective in presence of ulcerating lesions of intestinal mucosa.	Acute bacillary dysentery; preoperative and postoperative prophylaxis in colon surgery.	Intestinal form of canine distemper.
Succinylsulfathiazole (2-(N4-succinylsulfanilamido)-thiazole; sulfasuxidine)	Effective in intestinal tract infection due to poor absorption. Effective in presence of ulcerating lesions of intestinal mucosa.	Poorly absorbed—low blood levels.	Bacillary dysentery in acute and carrier stages; preoperative and postoperative prophylaxis in abdominal and intestinal surgery.	Bacillary white diarrhea of chickens.

Antibiotics.   The term literally means antigrowth substances **or** growth inhibitors.   Recently it has come to indicate bacteriostatic substances produced by microorganisms, and it is in this narrower sense that it is used here.

Mode of Action.   Antibiotics interfere with growth and metabolism of microorganisms, which gradually die.   According to Waksman, they may act in various ways:

(a) Substituting for essential nutrient.

(b) Interfering with vitamin utilization.

(c) Modifying intermediary metabolism of bacterial cell.

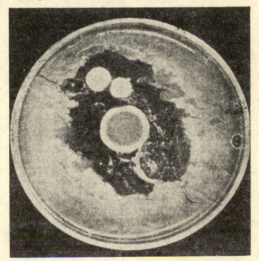

Courtesy, Merck & Co.

Fig. 23.   Colonies of *Penicillium notatum*, showing surrounding zone of bacterial inhibition.

(d) Combining with substrate or one of its constituents, which is thereby rendered inactive for bacterial utilization.

(e) Competing for enzyme required by bacteria.

(f) Interfering with respiratory mechanism, especially hydrogenase system.

(g) Inhibiting cellular oxidation directly.

(h) Acting as enzyme system and producing in the medium oxidation products such as peroxides injurious to bacterial cell.

(i) Favoring certain lytic mechanism in the cell.

(j) Affecting surface tension of bacteria, acting as a detergent.

All the antibiotics are much more active (bacteriostatic in higher dilutions) than the sulfonamides, but less active than the dyes.

TABLE OF ANTIBIOTICS

Antibiotic	Source	Action	Administration
Bacitracin	*Bacillus licheniformis*	Inhibits Gram-positive organisms. gonococci.	Local
Polymixin A, B, C, and D	*Bacillus polymyxa*	Intestinal antiseptic. Toxic.	Local
Tyrothricin	*Bacillus brevis* grown in soil	Inhibits Gram-positive bacteria and cocci.	Local
Carbomycin Erythromycin	*Streptomyces halstedii* *Streptomyces erythraeus*	Bacteriostatic to Gram-positive and some Gram-negative organisms that are resistant to penicillin. Nontoxic.	Oral
Chloramphenicol	*Streptomyces venezuelae*	Broad spectrum against bacterial and rickettsial infections, viral psittacosis.	Oral
Chlortetracycline (Aureomycin)	*Streptomyces aureofaciens*	Broad spectrum against Gram-positive and Gram-negative organisms, rickettsia, and some viruses.	Oral and intravenous
Neomycin Viomycin	*Streptomyces fradiae* *Streptomyces puniceus*	Resemble streptomycin, but toxic to kidneys and ears.	Oral and parenteral
Streptomycin	*Streptomyces griseus*	Active against acid-fast and Gram-negative bacilli, *H. influenzae*, *B. recurrentis*, salmonelloses, pulmonary tuberculosis. May injure 8th cranial or auditory nerve.	Many routes, parenteral

Antibiotic	Source	Action	Administration
Dihydrostreptomycin	Derivative of Streptomycin	Resembles streptomycin, but less toxic clinically.	Oral (never intravenous)
Oxytetracycline (Terramycin)	*Streptomyces rimosus*	Broad spectrum against bacteria, rickettsia, spirochetes, some viruses, typhoid, and amebiasis. Nontoxic.	Oral
Tetracycline (Achromycin)	*Streptomyces* Texas soil	Resembles the spectrum of chlortetracycline.	Oral and intravenous
Penicillin F, G, X, and K (dihydro F and G being most often used) Procaine penicillin G	*Penicillium notatum* and *P. chrysogenum*	Bacteriostatic and bacterial against streptococci, staphylococci, pneumococci, meningococci, syphilis. Some bacteria May be toxic. may produce penicillin-destroying enzyme penicillinase.	All routes, parenteral and oral
Penbritin	Derivative of Penicillin	Bactericidal to a wide range of Gram-positive and Gram-negative organisms in minimum inhibitory concentration (MID). Spectrum includes *Salmonella* and *Shigella* species, *H. influenzae*, *N. catarrhalis*, *E. coli*, *K. pneumoniae*, *S. paratyphi A* and *B*, *E. typhi*, septicemias, and wound infections. Not effective against penicillinase-producing staphylococci.	Oral
Fumagillin	*Aspergillus fumigatus*	Non-bactericidal, non-antiviral, non-fungicidal, but destructive to *Endamoeba histolytica*.	Oral
Fumigacin (Helvolic acid)	*Aspergillus fumigatus* strains	Bacteriostatic against Gram-positive organisms.	Oral
Interferon (Experimental)	Somatic or kidney cells	Apparently all virus diseases, including the common cold.	Various routes

Penicillin. This substance, produced by the mold *Penicillium notatum,* is the most widely investigated of the antibiotics. It was originally found by Alexander Fleming in England in 1929. Its successful isolation, clear proof of its clinical usefulness, its assay and dosage, and facts about its excretion from the body are the work of Howard Florey and coworkers in 1940. It was crystallized in 1946.

The formula for penicillin is $C_9H_{11}O_4SN_2R$. It has been produced synthetically; reports indicate that the antibacterial activity of the synthetic product may equal that of the original fungi strains. Penicillin filtrates contain several compounds, known as F, G, X, and K. Penicillin G is the type usually produced commercially, with one international unit equal to the activity of 0.6 micrograms of pure sodium penicillin G. The troublesome resistance of some microorganisms to penicillin is due to an enzyme, penicillinase, which destroys penicillin.

COMMERCIAL PREPARATIONS

Bicillin (Benzanthine penicillin G)
Declomycin (Demethylchlortetracycline) Erythrocin
Cosa-Signemycin (Tetracycline plus triacetyloleandomycin with glucosamine)
Cosa-Tetrastatin (Tetracycline with nystatin and glucosamine)
Erythromycin ester
Propionyl erythromycin ester lauryl sulfate

BIBLIOGRAPHY

Cowan, S. T., and E. Rowatt, eds. *The Strategy of Chemotherapy.* Eighth Symposium of the Society for General Microbiology. Cambridge: Cambridge University Press, 1958.

Fleming, Sir Alexander. *Penicillin, Its Practical Application.* Philadelphia: The Blakiston Company, 1946.

Jukes, T. H. *Antibiotics in Nutrition.* Antibiotics Monograph No. 4. New York: Medical Encyclopedia, Inc., 1955.

Karel, L., and E. E. Roach. *A Dictionary of Antibiosis.* New York: Columbia University Press, 1951.

Rahn, Otto. *Injury and Death of Bacteria by Chemical Agents.* Normandy, Mo.: Biodynamica, 1945. Pp. 14–168.

Reddish, G. F., ed. *Antiseptics, Disinfectants, Fungicides and Chemical and Physical Sterilization.* Second Edition. Philadelphia: Lea and Febiger, 1957.

Schnitzer, R. J., and E. Grumberg. *Drug Resistance of Microorganisms.* New York: Academic Press, 1957.

CHAPTER VII

YEASTS AND MOLDS

Although yeasts and molds are not in the same class (Schizomycetes) as bacteria, a brief study is presented here because they have many properties in common with bacteria and are often encountered together with them in nature and as culture contaminants.

IMPORTANCE

Many yeasts and molds are of great importance in industry, desirable because of their ability to produce fermentation products such as isopropyl alcohol and acetone, their usefulness in baking and cheese-making, etc., and undesirable because of their decaying and mildewing activities. They bring about significant changes in soil. Some cause plant or animal diseases. Several species of molds, like Penicillium and Aspergillus, produce powerful antibiotics.

GENERAL CHARACTERISTICS

Yeasts and molds are nonchlorophyllic, parasitic or saprophytic plants, reproducing generally by means of spores and infrequently by the sexual method. Some are unicellular (yeasts); most are multicellular filamentous types.

Some of the terms used in identifying yeasts and molds are:

Thallus. The actively growing vegetative portion of a fungus.

Hypha. A threadlike portion or individual filament, several of which make up the thallus.

Mycelium. A network or mass of branched hyphae.

Septa. Transverse partitions dividing a hypha into chains of cylindrical cells.

Fuseaux. Terminal, elongated, thick-walled, usually septate structures from which chlamydospores develop.

Stolon. A branch of the mycelium which extends outward, giving rise to a new plant body where it comes in contact with substratum.

Rhizoids. Branches along the stolons resembling plant roots.

Spirals. Terminal coils on hyphae, found in some species.

Pectinate Bodies. Comblike structures formed by some fungi.

Spores.   Cells which are set aside for reproduction.   They separate from the thallus, germinate after a period, and produce a new thallus.   Spores may be of the following types:

Conidia.   Asexual reproductive spores formed from the thallus by abstriction, budding, or septate division.

Endospores.   Spores formed within the membrane of the fungus.

Ascospores.   Spores, usually 2, 4, or 8 (the number limited to the species producing them), which are enclosed in a sac, the *ascus.*

Oöspores.   Spores that result from the fertilization of two dissimilar cells, namely a female cell and a male cell.

Zygospores.   Spores that result from the union of two similar cells.

Blastospores.   Spores which are formed by budding.

Arthrospores.   Spores which are formed by the segmentation of a hypha into a chain of cells which become separated.

Oidia.   Arthrospores which are cylindrical in shape.

Chlamydospores.   Large spores with a tough wall which are formed from the hypha by a collection of the protoplasm in a part of the filament which has become swollen.

Thallospores.   Spores formed directly from the hypha, as blastospores, arthrospores, and chlamydospores.

Basidiospores.   Sexual reproductive spores produced on a basidium, the number being limited and constant for a species.

Sterigma.   A short stalk which bears chains of conidia and is present in Aspergillus.

Conidiophore.   A hypha which bears one or a group of spores.

Coremium.   Group of conidiophores which occur in bunches.

Vesicle.   The enlarged end of a hypha upon which there are groups of spores.

Sporangium.   A sac situated at the end of a hypha and containing an indefinite number of asexual spores.

Sporangiophore.   A hypha having a sporangium.

Columella.   The end of a hypha, forming a centrally located support for the sporangium.

Basidium.   A special club-shaped stalk which bears basidiospores.

CLASSIFICATION

Classification of the yeasts and molds is still confused.   Relationships among groups are not clear; pleomorphism is common.   The following broad classification is fairly generally accepted and will serve to indicate the place of the yeasts and molds in the plant kingdom:

Phylum.   *Thallophyta.*   Plants without roots, stems, or leaves.

Subphylum I.   *Algae.*   Possess chlorophyll.

Subphylum II.   *Fungi.*   Do not contain chlorophyll.

Class 1.   *Schizomycetes.*   Bacteria.

Class 2.   *Phycomycetes.*   Alga-like fungi, with nonseptate mycelium.   Includes water and bread molds, mildews, blights, rots.

Class 3.   *Ascomycetes.*   Sac fungi, with septate mycelium if present.   Includes yeasts, Penicillium, Aspergillus.

Class 4. *Basidiomycetes.* Basidia fungi, with septate mycelium if present. Includes wheat rusts, smuts, true fungous plants as puff balls, mushrooms.

Class 5. *Fungi imperfecti.* Fungi as yet unclassifiable in other groups, with septate mycelium if present.

Classes 2, 3, 4, and 5 comprise the *Eumycetes* or true fungi.

LABORATORY STUDY

Cultivation. Fungi generally grow best at room temperature (20° to 25° C.) or at 37° C., usually in a medium of pH 5.5 to 6.5. Most fungi are aerobic. Sabouraud's test medium or conservation medium, and the Pennsylvania medium are most frequently used for isolation. (See p. 79.)

Examination. Morphological characteristics of yeasts and molds are observed best in unstained preparations prepared with the least amount of disruption.

1. *Petri dish cultures* may be examined under low-power lenses.

2. *Slide Cultures.*

(a) The simplest procedure is to seed a tube of melted and cooled agar, and spread a little in a thin film on a sterile slide. Place in a Petri dish containing moistened blotting paper to prevent drying, and incubate. Place cover-slip over growth and examine unstained. Slide may also be dried, fixed, and stained.

(b) Cement a sterile cover-slip about 1 mm. above a sterile glass slide, using bits of glass to keep the cover-slip up. Introduce seeded agar between slide and cover-slip, and incubate.

(c) A special type of slide culture vessel may also be used.

3. *Lesion Material.* Place pus, skin scrapings, sputum, or exudate on a slide. Place over it a drop of hot 20% NaOH. Cover with clean cover-slip. Examine after 15 to 30 minutes, which is usually sufficient time for the alkali to dissolve the tissue cells without injuring the fungus if present.

PHYCOMYCETES

The mycelium is composed of hyphae which are unicellular and multinucleated, up until the time of reproduction. The Phycomycetes reproduce by spores as well as by sexual reproduction. They are active enzyme-producers and are prominent in the decomposition and spoilage of fruits, vegetables, and other organic matter. Included in this class are the common, black, bread mold, *Rhizopus nigricans*, that is so frequently encountered as a contaminant of bacteriological plates and in moldy bread, and *Mucor corymbifer*, which is occasionally found in human infections.

ASCOMYCETES

The mycelium is composed of septate filaments, with sexual endogenous spores in sacs. Ascospores and usually conidia are present. The yeasts and common molds are found in this group. They produce such plant diseases as bitter rot of apples, peach leaf curl, and ergot on rye grain.

Yeasts. Yeasts reproduce vegetatively by budding and form spores and ascospores under certain conditions. The yeast cell contains a small but definite nucleus and intracellular granules which may be glycogen or oil. The usual method of reproduction of yeasts is by

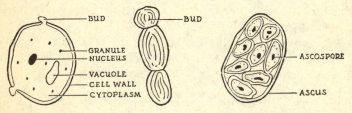

FIG. 24.   Left to right: A typical yeast cell; yeast showing budding; ascospores.

budding. Ascospores, when formed, germinate by bursting through the cell wall. The false yeasts or Mycodermata are included among the Fungi imperfecti. True yeasts are unicellular fungi.

Yeasts are noted for their high fermentative powers. They produce chiefly alcohol and carbon dioxide from sugar under anaerobic conditions. They are utilized in the manufacture of wines and beer and in the commercial production of alcohol. In the baking industry, yeasts are used to raise bread by the formation of bubbles of carbon dioxide. Some yeasts synthesize vitamins, such as riboflavin. Others, because of their own vitamin requirements, are used in vitamin assay.

Saccharomyces cerevisiae is used in brewing and baking; *Saccharomyces ellipsoideus*, in wine-making.

Penicillium. This genus is characterized by the formation of spores in chains at the head of the mold, giving it the appearance of a brush. They contribute to the spoilage of various organic materials, especially ripe fruits. Among the important varieties of Penicillium are: (a) *Penicillium expansum*—cause of common soft rot of apples in storage. (b) *Penicillium digitatum* and *Penicillium italicum*—cause of spoilage of citrus fruits, especially oranges. (c) *Penicillium camemberti*—produces a ripening and characteristic flavor in camembert cheese. (d) *Penicillium luteum var. pinophilum*—produces stains on wood and is used in production of gluconic acid. (e) *Penicillium brevicaule*—very active in proteolysis and causes spoilage of cheese and other dairy products. (f) *Penicillium notatum*—produces the antibiotic penicillin (see p. 92).

Aspergillus. On culture media Aspergilli appear felted. The conidia in this genus are borne on hyphae which end in a round enlarged head, and are so numerous that under the microscope the mold

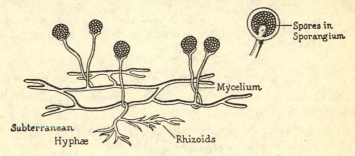

Rhizopus Nigricans (Bread Mold)

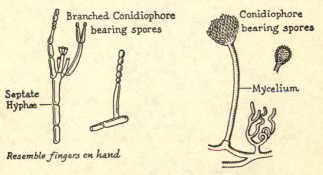

Penicillium Mold   Aspergillus Mold

Fig. 25.   Types of molds.

appears as a round mass of colored spores. They decompose a wide variety of waste matter, as well as causing spoilage of foods, wood, paper, and other organic commodities.

Among the Aspergilli are: (a) *Aspergillus flavus*—recognized by its spiny, septate conidiophores; widely distributed on grains and in soil. (b) *Aspergillus oryzae*—possesses smooth nonseptate conidiophores and is one of the most important of the industrial ferments.

(c) *Aspergillus nidulans*—of bright green color and often encountered as a contaminant in bacteriological work. It has been found in cases of otomycosis, an ear disease. (d) *Aspergillus niger*—recognized by its rather high and widely separated conidiophores terminating in a relatively enormous black, globular spore-head. It causes spoilage of foodstuffs. (e) *Aspergillus glaucus*—frequently found on decaying vegetables, jams, damp clothing, and as a contaminant in bacteriological work. (f) *Aspergillus fumigatus*—pathogenic, especially for birds, being the cause of aspergillosis of the air sacs in these animals. It also causes lesions of the lungs in domesticated mammals, especially horses and cattle.

BASIDIOMYCETES

The Basidiomycetes are characterized by septate mycelia, massed hyphae, and exogenous sporangia borne on stalks, or basidiospores. In this group are included smuts and weeping fungi, mushrooms, rusts, etc. No animal pathogens are found in this class.

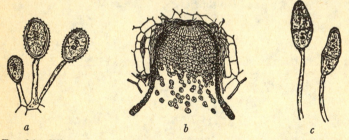

FIG. 26.   Wheat rust.   (*a*) Summer spores of the wheat rust which form the rusty lines on the wheat; notice the two nuclei in each spore. (*b*) A cluster-cup (on barberry) of the wheat rust containing rows of spores; each spore contains two nuclei. (*c*) Winter spores of the wheat rust; each spore has two cells, and each cell has two nuclei.

FUNGI IMPERFECTI

The Fungi imperfecti are so called because the "perfect" or sexual forms and structures resulting from conjugation have never been observed. The mycelium of this group of fungi is septate and varies greatly in diameter from bacillary size to a magnitude equal to that of other fungi (more than 1μ in diameter). Various types of spores are formed, except the sexual. Most of the fungi pathogenic for man are in this group, which includes the genera Cryptococcus, Pityrosporum, Torula, Mycoderma, Oöspora, Monilia, Oidium, Hormodendrum, Malassezia, Madurella, Indiella, and others. Some authorities place the genera Blastomyces and Coccidioides in this group. As there is no general agreement on the relationships among these organisms, classification is difficult and has not been standardized.

Classification by Types of Diseases Produced. Although many investigators have their own methods of classifying the Fungi imperfecti, a possible classification is offered here, based upon the characteristic diseases caused by certain genera.

Group	Diseases Produced	Fungus	Type of Involvement
Group A	Dermatomycoses Ringworm Favus	Microsporon Trichophyton Epidermophyton Achorion	Usually primary cutaneous lesions, with pus and no definite systemic involvement.
Group B	Ulcerative, granulomatous, exfoliative, suppurative, and vesicular types of dermatitis. Thrush.	Monilia Blastomyces Coccidioides Sporotrichum	Usually primary mucous membrane and skin infections, with frequent systemic involvement.
Group C	Cerebral and pulmonary granulomata and abscesses due to Torula. Otomycosis and focal pulmonary lesions due to Aspergillus.	Torula Aspergillus	Usually primary systemic involvement, with occasional instances of skin or mucous membrane infection.

Fungi and Diseases Produced by Them.*

Organism	Disease	Characteristics
Achorion schoenleinii	Favus, a disease of the scalp which occasionally attacks glabrous skin and nails. It is characterized by the formation of small, yellow, cup-shaped crusts at the mouths of hair follicles.	Central portion of crust is made up of rounded sporelike bodies. The hyphae are crooked and of irregular contour, often composed of chains of oval cells. Pear-shaped conidia are found on the sides of more delicate filaments. Chlamydospores are attached to clubbed tips of hyphae or occur in chains.
Microsporon audouini	Ringworm or tinea. Common form is tinea of the scalp which affects only children. It is highly contagious. Tinea of the body may occur secondary to scalp lesions in children or as a primary infection at any age.	In cultures, long, straight, coarse trunks radiate from center, giving off branches which form intricate tangle of threads; some terminal branches bear conidia; also form chlamydospores and fuseaux; characterized by a smooth colony growth.

* This table is necessarily incomplete and is intended to serve only as an outline for further study.

Fungi and Diseases Produced by Them (Cont.).

Organism	Disease	Characteristics
Trichophyton	**Tinea cruris,** ringworm on insides of thighs, groin, and genitals, characterized by brownish, eczematous patches. **Dhobie itch,** tropical groin eczema similar to tinea cruris but more intensely inflammatory.	Appear as chains of sporelike elements about 3μ in diameter. Possess conidia attached to hyphae by pointed tips, frequently arranged in clusters. Colony growth is rough.
Epidermophyton inguinale	**Eczema marginatum** or ringworm of the groin.	No conidia are formed, but the organism is characterized by innumerable blunt fuseaux which are borne on aerial hyphae, often in clusters, like a hand of bananas. Older cultures have chlamydospores. Colony is greenish-buff.
Mixed hyphomycetic infection	**Interdigital ringworm,** common name "Athlete's Foot." Interdigital exfoliation, with a contagious weeping exudate.	Various fungi have been isolated from infections of this type, including Monilia, Trichophyton, and Epidermophyton.
Torula histolytica	**Torula infection,** which rarely causes lesions of the skin, appears to have definite affinities for tissues of central nervous system and the lungs.	Yeastlike fungi; reproduce only by budding, do not produce mycelium or endospores, and do not ferment carbohydrates.
Monilia albicans	**Moniliases,** a group of clinicopathological conditions of the skin, mucous membranes, and internal viscera associated with the presence of Monilia fungi in the lesions, secretions, excretions, or exudates of affected tissues or organs. **Thrush,** a disease of the mouth characterized by creamy patches on an area of catarrhal inflammation and usually associated with malnutrition.	Yeastlike fungi; ferment sugars with gas formation, reproduce by budding, produce segmented, branched mycelium, and form spores or conidia. Oval or round cells possess budding forms and mycelial threads which are often long and branched. Blastospores are formed.
Blastomyces hominis	**Blastomycoses,** a group of closely allied clinical conditions characterized by the presence of certain granulomatous lesions in the cutaneous type of disorder and by pulmonary, osseous, meningeal, and other visceral manifestations in the systemic type of infection.	Round or oval, highly refractile bodies, $5-20\mu$ in diameter, containing granules of various sizes, and often vacuoles. They are surrounded by hyaline capsules. Budding forms are seen. On glucose agar the colony is round, white, glistening, and smooth (not unlike colonies of *Staphylococcus albus*).

Fungi and Diseases Produced by Them (Cont.).

Organism	Disease	Characteristics
Coccidioides immitis	**Coccidioidal granuloma,** which may be localized and relatively benign in character, or may be systemic, malignant, and rapidly fatal.	In tissues and pus the fungus appears as a sphere with contoured capsule, 5–50μ in diameter, containing a granular protoplasm. Reproduction is by endosporulation. Colonies are small, elevated, round, paraffin-like plaques covered with snowlike flakes. Sugars are not affected.
Sporotrichum schencki	**Sporotrichosis,** a chronic or subacute infectious disease. It is characterized by cutaneous and subcutaneous lesions resembling syphilitic gummata, which are clinically diagnostic. Occasionally, there is involvement of muscles, bones, and joints. In typical cases the lesions spread along the lymphatics which gradually soften and ulcerate.	In pus, Gram-positive, oval or cigar-shaped bodies, 2–10μ in length and 1–3μ in width are sometimes found. They are finely granular and basophilic, surrounded by a delicate capsule. The colony is a mass of tangled, creeping, branched, septate mycelia. Spores are 3–6μ long and 2–4μ wide, pear-shaped, and situated along the length of the filaments on slight stems.

BIBLIOGRAPHY

Alexopoulos, C. *Introductory Mycology.* New York: John Wiley and Sons, 1952.

Christensen, C. *The Molds and Man.* Minneapolis: University of Minnesota Press, 1951.

Conant, Norman F., *et al. Manual of Clinical Mycology.* Second Edition. Philadelphia: W. B. Saunders Company, 1954.

Cook, A. H., ed. *The Chemistry and Biology of Yeasts.* New York: Academic Press, 1958.

Flynn, J. E., and T. J. MacCreigh. *Biological Extracts,* pp. 1228–1269. Philadelphia: 1949.

Lindegren, G. C. *The Yeast Cell, Its Genetics and Cytology.* St. Louis: Educational Publishers, 1949.

Lodder, J., and N. J. W. Kreger-van Rij. *The Yeasts: A Taxonomic Study.* New York: Interscience Publishers, 1952.

Raper, K. B., and C. Thom. *Manual of the Penicillia.* Baltimore: Williams & Wilkins Company, 1949.

Roman, W., ed. *Yeasts.* New York: Academic Press, 1957.

Skinner, C. E., C. W. Emmons, and H. H. Tsuchiya. *Henrici's Molds, Yeasts and Actinomycetes.* Second Edition. New York: John Wiley and Sons, 1947.

Smith, G. M. *Cryptogamic Botany. I. Algae and Fungi.* New York: McGraw-Hill Book Company, 1955.

Thom, C., and K. B. Raper. *A Manual of the Aspergilli.* Baltimore: Williams & Wilkins Company, 1945.

Wolf, F. A., and F. T. Wolf. *The Fungi,* Vols. I and II. New York: John Wiley and Sons, 1947.

Chapter VIII

SOIL BACTERIOLOGY

Soil fertility is largely dependent upon bacterial activity. Hundreds of species of bacteria are present in different soils. These microorganisms vary in the different soils found in various parts of the world. The nature and extent of the soil bacterial population depend upon environmental conditions, such as favorable moisture, hydrogen ion concentration, temperature, available food supply, and aeration. Soil bacteria are absolutely essential to all life processes, for without putrefaction and decay there would be no decomposition of dead plant and animal matter. Soil microbes in the form of putrefactive bacteria decompose manure and plant or animal matter with the resultant liberation of simple chemical substances, such as sodium nitrate, calcium phosphate, sodium chloride, etc. Growing green plants can utilize these decomposition products and synthesize them in the process of photosynthetic food manufacture. The soil flora includes aerobic and anaerobic, proteolytic and carbohydrate-splitting organisms which are capable of bringing about decomposition and decay, and in addition bacteria which reduce nitrates to nitrites, bacteria which utilize the nitrogen of the air and ammonia, forming nitrates, and bacteria which utilize sulfur, iron, phosphorus, and manganese compounds, etc.

NITROGEN CYCLE

The constant withdrawal of nitrogenous substances from the soil by plants would soon lead to total depletion of soil nitrogen were it not for certain microorganisms continually at work replenishing the supply. Nitrogen-fixing bacteria convert atmospheric nitrogen to compounds utilizable by plants and nitrifying bacteria convert ammonia to utilizable nitrates.

Nitrates are used by plants to synthesize proteins which are returned to the soil as plant protein, animal protein, or urea. The putrefactive and urea bacteria act on these compounds and liberate the nitrogen as ammonia, which is then acted upon by the nitrifying bacteria.

Nitrogen-Fixing Bacteria. These organisms derive their energy from many of the simple carbohydrates, and their nitrogen from the atmosphere in the elementary or gaseous form. The nitrogen-fixing bacteria are usually found in the roots of various beans, peas, lupines, clovers, and other leguminous plants. The organisms gain entrance to the young roots to form a swelling or tubercle on the root. The bacteria (of the genus Rhizobium) and the leguminous plants have a

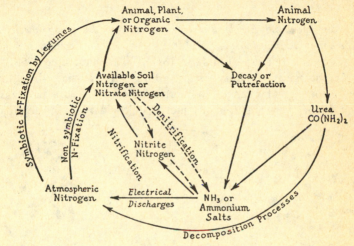

Fig. 27. The nitrogen cycle. (After Waksman.)

symbiotic relationship (living together of two kinds of organisms with mutual benefit). There are also nonsymbiotic nitrogen-fixing bacteria which include species of the genus Azotobacter (aerobic forms) and the genus Clostridium (anaerobic forms).

Symbiotic nitrogen-fixing bacteria do not fix nitrogen in culture media, but when grown in symbiosis with a plant, in neutral or nearly neutral soil, are able to fix atmospheric nitrogen to make it available to the host plant. The Azotobacter species operate in slightly alkaline soils, and must have trace amounts of molybdenum for their nitrogen-fixing activities. Nitrogen-fixation by *Clostridium pasteurianum* occurs in boggy acid soils.

Symbiotic nitrogen-fixation is utilized to increase soil fertility for nonleguminous crops in the following ways:

1. Inoculation with Rhizobium is successful when proper methods are used. However, the organisms do not persist indefinitely, usually disappearing or greatly decreasing in three years from legume-free soil.

2. Soil is enriched by cultivation and plowing under of legumes, because of the large amounts of combined nitrogen found in the root nodules, as well as the valuable organic material in the whole plants.

3. Nonleguminous crops have long been grown in association with legumes for greater yield of the former. The legumes apparently excrete fixed nitrogen into the soil.

RHIZOBIUM

Type Species. *Rhizobium leguminosarum* (*Bacillus radicicola, Pseudomonas radicicola*). 1. Habitat—soils in which legumes are grown. 2. Discoverers—Frank, in 1877, and Beijerinck, in 1888.

Morphological Characteristics. 1. Form—rod-shaped. "Life cycle" shows great changes in morphology. There are nonmotile coccoid forms; very small, highly motile, ellipsoidal forms; bacteroids

 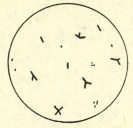 

Fig. 28. Bacteroid forms of *Rhizobium leguminosarum*. | Fig. 29. *Rhizobium leguminosarum*, showing common X and Y forms. | Fig. 30. *Clostridium pasteurianum*.

commonly irregular with X, Y, star- and club-shaped forms. Vacuolate forms predominate. It is suggested by Lewis that the "life stages" are merely periodic appearances due to the accumulation of food granules, fat vacuoles, etc. 2. Size—0.5 to 0.9μ by 1.2 to 3μ. 3. Nonsporing, may be encapsulated. 4. Staining properties—does not stain readily, the best stains being aniline gentian violet and carbol fuchsin. 5. Gram-negative and non-acid-fast. 6. Motility —motile with peritrichous flagella.

Cultural Characteristics. The organism is difficult to cultivate on ordinary culture media. The growth on mannitol agar is rapid, with a tendency to spread. The streak is raised, glistening, semi-translucent, white, slimy, and occasionally viscous.

Physiological Characteristics. 1. Optimum temperature, 25° C. 2. Strictly aerobic. 3. Slight acid production from dextrose, galactose, mannose, lactose, and maltose. 4. Forms root nodules on leguminous plants of the genera Lathyrus, Pisum, Vicia, and Lens, thus making it possible for the plants to utilize atmospheric nitrogen.

Related Species. 1. *Rhizobium trifolii* causes root nodule formation on species of Trifolium (clover). 2. *Rhizobium phaseoli* produces root nodules on species of Phaseolus (bean). 3. *Rhizobium lupini* causes nodule formation on the roots of members of the genera Lupinus, Serradella, and Ornithopus. 4. *Rhizobium japonicum* forms nodules on the roots of the soy bean. 5. *Rhizobium meliloti* produces nodules on the roots of the genera Melilotus, Medicago, and Trigonella (alfalfa and sweet clover).

AZOTOBACTER

Type Species. *Azotobacter chroococcum.* 1. Habitat—majority of field soils. 2. Discoverer—Beijerinck, in 1901.

Morphological Characteristics. 1. Form—short, thick rods with rounded ends; large ovoid forms, like giant diplococci; pear-shaped rods, and other forms. It tends to undergo cyclogenic changes from large yeastlike forms to coccoid or rod types, and amorphous stages. 2. Cell groupings—arranged singly or in pairs end to end; in old cultures appear in packets. 3. Size—2 to 3μ by 3 to 6μ. 4. Nonsporing; organisms surrounded by a slimy membrane of variable thickness, usually becoming brownish in older cultures. 5. Gram-negative and non-acid-fast. 6. Motility—motile by a polar flagellum.

Cultural Characteristics. 1. Gelatin stab—only slight growth, with no liquefaction. 2. Litmus milk—becomes clear in 10 to 14 days. 3. Potato—slight, glossy, slimy to wrinkled growth which may become yellowish, brownish-yellow, or chocolate brown. 4. Nutrient broth—no growth even in the presence of dextrose.

Physiological Characteristics. 1. Optimum temperature, 25° to 28° C. 2. Strictly aerobic. 3. Fixes atmospheric nitrogen, giving off CO_2, utilizing dextrose, maltose, lactose, starch, glycerol, ethyl alcohol, acetate, citrate, lactate, etc. 4. Forms dark brown or black pigment, which is insoluble in water, alcohol, ether, and chloroform.

Related Species. 1. *Azotobacter agile* also actively fixes atmospheric nitrogen and gives off CO_2, and is widely distributed in water and soil. 2. *Azotobacter vinelandii.*

Nitrification. This process involves the conversion of ammonia by autotrophic bacteria to the oxidized inorganic compounds of nitrogen such as nitrites and nitrates. Ammonia is liberated into the soil as a result of the decay of proteins of dead organisms. Two steps are recognized in the process of nitrification. The first, which is the oxidation of ammonia to nitrite, is carried out by bacteria of the **genera** Nitrosomonas and Nitrosococcus.

$$2 \text{ NH}_3 + 3 \text{ O}_2 \rightleftharpoons 2 \text{ HNO}_2 + 2 \text{ H}_2\text{O} + \text{energy liberated.}$$

The second, which is the oxidation of nitrite to nitrate, is performed by members of the genus Nitrobacter.

$$2 \text{ HNO}_2 + \text{O}_2 \rightleftharpoons 2 \text{ HNO}_3 + \text{energy liberated.}$$

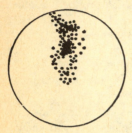

Nitrosomonas, Nitrosococcus, and Nitrobacter occur together in common soils, manure piles, sewage, and river water.

FIG. 31. *Nitrosococcus nitrosus.*

NITROSOMONAS

Type Species. *Nitrosomonas europaea.* 1. Habitat—soil. 2. Discoverer—Winogradsky, in 1892.

Morphological Characteristics. 1. Form —rod-shaped or spherical. 2. Cell groupings—arranged singly or aggregated into a mass by some slightly viscous substance; occurs rarely in chains of three to four. 3. Size— 0.9 to 1μ by 1.1 to 1.8μ. 4. Motility—motile by polar flagellum three to four times the rod length.

Cultural Characteristics. The organism grows readily in fluid medium without organic matter, but the medium should contain ammonium sulfate, potassium phosphate, and basic magnesium carbonate.

Physiological Characteristics. 1. Strictly aerobic. 2. Ammonia and ammonium salts are oxidized to nitrites.

Related Species. 1. *Nitrosomonas monocella*, and 2. *Nitrosomonas groningensis*, all of which are widely distributed in soil and convert ammonia and ammonium salts to nitrites. 3. *Nitrosococcus nitrosus* (*Nitrosococcus americanus*), and *Nitrosocystis javenensis* (*Nitrosomonas javenensis*), although of different genera, are related to these organisms and are widely distributed in soil, accomplishing the conversion of ammonium salts to nitrites.

NITROBACTER

Type Species. *Nitrobacter winogradskyi.* 1. Habitat—soil. 2. Discoverer—Winogradsky, in 1892.

Morphological Characteristics. 1. Form—rod-shaped, with gelatinous membrane. 2. Size—0.6 to 0.8μ by 1 to 1.2μ. 3. Staining properties—does not stain easily. 4. Gram-negative. 5. Nonmotile.

Cultural Characteristics. Can be cultivated on media free of organic matter. 1. Silicic acid jelly—small colonies, light brown, circular to irregular. 2. Mineral broth—flocculent sediment after 10 days.

Physiological Characteristics. 1. Optimum temperature, 25° to 28° C. 2. Strictly aerobic. 3. Nitrites are oxidized to nitrates.

Related Species. 1. *Nitrobacter flavum* grows on ordinary laboratory media and hydrolyzes cellulose. 2. *Nitrobacter agiles*. 3. *Nitrobacter opacum* grows on ordinary laboratory media, produces an orange pigment, hydrolyzes cellulose. 4. *Nitrobacter punctatum* hydrolyzes cellulose.

Denitrification. Denitrifying bacteria reduce nitrates to nitrites, nitrogen, or ammonia, e.g., *Flavobacterium denitrificans*, *Pseudomonas fluorescens*, *Thiobacillus denitrificans*, etc. Large quantities of unfermented organic matter stimulate the denitrification process; therefore excessive covering of soil with unrotted manure should be generally avoided. Denitrification occurs when the soil is sour, lightly packed, or water-logged so that anaerobic conditions prevail.

Decay and Putrefaction. Decay, the aerobic decomposition of protein, is brought about by a variety of bacteria including *Proteus vulgaris*, *Pseudomonas aeruginosa*, and many aerobic spore-formers like *Bacillus mycoides*. Putrefaction, the anaerobic decomposition of protein, is accomplished by a number of organisms, e.g., *Clostridium putrificum*.

The complete decomposition of protein to ammonia can be carried on by only certain microorganisms, e.g., *Bacillus mycoides*, *Proteus vulgaris*, and some molds and Actinomyces. The soil flora, however, comprises enough different kinds of organisms to effect the decomposition completely and quickly, one species attacking the decomposition products of another in a sort of chain reaction.

BACILLUS

A. **Bacillus Mycoides.** 1. Family—Bacillaceae. 2. Habitat—milk, water, soil, dust. 3. Discoverer—Flügge, in 1886.

Morphological Characteristics. 1. Form —large rod-shaped organisms with truncated

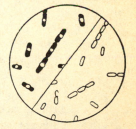

Fig. 32. *Bacillus mycoides*, showing equatorial spores, and young cultures without spores.

or slightly rounded ends. 2. Cell groupings—occurs singly, in pairs, small groups, and chains. 3. Size—0.8 to 1.3μ by 2 to 6μ. 4. Spores formed which are large, equatorial, and ellipsoidal, measuring 1 to 1.6μ by 2 to 2.25μ, and appearing within 24 to 48 hours. 5. Gram-positive and non-acid-fast, and stain with ordinary aniline dyes.

Cultural Characteristics. 1. Agar cultures—colonies are grayish, spreading, and rhizoid. 2. Agar slant—growth rhizoid, whitish, soft, glistening, becoming dull. 3. Gelatin stab—arborescent, filamentous growth, with saccate to stratiform liquefaction. 4. Litmus milk—not coagulated, slow peptonization to amber-colored fluid. 5. Potato—abundant grayish-brown growth of viscous nature, with granular surface. 6. Nutrient broth—vigorous flocculent growth. 7. Blood serum—not liquefied.

Physiological Characteristics. 1. Optimum temperature, 30° C. 2. Aerobic, and facultative anaerobe. 3. Voges-Proskauer positive. 4. Ammonia is produced. 5. Nitrites are produced from nitrates. 6. Acid is formed from dextrose, sucrose, xylose, and dextrin. 7. Some strains secrete a lysin which dissolves certain bacteria. 8. Resistance—spores resist steam at 15 pounds pressure for one hour, but are killed in one-half hour at 20 pounds pressure.

Pathogenicity. Nonpathogenic to man and animals.

B. **Bacillus Megatherium.** 1. Family—Bacillaceae. 2. Habitat—dust, soil, water, milk. 3. Discoverer—de Bary, in 1884.

Morphological Characteristics. 1. Form—large rod-shaped organisms with slightly rounded ends and curved axis. 2. Cell groupings—occurs singly, in pairs, and in chains. Long unsegmented forms are common, and shadow forms are observed early. 3. Size—1 to 1.5μ by 3 to 6μ. 4. Spores formed which are equatorial, oval, or ellipsoidal, and do not bulge; formed freely in 24 hours. 5. Nonencapsulated. 6. Gram-positive and non-acid-fast. 7. Motility— motile by four to eight peritrichous flagella.

Cultural Characteristics. 1. Agar cultures—round, raised colonies which are white to creamy, easily emulsified, and 3 to 5 mm. in diameter. 2. Agar slant—smooth growth, glistening, butyrous, cream-colored. 3. Gelatin stab—abundant filiform growth with infundibuliform or saccate liquefaction. 4. Potato—thick spreading growth which is grayish-white with greasy luster. 5. Litmus milk— peptonization without coagulation. 6. Bouillon—slight flocculent growth. 7. Blood serum—not liquefied, but abundant, moist, creamy, yellowish growth.

Physiological Characteristics. 1. Optimum temperature, 35° C. 2. Aerobic, and facultative anaerobe. 3. Ammonia is produced. 4. Nitrates are reduced to nitrites. 5. Acid, but no gas, is formed in glucose, maltose, and sucrose. 6. Resistance—spores withstand 18 pounds of steam pressure for one hour, but are killed by 19 pounds of pressure within the same period of time.

Toxic Substances Produced. The organism produces a powerful hemolysin which lyses the red blood corpuscles of man, monkey, and guinea pig. It appears in broth cultures at 37° C. on the second or third day, increases to a maximum on the sixth or seventh day, and then diminishes slowly. For its formation, oxygen is essential. It is destroyed by exposure to 56° C. for 30 minutes. When this hemolysin is injected into goats, it is possible to produce an antihemolysin.

Pathogenicity. The organism is generally nonpathogenic. The hemolysin is fatal to mice and guinea pigs injected with 1 to 2 ml. intraperitoneally.

C. Bacillus Subtilis (Hay Bacillus). 1. Family—Bacillaceae. 2. Habitat—hay, dust, milk, soil, water. 3. Discoverers—Ehrenberg, in 1838, and Cohn, in 1872.

Morphological Characteristics. 1. Form—rods, straight or curved, with rounded ends. 2. Cell groupings—occurs singly or in short chains. 3. Size—3 to 4μ by 1μ. 4. Spores formed which are oval, equatorial, subterminal, and germinate laterally. They are 1.2μ by 0.6μ, and appear on agar in 18 hours. 5. Gram-positive and non-acid-fast. 6. Motility—motile by eight to twelve peritrichous flagella.

Cultural Characteristics. 1. Agar cultures—small grayish colonies, with darker crumbly center surrounded by a lighter periphery with a curled edge. Emulsification difficult, and surface finely granular. 2. Agar slant—thin growth, spreading, grayish-white, raised, opaque, sometimes wrinkled. 3. Gelatin stab—filiform growth with rapid infundibuliform, saccate, crateriform, or stratiform liquefaction. A thick white membrane is formed on the surface, which adheres to the sides of the tube. 4. Litmus milk—partially clotted, peptonized and decolorized from above downwards. 5. Potato—luxuriant, yellowish-white, raised, creamy growth, becoming pinkish, with vesicles over surface; later becoming dry and mealy. 6. Broth—turbid growth becoming clear with coherent surface growth. 7. Blood serum—spreading growth, with slow liquefaction.

Physiological Characteristics. 1. Optimum temperature, 37° C. 2. Aerobic, and facultative anaerobe. 3. Voges-Proskauer positive.

4. Ammonia is produced. 5. Nitrates are reduced to nitrites.
6. Acid, but no gas, is produced in glucose, maltose, and sucrose.
7. No indol is produced. 8. Hydrogen sulfide production is slight.
9. Resistance—spores resist boiling for hours.

Pathogenicity. Most strains are nonpathogenic. *Bacillus subtilis* may give rise to conjunctivitis, iridochoroiditis, and panophthalmitis in man. It occasionally invades the bloodstream in cachectic diseases.

Related Organisms in the Soil. 1. *Bacillus mesentericus*, 2. *Bacillus ubicuitarius*, 3. *Bacillus agri*, 4. *Bacillus mutabilis*, 5. *Bacillus rufescens*, 6. *Bacillus agrestis*. 7. *Bacillus anthracis*, 8. *Bacillus cereus*, 9. *Bacillus macerans*, 10. *Bacillus brevis*, 11. *Bacillus maculatus*, 12. *Bacillus palustris*, 13. *Bacillus granularis*, and 14. *Bacillus fusiformis*.

PROTEUS

Proteus Vulgaris (Bacillus Proteus, Bacillus Proteus Vulgaris).
1. Family—Enterobacteriaceae. 2. Habitat—soil, intestinal contents of man, water rich in organic matter, putrefactive materials.
3. Discoverer—Hauser, in 1885.

Morphological Characteristics. 1. Form—small, thin, pleomorphic bacilli. 2. Cell groupings—occurs singly, in pairs, and frequently in long chains. 3. Size—0.5 to 1μ by 1 to 3μ. 4. Gram-negative.
5. Motility—motile with numerous peritrichous flagella, though slightly motile forms with four flagella, two at each end, may occur.
6. Non-spore-former.

Cultural Characteristics. 1. Agar cultures—opaque, luxuriant, thin, moist growth spreading over the entire surface. 2. Agar slant—uniform, slightly raised, bluish-gray, spreading growth. 3. Litmus milk—reduction of litmus, production of alkali, coagulation of casein and subsequent digestion. 4. Gelatin stab—filiform growth, with rapid cup-shaped liquefaction, which is complete in 10 to 12 days.
5. Potato—abundant growth, creamy to yellowish-gray, turning brown, usually referred to as café-au-lait growth. 6. Broth—marked turbidity, usually with thin pellicle.

Physiological Characteristics. 1. Optimum temperature, 37° C.
2. Aerobic, and facultative anaerobe. 3. Hydrogen sulfide is formed.
4. Ammonia is formed. 5. Nitrates are reduced to nitrites. 6. Acid and gas are produced from glucose, galactose, levulose, maltose, and sucrose. 7. Indol is produced. 8. Resistance—the organism is killed by moist heat at 55° C. in one hour.

Toxic Substances Produced. No soluble toxin is formed. *Proteus vulgaris* produces a hemolysin which acts on horse blood.

Serological Reactions. Certain strains, the so-called X strains isolated by Weil and Felix from the urine of patients with typhus fever, are important in the Weil-Felix reaction, which is diagnostic for typhus fever. After the fourth day of typhus fever the serum of patients contains agglutinins which react with certain varieties of Proteus organisms, which, though isolated from typhus cases, are not the cause of the disease. The reaction takes place in almost 100 per cent of those suffering from the disease. The X strains ferment maltose.

Pathogenicity. *Proteus vulgaris* is generally nonpathogenic, although it is frequently encountered in cystitis, infantile diarrhea, and suppurative lesions. It is one of the organisms of putrefaction, and cultures of it have an unpleasant odor.

Related Organisms. 1. *Proteus mirabilis*—is Gram-negative, more highly pleomorphic, and liquefies gelatin more slowly. It is found in dejecta of children with summer diarrhea and is a secondary invader in wound infection. 2. *Proteus rettgeri*—isolated from sporadic and epidemic gastroenteritis patients. Originally isolated from chickens afflicted with cholera-like epidemic.

SULFUR CYCLE

Sulfur, an element essential for all life, is found in the soil, free and combined in organic and inorganic compounds. It finds its way into the soil from decomposition of native rock, from organic manure, and from rain water. It is present in large quantities in sulfur springs. Volcanoes emit free sulfur and gaseous sulfur dioxide and hydrogen sulfide. Burning of coal also results in gaseous sulfur. These gases are dissolved in rain water and returned to the soil. "Sulfur bacteria" oxidize sulfur-containing gases to sulfates which are utilized by plants or removed from soil by drain water. The sulfur bacteria produce sulfuric acid which acts on insoluble soil compounds, e.g., calcium carbonate, magnesium carbonate, calcium silicate, tricalcium phosphate, making them soluble and thus available to plants.

Ectothiobacteria. These belong to the genus Thiobacillus in the autotrophic family Nitrobacteriaceae, order Eubacteriales. They are minute cells which do not store sulfur and which do not have red or green pigments for photosynthesis. They obtain their carbon from carbon dioxide or carbonates and bicarbonates in solution and their energy from the oxidation of sulfides, thiosulfates, or elementary sulfur, forming sulfur, persulfates, sulfuric acid, and sulfates.

Endothiobacteria. These are higher bacteria of the order Chlamydobacteriales and sub-order Rhodobacteriineae, containing granules of free sulfur or bacteriopurpurin or both. Including species of many sizes and shapes, they occur in hydrogen-sulfide-containing waters, and utilize and grow best in the presence of this compound. Endothiobacteria are divided as follows:

1. Those containing the pigment system of red bacteriopurpurin and green photosynthetic* bacteriochlorophyll. These organisms comprise the sub-order Rhodobacteriineae. Under anaerobic conditions they obtain the energy necessary for assimilation of carbon dioxide from sunlight. In this case, hydrogen sulfide is not used as an energy source but as a hydrogen donor in the photosynthetic reduction of carbon dioxide. The hydrogen sulfide is changed to sulfur which may be stored in the cell (by Thiorhodaceae; the sulfur is eventually oxidized to sulfuric acid) or liberated (by Athiorhodaceae).

2. Those not containing this pigment system. These organisms utilize hydrogen sulfide and sulfur as sources of energy for the assimilation of carbon dioxide and the synthesis of cell substances. The filamentous forms belong to the family Beggiatoaceae; the nonfilamentous forms, to the family Achromatiaceae of the order Chlamydobacteriales.

CARBON CYCLE

Removal of carbon from the atmosphere is brought about by the utilization of carbon dioxide in weathering rocks, or by photosynthesis where carbon dioxide is taken in and oxygen is liberated. This removal of carbon from the atmosphere is compensated for by the processes of decay, combustion, and animal respiration which are continually occurring. The decay process involves the liberation of carbon and hydrogen as carbon dioxide, methane, water, etc. Enormous quantities are evolved from mineral springs and volcanoes. The ratio of carbon to nitrogen in soils indicates the extent of decomposition and the nature of organic matter contained in them. A wide carbon-nitrogen ratio is indicative of a fertile soil; a narrow ratio suggests a poor soil. Simple carbohydrates and many proteins are decomposed by many soil microorganisms, but cellulose and nucleoproteins are difficult to decompose and relatively few soil organisms can do it. Cellulose may be digested by members of the genera Cellulomonas, Cellvibrio, Clostridium, Pseudomonas, Actinomyces, some molds, etc. This decomposition of cellulose, which is found as plant residues in large quantities in soil, results in the production of (1) acids which react with insoluble material rendering them available as plant foods, and (2) gases such as CO_2, H_2, CH_4, etc., some of which pass into the atmosphere and the remainder furnish energy to other soil

---

* Photosynthesis is the biological reduction of carbon dioxide to organic matter by means of light energy. It is carried on by all green plants by means of the pigment chlorophyll.

bacteria. Cellulose ferments are used in waste disposal, water purification, and soil humus formation, and are being investigated as a means of commercial production of acetic acid.

Those bacteria which utilize carbon are regarded as intermediate between the true autotrophic and the heterotrophic forms. Among

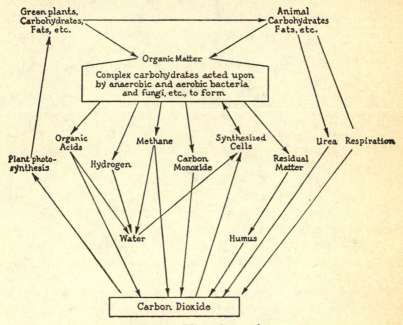

Fig. 33. The carbon cycle.

the bacteria which oxidize carbon oxides, methane, and other hydrocarbons are both aerobic and anaerobic forms which may be classified as follows:

1. Bacteria which oxidize carbon monoxide to carbon dioxide, e.g., *Carboxydomonas oligocarbophila.*

2. Bacteria which oxidize methane, but not other hydrocarbons, to form carbon dioxide and water, e.g., *Methanomonas methanica.*

3. Bacteria which can utilize paraffin and certain petroleum oils, as well as methane, as their source of carbon and energy, e.g., *Methanomonas aliphatica.*

4. Strictly anaerobic, purple, sulfur-free bacteria which require simple fatty acids in order to take up carbon dioxide. When subjected to infrared rays, they can use hydrogen, thus becoming true autotrophic forms.

PHOSPHORUS CYCLE

Next to nitrogen, phosphorus is the limiting element of crop production in most soils. The mineral phosphates of the soil are changed by bacterial action to soluble phosphorus which is utilized by living plants and transformed into organic and inorganic phosphorus in the plant tissues. If the plant is eaten by an animal the phosphorus is found in the body and in the excreta as organic or inorganic compounds. Excreted organic phosphorus must be acted upon by bacteria to be available to plants; the inorganic compounds are utilized directly.

OTHER SOIL BACTERIA

Iron Bacteria. These bacteria, which are found in iron-bearing waters, derive their energy from the oxidation of ferrous compounds. Members of the genus Gallionella, sub-order Caulobacteriineae, are stalked bacteria, the stalks being slender, twisted bands composed of ferric hydroxide, which completely dissolves in dilute hydrochloric acid. The most outstanding member of this group is *Gallionella ferruginea*. Members of the order Chlamydobacteriales also utilize iron compounds as a source of energy.

Hydrogen-Oxidizing Bacteria. Members of the genus Hydrogenomonas are short rods which can grow in the absence of organic matter and obtain energy by the oxidation of hydrogen to form water. These forms commonly occur in canal mud or swamps, or in soils where large amounts of hydrogen are produced by anaerobic bacteria. The type species is *Hydrogenomonas pantotropha*. Other species as *Hydrogenomonas vitrea* and *Hydrogenomonas flava* are also widely distributed in soil.

Miscellaneous Bacteria. These include a variety of bacteria which obtain their energy from the metabolism of manganese, selenium, arsenic, and other metallic compounds.

BIBLIOGRAPHY

Barnett, A. J. G. *Silage Fermentation.* New York: Academic Press, 1954.

Fry, B. *The Nitrogen Metabolism of Microorganisms.* New York: John Wiley and Sons, 1955.

Hofer, A. W., and J. K. Wilson. *Inoculation for Legumes.* Geneva, N. Y.: New York State Agricultural Experiment Station, Circular No. 179, 1938.

Löhnis, Felix, and Edwin Broun Fred. *Textbook of Agricultural Bacteriology.* New York: McGraw-Hill Book Company, 1923.

McElroy, W. D., and B. Glass, eds. *Inorganic Nitrogen Metabolism.* Baltimore: Johns Hopkins Press, 1956.

Nicol, H. *Microbes and Us*. Harmondsworth, England: Pelican Books, 1955.

Porter, J. R. *Bacterial Chemistry and Physiology*. New York: John Wiley and Sons, 1946. Pp. 663–670.

Stephenson, Marjory. *Bacterial Metabolism*. Third Edition. New York: Longmans, Green and Company, 1949.

Swingle, D. B. *General Bacteriology*, rev. by G. W. Walter. New York: D. Van Nostrand Company, 1947. Pp. 70–89.

Umbreit, W. W. *Modern Microbiology*. San Francisco: W. H. Freeman and Company, 1962. Pp. 327–341.

Waksman, S. A. *Humus*. Baltimore: Williams & Wilkins Company, 1938.

———. *Principles of Soil Microbiology*. Second Edition. Baltimore: Williams & Wilkins Company, 1932.

———. *Soil Microbiology*. New York: John Wiley and Sons, 1952.

Walter, W. G., and R. H. McBee. *General Microbiology*. Second Edition. Princeton: D. Van Nostrand Company, 1962. Pp. 204–222.

Wilson, Perry W. *The Biochemistry of Symbiotic Nitrogen Fixation*. Madison, Wis.: University of Wisconsin Press, 1940.

Yearbook, U. S. Department of Agriculture. Washington. D. C.: U. S. Government Printing Office, 1947. "Soil Organisms," pp. 511–527.

Yearbook: Soils and Men, U. S. Department of Agriculture. Washington, D. C.: U. S. Government Printing Office, 1938.

Chapter IX

AIR BACTERIOLOGY

Because of lack of moisture and nutrients, and at times the presence of sunlight, air cannot be considered a natural bacterial environment. It has, therefore, no normal bacterial flora. However, large numbers of microorganisms do find their way into the air from dust, soil, excreta, etc. They are important as contaminants in food and other industries and in the laboratory, and as a means of infection.

Droplet infection may be the source of pathogenic organisms in the air. Bacteria in the mouth may be coughed or sneezed in droplets sometimes for a distance of 12 feet. Tubercle bacilli may remain viable for months in dust, and be carried into the air. Droplets carrying bacteria and viruses evaporate rapidly while falling and leave droplet nuclei of bacteria which may remain in circulation in rooms for hours or even days. Common cold, whooping cough, measles, diphtheria, tuberculosis, and other diseases may be disseminated by droplet infection.

BACTERIA FOUND IN AIR

Kinds. Only species resistant to desiccation and exposure to sunlight can persist in the air for long. These are largely saprophytic. Most common are the spore-forming bacilli such as *Bacillus subtilis*, micrococci such as Sarcina, nonpathogenic Corynebacteria, chromogenic non-spore-forming rods such as Serratia, spores of yeasts and molds, and Actinomyces. Indoors, in highly populated places such as schools and theaters, the organisms common in the normal mouth and throat—Streptococcus, pneumococcus, and Staphylococcus—find their way into the air through coughing, sneezing, talking, and laughing, although gently expired air itself is reputedly sterile.

Number. The number of bacteria in the air depends largely on the locale and the activity in the environment. Free, unattached organisms are slightly heavier than air and will settle out very slowly in a quiet atmosphere. A gentle current, however, can keep them in suspension almost indefinitely. Bacteria-laden dust particles settle out rapidly. Droplets do not necessarily fall to the ground immediately, falling more slowly as they get smaller.

Moisture decreases the number of air bacteria. Air in summer carries more bacteria than in winter. In general, the number of bacteria will vary in inverse proportion to the distance from the earth.

Dust harbors an enormous number of bacteria. A dusty schoolroom in an old, poorly ventilated building averaged 700 bacteria per cubic foot of air space, whereas another classroom with windows opening on a tree- and grass-covered campus had only one or two bacteria per cubic foot. Sea, mountain, and desert air is extremely free from bacterial contaminants.

BACTERIOLOGICAL EXAMINATION OF AIR

Air may be examined to determine both the kind and the number of microorganisms present, although generally accepted, efficient methods have not yet been fully attained. The quantitative results vary with the method of collection, sampling time, culture media used, incubation time, etc. The methods most commonly employed are:

1. *Exposing Petri dishes* containing suitable culture media for a definite period of time. The dishes are then incubated and the colonies counted. Although this method is generally used for qualitative rather than quantitative evaluation, it affords a rough idea of the numbers and types of organisms present.

2. *Sand Filtration.* A measured amount of air is drawn through a layer of sterile sand in a small glass tube. The sand is then placed in sterile isotonic saline and shaken, and aliquots of the supernatant fluid are plated. The method has the advantages of simple construction, and portability. The sand, however, must be screened to a certain size and carefully sterilized to avoid caking. Another factor to be reckoned with is that only those organisms which survive filtration and the time lapse until the culture is plated will be observed.

3. *Bubbling* a known amount of air through a liquid medium, such as isotonic saline, aliquots of which are then plated.

4. *Wells's Air Centrifuge.* By use of this apparatus, bacteria are precipitated onto the culture medium. The centrifuge creates the air flow, measures the amount of air drawn through it, and collects the bacteria on suitable media in sterile bottles. These are incubated and the colonies counted and studied without the need of transfer to Petri dishes. The Wells centrifuge has met with much favor. Its operation is simple and rapid, it is portable, the testing is done in one step, and it covers a wide range of applications because any type of solid medium can be used.

5. *Atomization,* used by Moulton, Puck, and Lemon to produce

a layer of liquid around each bacterial particle. The bacteria-laden mist is absorbed in a chamber containing a sterile mixture of broth and olive oil. The atomizing and collecting chambers are rinsed with the collecting fluid, and aliquots of this are plated.

6. *Funnel device* of Hollaender and Dalla Valle, consisting of a brass container with removable bottom in which a funnel is inverted about 1 cm. above the bottom half of a Petri dish containing suitable medium. A measured amount of air is drawn through the funnel stem, and the organisms present are collected on the medium in the dish, which is subsequently incubated. The apparatus is simple, efficient, and portable, carrying on the test in one operation.

PURIFICATION OF AIR

For the killing of microorganisms in the air, two general methods have been developed:

1. *Irradiation* with ultraviolet light. This is being increasingly applied in schools, factories, storage warehouses, and hospitals.

2. *Aerosols*, sprays, or very fine mists of bactericidal substances. These are more readily available, cheaper to install, and easier to improvise in an emergency than irradiation. They are, however, more difficult to control and apt to be toxic or cause irritation. Care must be taken to choose a substance which is not toxic on prolonged inhalation, does not vaporize readily but remains as droplets, and is not inflammable. Sodium hypochlorite, resorcinol, hexylresorcinol, propylene glycol, and triethylene glycol have been used most successfully.

Water spraying has also been employed. The air is supersaturated with water, which carries along the dust nuclei as it falls to the ground. Before and after spraying, a disinfectant solution is applied to the ground.

BIBLIOGRAPHY

American Public Health Association Yearbooks. New York, 1940 Supplement, 1941, 1942, 1942 Supplement.

Jawetz, E., J. L. Melnick, and E. A. Adelberg. *Review of Medical Microbiology.* Los Altos: Lange Medical Publications, 1962. Pp. 80–81.

Moulton, Forest Ray, ed. *Aerobiology, a Symposium.* Washington, D. C.: American Association for the Advancement of Science, Publication No. 17, 1942.

O'Hara, Dwight. *Air-borne Infection.* New York: The Commonwealth Fund, 1943.

Rosebury, T. *Experimental Air-borne Infection.* Baltimore: Williams & Wilkins Company, 1947. Pp. 137–207.

Winslow, Charles-Edward A. *The Conquest of Epidemic Disease.* Princeton, N. J.: Princeton University Press, 1943.

CHAPTER X

WATER BACTERIOLOGY

The suitability of a water supply is determined by four types of analyses: (1) The *chemical analysis* determines total solids, hardness, etc., and detects any harmful chemical ingredients, such as poisonous lead or zinc salts. (2) The *physical examination* determines if the water has any objectionable turbidity, color, taste, or odor. (3) The *biological analysis* detects algae, fungi, protozoa, nematode worms, the smaller species of Crustacea, and the larvae of aquatic insects. (4) The *bacteriological analysis* is the most valuable examination and is vital in preventing epidemics as a result of water pollution.

The bacteriological examination of water usually involves (1) estimation of the number of bacteria as determined by total plate count, and (2) the more significant detection of the presence or absence of members of the coliform group. Finding these organisms in a water supply is taken as evidence of fecal contamination and, therefore, of possible presence of the intestinal pathogens causing typhoid fever, paratyphoid fever, dysentery, and cholera.

The methods discussed in this chapter are those set forth in the latest edition of *Standard Methods* of the American Public Health Association.

PREPARATIONS FOR TESTING

Collection of Specimen. The recommended container is a colorless, wide-mouthed glass bottle with ground-glass stopper. The specimen should be a representative sample and should be examined with the least possible delay. For impure waters, the maximum time lapse is 6 hours; for relatively pure waters, 12 hours. Samples should be stored at 6–10° C. In testing water from *swimming pools*, 20–50 mg. sodium thiosulfate is added to the sample bottle to neutralize the residual chlorine and prevent it from destroying bacteria in the interval between collection and testing.

Materials. All glassware should be sterilized in a dry heat oven at 170° C. for 1 hour unless otherwise indicated.

1. One ml. pipettes, plugged with nonabsorbent cotton.
2. Test tubes. These are $5 \times \frac{5}{8}$ inch test tubes, to which 9.5 ml. of tap water are added, and which are sterilized in the autoclave at 15 pounds pressure for 30 minutes and then cooled to room temperature. After sterilization, each tube should have exactly 9.0 ml. in it. If the water is below or above the mark, either add sterile water or remove a portion.
3. Sterile corks. If $5 \times \frac{5}{8}$ inch test tubes are used, a number 5 cork will exactly fit. Prepare one cork for each test tube. The corks may be sterilized by boiling for 20 minutes in water, or by wrapping each cork separately and sterilizing in the hot air oven at 170° C. for two hours.
4. Dilution bottles or test tubes with close-fitting ground-glass or rubber stoppers.
5. Petri dishes, 10 cm. in diameter.
6. Fermentation tubes.
7. Formulae for media will be found in Chapter V.

Dilutions. All dilutions of the water specimens must be accurate and made with sterile precautions. A sample dilution scheme follows:

1. Shake the specimen bottle vigorously 25 times.
2. Make a *1:10 dilution of the water* by adding 1 ml. of the water specimen to 9 ml. of sterile water in a test tube.
3. Then place a sterile cork carefully in the test tube to replace the cotton plug. Shake vigorously 25 times.
4. Make a *1:100 dilution of the water* by adding 1 ml. of the water specimen diluted 1:10 to 9 ml. of sterile water.
5. Replace the cotton plug with a sterile cork and mix as in 3.
6. Make a *1:1000 dilution of the water* by adding 1 ml. of the water specimen diluted 1:100 to 9 ml. of sterile water.
7. Replace the cotton plug with a sterile cork and mix as in 3.
8. Dilutions may also be made by adding 10 ml. of the water sample or previous dilution to 90 ml. of sterile water in dilution bottles.
A fresh pipette is used for each successive dilution.

PLATE COUNT

Preparing the Plates. The procedure for preparing plates to be used for counting is as follows:

1. Within 20 minutes after dilutions are made, transfer 0.5 ml. or 1 ml. of the sample and each of the dilutions to appropriately marked Petri dishes.
2. Add 10 ml. liquefied and cooled nutrient agar, tryptone glucose extract agar, or gelatin. Mix thoroughly. Allow to solidify.
3. Invert and incubate in dark, well-ventilated incubators practically saturated with moisture, at 20° C. for 48 hours and 37° C. for 24 hours. Gelatin plates are incubated at 20° C. only.

Counting. After the incubation period, the colonies are counted with a lens having a magnification of $1\frac{1}{2}$ diameters. Select the plates in which the colonies of bacteria number between 30 and 300.

(It may sometimes be necessary to count plates with fewer than 30 colonies, as 1 ml. is set as the maximum amount of water to be used in preparing plates.) At least two plates should be counted. Multiply the number of colonies counted, rounded out to two significant figures, by the dilution of the water. Take the average and report the result as the number of bacteria in each ml. of undiluted water, for each incubation temperature.

Suppose, for example, that 343 colonies are counted on the plate containing 1 ml. of the water specimen diluted 1:10, 45 on the plate containing the 1:100 dilution, and 6 on the plate containing the 1:1000 dilution. The number of bacteria reported (each colony being counted as one bacterium) per ml. of undiluted water will be:

$$340 \times 10 = 3400$$
$$45 \times 100 = \underline{4500}$$
$$\frac{7900}{2} = 4000 \text{ bacteria per ml. of undiluted water.}$$

Interpretation. The maximum number of bacteria allowable for safe water is not set in *Standard Methods*, but is generally accepted to be 100 per ml. at 37° C. The organisms normally found in water grow best at 20° C.; the coliform organisms, at 37° C. If the total counts at the two temperatures approximate each other, there is a strong suspicion that sewage bacteria are present. One of the greatest values of routine total counts is the rapid spotting of sudden pollution as indicated by a sharp rise above the count usually obtained from the same source.

TEST FOR PRESENCE OF MEMBERS OF THE COLIFORM GROUP

Preliminary Considerations. 1. The coliform group includes all aerobic and facultative anaerobic Gram-negative, non-spore-forming bacilli which ferment lactose with gas formation.

2. Formation of gas in a standard 0.5% lactose broth fermentation tube within 24 hours at 37° C. is presumptive evidence of the presence of members of this group.

3. The appearance of aerobic, lactose-splitting, typical *Escherichia coli* colonies on Endo or eosin methylene blue plates made from a lactose broth fermentation tube in which gas has formed, or the formation of gas within 48 hours at 37° C. in a fermentation tube containing brilliant green lactose bile broth which has been seeded from a lactose broth fermentation tube in which gas has formed. constitutes a confirmatory test.

4. Demonstration that one or more aerobic plate colonies consist of Gram-negative, non-spore-forming bacilli which form gas when inoculated into a lactose broth fermentation tube is a completed test.

Presumptive Test.

1. Inoculate a series of lactose broth fermentation tubes with appropriate graduated quantities of the water to be tested.   The concentration of ingredients in the final mixture of medium and sample must be maintained as is indicated in the formula for standard lactose broth, except that 1 ml. or less of sample may be added to 10 ml. single-strength lactose broth.   Multiple strengths of the media may be employed as necessary.

2. Incubate the tubes at 37° C. for 48 hours.

3. Examine each tube at 24 and 48 hours for gas formation.

A positive presumptive test is indicated by the formation of gas within 24 hours.

Besides the coliforms, a positive presumptive test may be attributable to *Clostridium perfringens*, yeasts, or the synergistic combined action of two organisms.

A doubtful test is indicated if gas is observed at the end of 48 but not 24 hours.   In this case the confirmed and completed tests should be performed.

A negative test is the absence of gas formation after incubation for 48 hours.

Confirmed Test.

1. For this test, use all of the lactose broth fermentation tubes showing gas or the ones made from the two highest dilutions.

2. From each of these, streak one or more Endo or eosin methylene blue plates and incubate at 37° C. for 24 hours, or inoculate brilliant green lactose bile broth fermentation tubes and incubate at 37° C. for 48 hours.

Plates: If, within the incubation period, typical *Escherichia coli* colonies develop (distinct red on Endo, metallic green on eosin methylene blue), the confirmed test is considered positive.   If only atypical colonies develop, the test cannot be considered definitely negative, since it often happens that members of the coliform group fail to form typical colonies on these media, or that colonies develop slowly.   In this case, proceed with the completed test.

Tubes: Gas formation constitutes a positive confirmed test.   If decolorization occurs (probably due to *Clostridium welchii*), proceed with completed test.

Completed Test.

1. This test may be performed on lactose broth fermentation tubes showing gas, colonies found on plates for confirmed test, or brilliant green lactose bile broth tubes showing gas for confirmed test.

2. If using one of the broths, streak one or more Endo or eosin methylene blue plates and incubate at 37° C. for 24 hours.

3. From each of the plates used for the confirmed test or from those prepared in step 2 above, fish typical colonies or those most likely to be coliform organisms and inoculate agar slant and lactose broth fermentation tube. Incubate at 37° C. for 24–48 hours.

The formation of gas in the lactose broth tube and the demonstration of typical Gram-negative, non-spore-forming bacilli in the agar culture constitute a satisfactory completed test, demonstrating the presence of a member of the coliform group. The absence of gas formation in the lactose broth tube, or the failure to demonstrate Gram-negative, non-spore-forming bacilli in a gas-forming culture, constitutes a negative test.

If spore-forming, lactose-fermenting organisms are found, transfer to formate ricinoleate broth and incubate at 37° C. for 48 hours. If no gas is produced, the conclusion is that only spore-formers are present. If gas is produced, coliform organisms are probably present. In this case, inoculate standard lactose broth fermentation tubes and agar slants from the formate ricinoleate. If gas is produced in the lactose broth and no spores are formed in the agar slant, the test is considered positive. If spores are present, coliform organisms are considered absent.

DIFFERENTIATION OF COLIFORM ORGANISMS

The scheme for differentiation of the coliform organisms into those generally designated as of fecal origin (*Escherichia coli*) and those generally designated as of nonfecal origin* (*Escherichia freundii* or intermediates and *Aerobacter aerogenes*) is included in the Appendix to *Standard Methods*, which comprises certain provisional or nonstandard methods not yet included in routine standard procedure. The four tests used are indol production (see p. 40), methyl red test (see page 40), Voges-Proskauer test (see p. 41), and utilization of sodium citrate as the sole source of carbon. The reactions are tabulated as follows:

	Indol	Methyl Red	Voges-Proskauer	Citrate Utilization
Escherichia coli				
Variety I	+	+	−	−
Variety II	−	+	−	−
Escherichia freundii (Intermediates)				
Variety I	−	+	−	±
Variety II	+	+	−	+
Aerobacter aerogenes				
Variety I	−	−	+	±
Variety II	−	−	−	+

Typical reactions in the mnemonic IMViC are:
Escherichia coli, + + − −; *Aerobacter aerogenes*, − − + +.

---

* It must be remembered, however, that all types of coliform organisms may occur in feces.

DRINKING WATER STANDARDS

The presence of a single coliform organism is not a ground for condemning a water as unfit for human consumption. It is the relative abundance of these organisms that is important. For this reason, *Standard Methods* includes estimation of coliform density and the most probable number of coliform organisms present.

In this country, the only national standard available is the one set by the United States Public Health Service for drinking waters transported interstate on common carriers. This establishes that not more than 10% of all the standard 10 ml. portions examined in a month shall show the presence of coliform organisms.

WATER PURIFICATION

It is estimated that 10–40% of the total cases of typhoid are a result of water contamination. In cities with efficient water purification systems, however, typhoid and other diseases such as cholera resulting from contaminated water have practically vanished. The methods of purifying water are noted briefly below.

Natural Purification. Natural or self-purification of water is brought about by distillation, sedimentation, oxidation including the action of light, and filtration.

Artificial Purification. The methods of treating a water which is unsafe for human consumption may be summarized as follows (after Jordan and Burrows):

1. Mechanical.
(a) Storage.
(b) Filtration.
(1) Slow sand filtration.
(2) Coagulation and rapid sand filtration.
2. Chemical—almost exclusively with chlorine-yielding substances.
(a) Large scale—hypochlorites and liquid chlorine.
(b) Small scale—hypochlorites, ultraviolet light, ozone, etc.

Frequently a combination of filtration and chlorination is employed.

For the family or individual, when using an impure water supply or when water-borne disease is prevalent, boiling for 5 minutes is a simple and efficient means of destroying all the intestinal pathogens.

BIBLIOGRAPHY

British Ministry of Health. *The Bacteriological Examination of Water Supplies.* Revised Edition. Bulletin No. 71. London: His Majesty's Stationery Office, 1934.

Bryan, C. A., and A. H. Bryan. "Marine Microbiology," *Journal of Ecology.* Baltimore: Williams & Wilkins Company, October, 1959.

Dubos, R. J., ed. *Bacterial and Mycotic Infections of Man.* Third Edition. Philadelphia: J. B. Lippincott Company, 1958. Pp. 662–663.

Gainey, P. L., and T. H. Lord. *Microbiology of Water and Sewage.* New York: Prentice-Hall, 1952.

Hopkins, E. S., and W. H. Schulze. *The Practice of Sanitation.* Third Edition. Baltimore: Williams & Wilkins Company, 1958.

Horwood, M. P. *The Sanitation of Water Supplies.* Springfield, Ill.: Charles C. Thomas, 1932.

Jawetz, E., J. L. Melnick, and E. A. Adelberg. *Review of Medical Microbiology.* Los Altos: Lange Medical Publications, 1962. Pp. 74–75.

Maxcy, K. F., ed. *Rosenau Preventive Medicine and Public Health.* Eighth Edition. New York: Appleton-Century-Crofts, 1956.

Medical Research Council. *A System of Bacteriology in Relation to Medicine* Vol. III. London: His Majesty's Stationery Office, 1929.

Prescott, Samuel Cate, Charles-Edward A. Winslow, and Mac Harvey McGrady. *Water Bacteriology.* Sixth Edition. New York: John Wiley and Sons, 1946.

Standard Methods for the Examination of Water, Sewage and Industrial Wastes. Tenth Edition. New York: American Public Health Association, 1955.

Suckling, E. V. *The Examination of Water and Water Supplies.* Fifth Edition. Philadelphia: The Blakiston Company, 1943.

Tanner, Fred Wilbur. *The Microbiology of Foods.* Second Edition. Champaign, Ill.: Garrard Press, 1944. Pp. 156–250.

Taylor, E. W. *The Examination of Waters and Water Supplies.* Seventh Edition. Boston: Little, Brown and Company, 1958.

Wadsworth, A. B. *Standard Methods.* Baltimore: Williams & Wilkins Company, 1947. Pp. 257–538.

Walter, W. G., and R. H. McBee. *General Microbiology.* Second Edition. Princeton: D. Van Nostrand Company, 1962. Pp. 234–237.

Chapter XI

SEWAGE AND SHELLFISH BACTERIOLOGY

SEWAGE BACTERIOLOGY

Sewage consists of the human excreta and other wastes found in sewers. Health standards are dependent upon efficient waste disposal, for a number of serious diseases may be transmitted by discharges from the gastrointestinal tract, such as typhoid fever, paratyphoid fever, cholera, dysentery, and certain nematode infections. Most bacteria present in sewage, however, are saprophytes which not only are harmless but actually perform vital functions in the nitrogen and sulfur cycles, etc.

Purification of sewage in cesspools and municipal disposal plants involves the usual aerobic decay and anaerobic putrefactive processes. There are two general methods of disposal, "dry" and "wet."

Dry Sewage Disposal. This is of two kinds:

1. The *privy* is placed over an open pit as in U. S. army latrines and farm outhouses.
2. The *chemical closet*, an iron tank containing a chemical disinfectant, is placed beneath the toilet seat.

Wet Sewage Disposal. There are various methods whereby sewage may be purified and separated from suspended particles.

1. The *cesspool*, in which the solids accumulate in the bottom and are greatly reduced in volume through the septic action of putrefactive anaerobic bacteria.
2. The *septic tank*, which is also based on the principle of anaerobic putrefaction.
3. The *sprinkling filter*, in which the sewage is sprayed through the air for activation with oxygen.
4. *Chemical purification;* the sewage is purified through the disinfecting action of chemical agents.
5. *Direct disposal of sewage* by discharging it into a body of water.

Bacteriological Examination. Sewage may be examined bacteriologically in a like manner to water (see Chapter X), but dilutions must be much greater because so many more organisms are present. The total colony count is important in the examination of sewage.

SHELLFISH BACTERIOLOGY

Because shellfish, such as oysters and clams, which are eaten raw are often the cause of epidemics of typhoid fever and other enteric infections, certain standardized procedures have been set forth for their sanitary supervision.

Selection of Sample. The oysters selected should represent the average sizes of the lot under examination, care being taken to choose those with deep bowls, short lips, and tightly closed shells. When the oysters are sent to the laboratory, a statement should be attached as to the conditions under which they were taken from the water. The oysters should be packed in a suitable container and should be properly iced if sent for a long distance.

Technical Procedure. At the laboratory a record is first made of the general appearance of the oysters, giving the condition in which they were received, the presence of abnormal odors, and the temperature of the stock.

1. The oysters are thoroughly cleaned and washed.

2. The shells are opened by means of a sterile oyster knife in the usual manner, or by drilling through a flamed portion of the shell near the hinge with a sterile drill.

3. The shell liquor of at least five oysters from each lot is pooled and diluted with sterile tap water containing 2% sodium chloride.

4. The shell liquor is tested for members of the coliform group by adding 1 ml. amounts of it to five lactose broth fermentation tubes in the undiluted state and in dilutions of 1:10 and 1:100.

5. Incubate for two days at 37° C. and note gas formation.

6. *Interpretation of the Results.* The presence of *Escherichia coli* in each fermentation tube, as evidenced by the production of gas, is given the following values, which represent the reciprocals of the greatest dilution in which the test is positive.

The value of 1, when the test is positive in undiluted, but not in 1:10.

The value of 10, when the test is positive in 1:10, but not in 1:100.

The value of 100, when the test is positive in 1:100.

The values for the five fermentation tubes are added and this gives the score. A score over 50 is commonly adopted as a basis for exclusion from market.

Example:

RESULTS OF TESTS FOR ESCHERICHIA COLI IN DILUTIONS INDICATED

Sample	Undiluted	1:10	1:100	Numerical Value
1	+	+	0	10
2	+	+	0	10
3	+	0	0	1
4	+	+	0	10
5	+	+	0	10

Total or coliform score...................... 41

Shellfish Contamination. The degree of contamination is dependent upon the degree of pollution of the water in which they are cultivated. During the "drinking" season the content of bacteria in oysters is high, but this falls rapidly after hibernation begins. Oysters are purified by transferring them from polluted waters to clean waters.

When mussels are subject to sewage pollution, they may frequently give rise to gastrointestinal disturbances which sometimes end in paralysis and even in death.

BIBLIOGRAPHY

Babbitt, H. E. *Sewage and Sewage Treatment.* Seventh Edition. New York: John Wiley and Sons, 1953.

Ehlers, V. M., and E. W. Steel. *Municipal and Rural Sanitation.* Fifth Edition. New York: McGraw-Hill Book Company, 1958.

Fair, G. M., and J. C. Geyer. *Water Supply and Waste-Water Disposal.* New York: John Wiley and Sons, 1954.

Gainey, P. L., and T. H. Lord. *Microbiology of Water and Sewage.* New York: Prentice-Hall, 1952.

Gradwohl, R. B. H. *Clinical Laboratory Methods and Diagnosis.* St. Louis: C. V. Mosby Company, 1948.

Hardenberg, W. A. *Sewerage and Sewage Treatment.* Third Edition. Scranton, Pa.: International Textbook Company, 1950.

Hopkins, E. S., and W. H. Schulze. *The Practice of Sanitation.* Third Edition. Baltimore: Williams & Wilkins Company, 1958.

Imhoff, K., and G. M. Fair. *Sewage Treatment.* Second Edition. New York: John Wiley and Sons, 1956.

Maxcy, K. F., ed. *Rosenau Preventive Medicine and Public Health.* Eighth Edition. New York: Appleton-Century-Crofts, 1956.

Park, William Hallock, and Anna Wessels Williams. *Pathogenic Microörganisms.* Eleventh Edition. Philadelphia: Lea and Febiger, 1939. Pp. 944–949.

Rudolfs, Willem. *Principles of Sewage Treatment.* Washington, D. C.: National Lime Association, Bulletin 212.

Standard Methods for the Examination of Water Sewage and Industrial Wastes. Tenth Edition. New York: American Public Health Association, 1955.

Swingle, D. B. *General Bacteriology,* rev. by G. W. Walter. New York: D. Van Nostrand Company, 1947. Pp. 243–249.

Tanner, Fred Wilbur. *The Microbiology of Foods.* Second Edition. Champaign, Ill.: Garrard Press, 1944. Pp. 156–250.

ZoBell, C. E. *Marine Microbiology.* Waltham, Mass.: Chronica Botanica, 1946.

Chapter XII

DAIRY BACTERIOLOGY

Milk secreted into the udders of healthy cows is sterile, but by the time it finds its way into the milking pail, it has acquired a sizable bacterial population. Some of the organisms come from the cow's udder or skin, others from dust, fodder, milk utensils, and also from the hands, clothing, etc., of the person doing the milking. These bacteria are normally nonpathogenic. Many are useful and are considered to be the common bacteria of milk.

BACTERIA FOUND IN MILK

Milk bacteria may be classified and studied according to the changes brought about in the milk, and include the following groups: (a) acid-forming, (b) gas-forming, (c) proteolyzing, (d) alkali-forming, and (e) inert.

Acid-Forming Bacteria. These bacteria split lactose and form acids causing the souring and later the curdling of milk. Warm climate tends to increase the acidity of milk, and the pH varies with the number of bacteria present. The principal acid-formers present in milk are:

1. *Streptococcus lactis* (*Streptococcus acidi lactici, Bacillus acidi lactici*), first isolated by Lister in 1873 from milk and present in milk and milk products, especially cheese, and on certain plants. This organism is largely responsible for the normal souring of milk.

2. *Streptococcus agalactiae* (*Streptococcus mastitidis*), isolated by Lehmann and Neumann in 1896. It causes mastitis of cows and sometimes sore throat in the human being, and summer complaint in children.

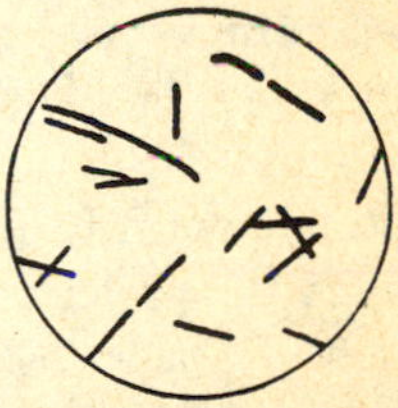

Fig. 34. *Lactobacillus bulgaricus*.

3. *Lactobacillus bulgaricus*, present in fermented milks. It normally occurs in ensilage and is carried to milk through ensilage.

4. *Lactobacillus acidophilus*—has gained importance in sour milk therapy. The oral administration of living cultures of *Lactobacillus*

131

acidophilus often will result in its establishment within the human intestine and the subsequent suppression of putrefactive microbic types.   Based upon extensive feeding experiments, investigations have called attention to the clinical value of a simplified flora dominated by *Lactobacillus acidophilus*.   Reports have been made that the ingestion of living cultures of *Lactobacillus acidophilus* resulted in relief from chronic constipation as well as chronic diarrhea and colitis.

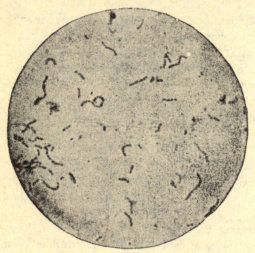

Fig. 35.   *Lactobacillus acidophilus.*

Gas-Forming Bacteria.   These belong largely to the coliform group, some of which ferment lactose with the formation of acid and gas, and exert a proteolytic action decomposing milk proteins, particularly casein.   Examples are: *Escherichia coli, Salmonella paratyphi,* and *Aerobacter aerogenes.*

Proteolyzing Bacteria.   These bacteria produce an alkaline reaction and, in addition, hydrolyze the milk proteins.   Two groups of enzymes are involved: a rennin-like enzyme, which precipitates the protein with the formation of a soft curd, and casease, which hydrolyzes the protein, transforming the milk to a clear fluid (peptonization).   The bacteria responsible for these changes include: 1. *Bacillus subtilis,* the hay bacillus. 2. *Bacillus mesentericus* (*Bacillus vulgatus*), the potato bacillus. 3. *Proteus vulgaris.*   (See p. 112, Chapter VIII.)

Alkali-Forming Bacteria.   Some of the bacteria commonly found in milk render it alkaline without producing any change in its appearance, taste, or odor. These include *Shigella alkalescens* and *Pseudomonas fluorescens*.

Inert Bacteria. A large number of the common milk bacteria, as well as most of the pathogens which may be found in milk, produce no change in the appearance, odor, taste, or reaction of milk, and are consequently said to be inert.

Anaerobic Bacteria. These bacteria, which may be contaminants of milk, ferment lactose, forming gas and producing strong-smelling acids, like butyric, valerianic, and propionic. The organisms may be present ordinarily in small numbers and their development is usually prevented by the acid-forming bacteria. When the anaerobic bacteria grow in milk in large numbers, a curd containing many gas bubbles is usually formed. These organisms include certain members of the Clostridium group, such as *Clostridium perfringens, Clostridium pruchii, Clostridium butyricum,* and *Clostridium multifermentans*.

Chromogens in Milk. Certain pigment-producing bacteria cause color changes in milk. 1. *Pseudomonas syncyanea (Bacillus syncyaneus)* causes blue milk. 2. *Serratia marcescens (Bacillus prodigiosus)* imparts a red color to milk. 3. *Pseudomonas aeruginosa (Bacillus pyocyaneus)* causes blue-green milk. 4. *Flavobacterium synxanthum* causes yellow milk.

Ropiness. Sometimes milk on standing becomes slimy or ropy, due mainly to the capsule-forming activities of *Alcaligenes viscosus*.

Number and Kinds of Bacteria in Milk. In one milk sample, the bacteria may be largely of the inert forms, and in another the bacteria of the gas-forming group may predominate. Market milk containing not more than 50,000 bacteria per ml. is considered of good quality. Certified milk is limited to 10,000 bacteria per ml. The number of bacteria present in market milk depends upon the original contamination, the temperature at which the milk has been kept, and the period for which the milk has been held.

DISEASES TRANSMITTED THROUGH MILK

Diseases Primarily of Bovine Origin. Pathogenic organisms from diseased cows may find their way into milk and be the means of transmission of the following: (1) tuberculosis (*Mycobacterium tuberculosis var. bovis*), (2) mastitis (*Streptococcus agalactiae*), (3) undulant fever or contagious abortion (*Brucella abortus*), (4) foot-and-mouth disease (virus), (5) cowpox (virus), and (6) anthrax (*Bacillus anthracis*).

Diseases Primarily of Human Origin.   A large number of diseases are transmissible through milk contaminated by persons collecting and handling it.   These include:

1. Typhoid fever caused by *Salmonella typhosa*.   The organism gets into milk through some person who is a carrier or has been in contact with a case of typhoid fever, through washing milk cans in water containing the germs, or from the cow which may carry the organism on its teats consequent to walking in water containing typhoid bacilli.

2. Enteric fevers caused by *Salmonella paratyphi*, *Salmonella schottmuelleri*, *Salmonella enteritidis*, and *Salmonella typhimurium*.

3. Scarlet fever caused by *Streptococcus scarlatinae*.

4. Septic sore throat caused by *Streptococcus epidemicus*.

5. Diphtheria caused by *Corynebacterium diphtheriae*.

6. Cholera caused by *Vibrio comma*.

BACTERIOLOGICAL EXAMINATION OF MILK

Collection of Samples.

1. Collect samples from thoroughly mixed milk.   The samples should be representative of the lot from which they are taken.

2. Transport to a laboratory as soon as possible in ice-packed sample cases. The temperature should be below 10° C. for transportation.

Plate Count Method.

1. Shake sample bottles 25 times, with an up and down motion.

2. Prepare dilutions of the milk as follows:

(a) Add 10 ml. of milk sample to bottle containing 90 ml. sterile water (1:10 dilution).   Mix thoroughly.

(b) Transfer 10 ml. of the 1:10 dilution to 90 ml. sterile water (1:100 dilution).

(c) Transfer 1 ml. of the 1:10 dilution to 99 ml. sterile water (1:1000 dilution).

(d) If high bacterial counts are expected (300,000 or over), prepare 1:10,000 and 1:100,000 dilutions.

(e) Shake each dilution 25 times.

(f) With a sterile 1 ml. pipette, transfer 1 ml. amounts from desired dilutions to each of two sterile Petri dishes, selecting dilutions that will produce from 30 to 300 colonies per plate.

(g) Pour 10–12 ml. of melted and cooled standard agar* into each plate and mix with sample thoroughly.

---

* Formula for the standard agar is as follows:

Agar	15 gm.
Beef extract	3 gm.
Tryptone	5 gm.
Glucose	1 gm.
Distilled water	1000 ml.

The desired reaction is pH 6.6–7.0, a pH of 7.0 being preferred.   1% skim milk is added where the dilutions are greater than 1:10.

3. After incubating the plates at 37° C. for 48 hours, with a lens magnifying 2½ diameters, count the colonies on plates showing between 30 and 300 each. The number of colonies in these plates should be multiplied by the dilution and reported as the final count. If no plate is within these limits the one that comes nearest to 300 is counted, only a part of the plate being counted, and the total is estimated from this.

Although it is one of the two widely used methods for the determination of the bacterial count of milk, the agar plate count involves certain disadvantages which may be summarized as follows:

1. Pathogenic organisms are not detected. Many pathogens (e.g., *Mycobacterium tuberculosis*) do not grow; others (e.g., *Salmonella typhosa*) are not distinguishable.

2. Counts are not accurate. Many organisms do not find the conditions provided favorable for development: the medium lacks nutrients for some, the air supply prevents growth of anaerobes, the incubation temperature is unsuitable for some. Another reason for inaccuracy is that many colonies come from clumps and not individual cells as is assumed in the calculation.

3. Period required before the results are available is too long.

4. Specific information about species is not furnished.

Microscopic or Breed Count. The counting of the actual bacteria found on a dried film of milk is recognized in *Standard Methods* as a technic of equal standing with the colony count from agar plates. The Breed count is of special value in judging the quality of fresh milk at milk-receiving stations.

1. For this method the following equipment is required:

(a) A pipette which will deliver 0.01 ml. of milk as tested by weighing the amount of milk discharged—0.0103 gm.

(b) Slides bearing 1 square centimeter marking or cards with square centimeter divisions which can be placed under the slide.

2. The procedure consists of the following steps:

(a) Deliver 0.01 ml. of milk on the slide and spread uniformly over the area of 1 sq. cm.

(b) Dry in a warm place on a level surface within 5 minutes, avoiding excessive heat.

(c) Dip in xylol for one minute to remove fat.

(d) Drain and dry.

(e) Fix in 90% ethyl alcohol for 1 minute.

(f) Stain with Löffler's methylene blue solution.

(g) Rinse with water.

(h) Decolorize in alcohol to a faint blue tinted background. This requires from 3 seconds to several minutes.

(i) Dry and examine under the microscope.

The number of bacteria per ml. of milk is estimated by counting all the organisms within a given area in a microscopic field. This field is carefully measured and its ratio to a square centimeter is determined. With the Breed method, at least 1/100,000 ml. of milk should be examined. With milk of a high grade, proper precautions should be taken to obtain a fair estimate of the number of bacteria present. The microscope should be so adjusted that each field covers a certain known fraction of the area of a square centimeter. The microscope may be standardized by placing a micrometer slide on the stage and by adjusting the draw tube or by selecting proper oculars, so that the diameter of the microscope field equals exactly 0.205 mm. Thus, each field of the microscope covers an area of approximately 1/3000 sq. cm., or measures the number of bacteria in 1/300,000 ml. of milk. About 30 fields should be counted and multiplied by 300,000. The average should be reported as the number of bacteria present per ml., with a notation as to whether clumps or individual cells were counted.

The advantages of the direct microscopic count over the plate count are:

1. Results are obtained more quickly.
2. Less work is required.
3. Less equipment is required.
4. A permanent record of the test is obtained.
5. Some idea may be had of morphological types.
6. Presence of leucocytes and other body cells is observed.
7. A more accurate quantitative estimation is obtained.

The disadvantages of the method are:

1. It cannot be used on pasteurized milk.
2. Dead cells are not distinguished from living cells.
3. The large conversion factor used in the calculation limits its accurate use to high-count milk.

The ratio between the results of the microscopic and plate counts is generally set as 4 : 1.

Frost Microscopic Colony Count. This method is carried out as follows:

1. Milk is mixed with agar on a sterile glass slide and the mixture spread over a definite area. (To 0.05 ml. of milk an equal amount of liquefied agar at 42° to 45° C. is added.)
2. After hardening, the preparation is incubated in a moist chamber, usually for 8 hours at 37° C.
3. It is then quickly and thoroughly dried at a temperature which is a few degrees below the boiling point of water.
4. The preparation is then stained, by treating it first with a 10% solution of glacial acetic acid in 95% alcohol and then with methylene blue or carbol thionine solution which is about one-fourth the usual strength.

The colonies per microscopic field are determined and the results calculated. Roughly, when a 10 X eyepiece is used, the colony under a 16 mm. lens means 4000 colonies per ml. of milk; the colonies in an average field of a 4 mm. lens should be multiplied by 100,000; those in an average field of a 1.9 mm. or oil immersion lens should be multiplied by 400,000. At least five representative fields should be counted and averaged.

The Frost Microscopic Colony Count method has the following advantages:

1. It is rapid, not over 7 or 8 hours being required.
2. It is reliable in that failure due to spreaders or improper dilutions is unlikely.
3. A relatively large amount of milk, usually 0.05 ml., is studied.
4. A permanent record of the count is provided.
5. Only a small amount of glassware and medium is required.

OTHER METHODS OF TESTING MILK

Examination for Streptococci. 1. If long-chain streptococci are found in the milk sediment, in association with an excess of leucocytes, and the latter cells are clumped together and consist largely of the polymorphonuclear type, the streptococci are very likely pathogenic. 2. If the plates contain numbers of extremely minute brownish colonies which upon microscopic examination are found to consist of streptococci in rather long chains, an examination of the herd from which the milk came may result in the discovery of cows affected with catarrhal mastitis. Market milk should be excluded or condemned when these three tests agree:

1. Microscopic examination of the sediment shows streptococci, diplococci, or cocci.
2. The plate from the same sample shows colonies exceeding a count of 100,000 per ml.
3. Bouillon culture from these colonies shows long-chain streptococci only or in great excess compared with the other bacteria present.

A chemical test for mastitis in cows is as follows:

1. Four drops of bromthymol blue solution (0.04%) are added to 3 ml. of freshly drawn milk.
2. These are well mixed. Scrupulous cleanliness should be observed since any acid or alkali added will influence the test.
3. Normal milk is greenish yellow when the indicator is added to it. A light green milk indicates a slightly infected milk, while a greenish blue to blue color in the milk indicates a highly infected specimen.

This test is based on the increased alkalinity of milk from cows suffering from mastitis, due to the secretion of inflammatory exudate.

The direct microscopic method may be used on freshly collected milk. Newman has devised a rapid stain for milk films. The stain consists of:

1% alcoholic solution methylene blue...........	30 ml.
Xylene......................................	30 ml.
95% alcohol.................................	100 ml.
Glacial acetic acid..........................	1 ml.

The technic is as follows:

1. Place a loopful of the milk to be examined on a clean slide. Avoid butterfat, if possible.

2. Spread the milk into a uniform film with a straight needle.

3. Dry film by placing the slide on a warm surface, but *not passing through a flame*. Drying must not be too rapid or the film will crack and peel off.

4. Immerse the slide in a Coplin jar containing the stain and allow to remain for 2 to 3 minutes.

5. Dry without washing.

6. After drying, wash by immersing the slide in a beaker of water. Washing must be done very carefully to avoid washing off the film.

7. Allow the slide to dry in air.

With this method leucocytes and bacteria will be stained deep blue against a pale blue background.

Methylene Blue Reduction Test. This method is useful for making a rapid inspection of a large number of raw milk samples. Methylene blue is decolorized by reduction. When mixed with milk, the rate of this change is an indication of the bacterial oxygen consumption, and roughly of the number of bacteria present.

1. Place 10 ml. samples of milk in sterile test tubes and add 1 ml. of a standard certified methylene blue solution which is prepared by dissolving 1 tablet (which is purchased commercially) in 50 ml. of boiling distilled water and adding 150 ml. of cold distilled water. Prepare control tubes by adding the methylene blue to tubes of sterile milk. Mix until the color is uniform.

2. Incubate the tubes at 37° C. in a water bath.

3. Inspect the tubes at specified intervals and determine the time, after placing in water bath, that is required for discoloration.

4. Examine the tubes after 24 hours and 48 hours of incubation at 37° C. Note carefully the changes in texture, consistency, and composition of the milk at those intervals.

Milk, on the basis of this test, may be graded as follows: Class I. Excellent, not decolorized in 8 hours. Class II. Good, decolorized in less than 8 hours, but not less than 6 hours. Class III. Fair, decolorized in less than 6 hours, but not less than 2 hours. Class IV. Poor, decolorized in less than 2 hours.

Production of Gas. The organisms of most importance in producing gas in dairy products include the Escherichia- Aerobacter group, certain non-spore-forming yeasts that ferment lactose, and certain Clostridia.

1. Add 10 ml. each of *pasteurized* and *raw* milk to sterile plugged test tubes by means of sterile pipettes.
2. Pour into each a column of melted vaseline, and permit to harden.
3. Incubate the tubes at 30° C.
4. Examine at regular intervals for gas production and record the differences in location of the vaseline column in millimeters.

PASTEURIZATION

Pasteurization is a process of sterilizing or partially sterilizing organic solutions without altering their chemical properties. This method was devised by Pasteur and was first applied to beer and wines, and later to milk. The usual, or *holder*, method involves heating every particle of milk or milk product to a temperature of 143°–145° F. and holding at such temperature for not less than 30 minutes in an approved apparatus. This should bring about a reduction of 90% or more in the bacterial content, including destruction of all pathogenic organisms. After heating, the milk must be rapidly cooled to 50° F. or below, to prevent the growth of the surviving organisms (largely proteolytic spore-formers).

Flash pasteurization involves heating milk to 160°–165° F. for 15 seconds and rapid cooling. Its use is limited, however, because the time and temperature are hard to control accurately and because a "cooked" flavor is sometimes imparted to the product.

The advantages of proper pasteurization are twofold: a reduction in the spread of pathogens through milk, and an improvement in its keeping quality. That these advantages have been recognized is indicated by the results of a recent survey which showed that in cities in the United States with a population over 1000, 74% of the market milk was pasteurized.

Pasteurization does not affect the digestibility of milk. Excessive heating (above present pasteurization conditions), however, does alter the size of the butterfat globules and prevents formation of a distinct cream line.

Phosphatase Test. This test serves as a means of determining the efficiency of pasteurization. It is based on the power of the heat-sensitive enzyme phosphatase, normally present in milk, to liberate phenol from phosphoric-phenyl esters. *Mycobacterium tuberculosis,* most resistant of the non-spore-forming pathogens commonly found in milk, is always destroyed more quickly by heat than is phosphatase. The enzyme is 96% destroyed if subjected to 143° F. for 30 minutes. The test can detect 1° F. or 5 minutes underheating, or the addition of 0.5% raw milk. The Scharer Rapid Field Modification is performed as follows:

   1. The reagents required are:
   (a) 2,6-dibromo-quinone-chloroimide (BQC) dissolved in 95% ethyl alcohol.
   (b) Buffered substrate containing disodium phenyl phosphate, sodium borate buffer, and magnesium, dissolved in water and rendered phenol-free by treatment with BQC and extraction with butyl alcohol.
   Both these reagents may be purchased ready prepared.
   2. The technic is as follows:
   (a) Add 5 ml. buffered substrate to 0.5 ml. milk. Shake briefly.
   (b) Incubate 10 minutes in water bath at 37° C.
   (c) Remove and add 6 drops BQC solution. Shake well.
   (d) After 5 minutes, compare color with opaque standards.

After proper pasteurization, tested milk is gray or brown; cream, gray or white. Raw milk or cream is an intense blue. The intensity of color is proportional to the degree of underpasteurization or the amount of raw milk added.

MILK PRODUCTS

The subject of milk products is of vast importance from both an economic and a nutritional standpoint, and one in which bacteria play a vital role. We can, however, touch on it here only in the briefest manner.

Butter. Butter is composed of a minimum of 80% fat and 10–16% water, the remainder comprising milk solids-not-fat, and sometimes salt. Cream is separated from milk and is then usually pasteurized to destroy the organisms present. A bacterial culture ("starter") is added consisting of lactic acid bacteria like *Streptococcus lactis,* which curdle and coagulate the milk, and also organisms like *Leuconostoc citrovorum* and *Leuconostoc dextranicum* (*Streptococcus paracitrovorus*), which attack principally the citric acid in milk yielding products such as diacetyl to impart desirable flavor and aroma. At the required point, the ripening is checked by cooling and the product churned to form butter.

Cheese. Cheese is a concentration of the constituents of milk, principally fat, casein, and insoluble salts. To retain these in concentrated form, milk is coagulated either by bacteria-produced lactic acid or by the addition of rennet.

There are in general three standard types of cheese as follows:

1. Soft, acid-curd, or cottage type cheese and cream cheese, eaten in fresh or unripened state.

2. Hard, rennet-curd cheese, ripened by the growth of bacteria or molds or both, which do not cause extensive proteolysis. Included in this group are American, Roquefort, Cheddar, and Swiss cheeses.

3. Soft or semisoft rennet-curd cheese, ripened by more or less proteolytic bacteria (as in Limburger) or molds (as in Camembert), which soften the curd.

Ice Cream. Ice cream is made principally from cream and sugar, and contains 8–14% butterfat. The ingredients, particularly unpasteurized cream, and the utensils are the main sources of bacteria. Although the types of bacteria present in ice cream are generally the same nonpathogens usually found in milk and cream, ice cream ranks next to milk among dairy products as a cause of epidemics. Low holding temperatures do not free ice cream of pathogens, as is evidenced by epidemics of typhoid and dysentery traced to it.

Fermented Milks. Lactobacilli in combination with certain yeasts and Streptococci have been used for hundreds of years to produce fermented beverages from the milk of cows, mares, goats, etc. Some of these are the yoghurt of central Europe, busa of Turkestan, kefir and koumiss of Russia, leben of Egypt, and, in this country, acidophilus milk and "cultured buttermilk." The latter is prepared principally from fresh, sweet milk by the addition of ordinary butter starters. (See also pp. 131–132.)

Evaporated Milk. This is whole milk concentrated to slightly less than one-half its original volume. It is processed under steam pressure after canning and should be sterile.

Sweetened Condensed Milk. This is concentrated whole milk to which approximately 40% cane sugar has been added. The product is not sterile, but the high concentration of sugar inhibits bacterial multiplication.

BIBLIOGRAPHY

Adams, H. S. *Milk and Food Sanitation Practices.* New York: The Commonwealth Fund, 1947.

Eckles, Clarence H., Willes B. Combs, and Harold Macy. *Milk and Milk Products.* Fourth Edition. New York: McGraw-Hill Book Company, 1951.

Elliker, P. R. *Practical Dairy Bacteriology.* New York: McGraw-Hill Book Company, 1949.

Enock, Arthur Guy. *This Milk Business: A Study from 1895 to 1943*. London: H. K. Lewis and Company, 1943.

Foster, E. M., *et al*. *Dairy Microbiology*. Englewood Cliffs, N. J.: Prentice-Hall, 1957.

Grant, F. M. *Cleaning and Sanitizing Farm Milk Utensils*. Farmers' Bulletin No. 2078. Washington, D. C.: Superintendent of Documents, 1955.

Hammer, B. W., and F. J. Babel. *Dairy Bacteriology*. Fourth Edition. New York: John Wiley and Sons, 1957.

Methods and Standards for the Production of Certified Milk. New York: American Association of Medical Milk Commissions, 1957.

Prescott, S. C., and C. C. Dunn. *Industrial Microbiology*. Third Edition. New York: McGraw-Hill Book Company, 1959.

Rahn, O. *Microbes of Merit*. New York: Ronald Press, 1945.

Rettger, Leo F., Maurice N. Levy, Louis Weinstein, and James E. Weiss. *Lactobacillus Acidophilus and Its Therapeutic Application*. New Haven: Yale University Press, 1935.

Rogers, Lore A., and associates. *Fundamentals of Dairy Science*. Second Edition. New York: Chemical Catalog Company, 1939.

Sammis, J. *Cheese Making*. Madison, Wis.: Cheesemaker Book Company, 1942.

Schaub, I. G., and M. K. Foley. *Diagnostic Bacteriology*. Fourth Edition. St. Louis: C. V. Mosby Company, 1952.

Standard Methods for the Examination of Dairy Products. Tenth Edition. New York: American Public Health Association, 1953.

Swingle, D. B. *General Bacteriology*, rev. by G. W. Walter. New York: D. Van Nostrand Company, 1947. Pp. 231–241.

Tanner, Fred Wilbur. *The Microbiology of Foods*. Second Edition. Champaign, Ill.: Garrard Press, 1944. Pp. 250–594.

U. S. Department of Agriculture. *Cheese Varieties and Descriptions*. Handbook No. 54. Washington, D. C.: Superintendent of Documents, 1953.

U. S. Public Health Service. *Milk Ordinance and Code*. Public Health Service Publication No. 229. Washington, D. C.: Superintendent of Documents, 1953.

Van Slyke, L. L., and W. V. Price. *Cheese*. New York: Orange Judd Publishing Company, 1938.

Wadsworth, A. B. *Standard Methods*. Baltimore: Williams & Wilkins Company, 1947. Pp. 257–538.

Walter, W. G., and R. H. McBee. *General Microbiology*. Second Edition. Princeton: D. Van Nostrand, 1962. Pp. 255–276.

Chapter XIII

FOOD BACTERIOLOGY

In this chapter we shall concern ourselves mainly with a discussion of the methods of food preservation and the illnesses spread through food, excluding dairy products. (See Dairy Bacteriology, Chapter XII.) The useful role of microorganisms in the production of foods, such as bread, vinegar, etc., will be taken up in Chapter XIV.

FOOD PRESERVATION

With few exceptions, food would remain edible indefinitely if kept free from microorganisms or if the organisms were prevented from multiplying. Since this is scarcely practical, adequate methods for preventing spoilage of food have long been the goal of both scientific workers and the food industry.

Spallanzani, in 1775, proved that meat broth could be preserved if heated in corked glass bottles. Early in the nineteenth century, a Frenchman, François Appert, discovered that when perishable food was placed in proper containers, sufficiently heated, and then hermetically sealed, it kept without spoiling. Since these early beginnings, the ensuing years have seen the development of the many methods used today for food preservation. These methods, exhaustively discussed by Tanner, are reviewed below.

Asepsis. Food should be prepared by clean, healthy people in clean plants. Before further processing, food is washed wherever possible to remove surface dirt.

High Temperatures. The marked susceptibility of bacteria to heat is the basis of this most reliable method of food preservation. Care must be exercised to insure heat penetration into the entire foodstuff, with special consideration of the size of the food mass, altitude, reaction, etc.

1. *Pasteurization.* Food material is heated to a temperature below the boiling point, held there for a specified length of time, and rapidly cooled. Examples: milk (see p. 139), fruit juices, beer, wine.

2. *Boiling.* Examples: most fruits and some acid vegetables in glass or tin containers.

3. *Canning*.   Destruction of all or nearly all the living bacteria is obtained by subjecting the food to high temperatures and holding it under conditions which prevent the ingress of other bacteria.   The process usually involves these steps: cleansing, blanching, exhausting or preheating, hermetically sealing the container, heat-processing usually by steam under pressure, and cooling.   Examples: vegetables, milk, meat, in tin cans or glasses.

Low Temperatures.

1. *Refrigeration or Chilling*.   Bacterial action is retarded but not prevented.   Examples: vegetables, fruits, beverages, meat, dairy products, etc.

2. *Cold Storage.*

(a) *Slow-freezing*.   This method usually yields undesirable products.

(b) *Quick-freezing*.   This method, rapidly growing in importance, generally involves direct immersion of the food in a liquid refrigerant, such as calcium chloride brine.   Investigations have shown that the product is not sterile but contains a few living organisms.   After processing, the food must be carefully handled and kept frozen.   Examples: meat, fruits, vegetables, ice cream.

Chemical Preservatives.   Food preservation by the use of chemical preservatives is gradually losing favor because (1) other and better methods are available, and (2) the preservatives have been demonstrated to be harmful in many instances.

Benzoic acid and benzoates, salicylic acid and salicylates, boric acid, sulfurous acid and sulfites, nitric acid and nitrates, nitrous acid and nitrites, formaldehyde, and creosote have at one time or another been employed.

Sulfur dioxide, one of the few preservatives now legally allowed, is used in dried fruits to preserve the natural color and for its probable antiseptic value.   Sulfurous acid and sulfites are added to alcoholic liquors, especially wines.

Boric acid, once widely used, is now prohibited in most countires.   It cannot kill organisms, but prevents the growth of many.   It has been found to be poisonous if ingested in large quantities.

Sodium nitrate (saltpeter) and small amounts of sodium nitrite are usually added to salt solutions in pickling meats.   *Benzoic acid* and *benzoates* are used for the preservation of vegetables.   Small amounts of sodium benzoate are sometimes added to tomato catsup after the bottle has been opened.

The practice of adding *formaldehyde* to milk and cream, once fairly common, is now prohibited by law.

In the smoking of meat, small amounts of *creosote* are furnished by burning wood or by injection of the creosote itself.

Spices and essential oils have been used in fruit cake, mince meat, condiments, etc. Spices have been known to harbor organisms which cause spoilage, although a few show appreciable germicidal properties.

Acids naturally present in foods like vinegar and sour milk serve as their preservative.

Drying. This method involves the abstraction of moisture to a point intolerable to bacterial growth or enzyme activity. The organisms themselves also lose water to the dehydrated foodstuff.

The advantages peculiar to dried foods made them a valuable weapon in the recent war. Food can be kept in this manner for long periods of time; bulk is greatly reduced, saving shipping space and facilitating handling; dried foods are good substitutes for fresh foods; they do not require sterilization or maintenance of sterile conditions during preparation; they are economical to use.

Certain disadvantages, however, cannot be overlooked. A long soaking period and proper care is required for proper rehydration; if improperly packaged or handled, the product may become infested with insects; if the food is accidentally moistened, it may be attacked by bacteria, yeasts, and molds.

For civilian use, raisins, prunes, etc. are marketed dried in their natural state. Gelatin, eggs, and milk are sold in powdered form, the latter two being widely used in the manufacture of baked goods, ice cream, etc.

Concentration. Development of microorganisms is repressed by high osmotic pressure, usually obtained by concentrated solutions of salt (meat, fish, and pickling of cucumbers and cabbage), sugar (fruits, jellies, milk), or both.

Fermentation. Carbohydrates in foods may be fermented to organic acids like lactic acid, checking the development of putrefactive bacteria. Examples: sauerkraut, pickles, fermented milks.

PROBLEMS OF SPECIFIC FOODS

Meat. Contrary to long-held belief, it has recently been shown that meat from healthy, freshly slaughtered animals is seldom sterile, but contains some bacteria. By the time meat is purchased on the open market, it may contain hundreds of thousands or millions of organisms per gram. There appears, however, to be no correlation between the bacterial count and the sanitary quality, as the kind rather than the number of bacteria is important.

In the spoilage of high protein foods such as meat and fish, both aerobes and anaerobes are involved, but the latter are chiefly responsible for the changes produced. First the aerobic organisms act and create an environment favorable to the growth of the anaerobes. Then these attack proteins and decomposition products released by the aerobes, with resultant liberation of foul-smelling compounds.

Salle classifies the organisms involved as follows:

1. Gram-positive, aerobic, spore-forming saprophytic rods, which liquefy gelatin rapidly; e.g., *Bacillus subtilis, Bacillus albolactis.*

2. Gram-negative, aerobic, non-spore-forming rods; e.g., *Escherichia coli, Proteus vulgaris, Aerobacter cloacae, Pseudomonas fluorescens.*

3. Cocci; e.g., *Micrococcus pyogenes* var. *aureus, Micrococcus candidus.*

4. Anaerobes; e.g., *Clostridium perfringens, Clostridium tertium.*

5. Molds and yeasts, particularly Penicillium and Mucor.

The most common bacterial infections transmitted to man through meat are tuberculosis, anthrax, actinomycosis, paratyphoid fever, puerperal fever, pleuropneumonia, glanders, mucous diarrhea, salmonellosis, and brucellosis.

Eggs.   From 5 to 18% of fresh eggs contain living bacteria within the shell; all eggs have organisms upon the shells. The shells are porous, and damp soiled eggs soon become contaminated, the moisture aiding in carrying the organisms through the pores. Eggs may be preserved by cold storage, freezing, drying, immersion in sodium silicate solution (water glass), and various methods designed to prevent the passage of air through the shell pores. These include packaging in brine or sawdust, coating with vaseline or paraffin, wrapping in oiled paper, or immersing in lime water.

Bread and Pastry.   There are large numbers of organisms on unwrapped bread, but these are primarily of commercial rather than public health importance. Molds cause extensive deterioration. Organisms like *Serratia marcescens* produce a disagreeable red tinge. Cream puffs and éclairs are an ideal medium for staphylococci and are the cause of a number of outbreaks of food poisoning (see below).

Shellfish.   Oysters and other shellfish grown in fecal contaminated water may spread typhoid fever.   (See also Chap. XI.)

FOOD POISONING AND FOOD INFECTION

Food poisoning, as used in this textbook, refers to the conditions caused by the ingestion of food contaminated with harmful bacteria or their exotoxins. It includes toxemia from ingestion of foods containing poisons produced by *Clostridium botulinum* and certain Micrococci, and food infections or, more accurately, food-borne infections. These latter are of two general types: (1) Salmonella infections, in which the symptoms are typical of food poisoning, and (2) those diseases, such as typhoid fever, cholera, and dysentery, which are caused by agents transmitted by many vectors of which food is but one, and whose symptoms are not those usually associated with food poisoning.

The term "ptomaine poisoning," often used by the layman to indicate food poisoning, is misleading and inaccurate. Ptomaines are protein decomposition products such as cadaverine, putrescine, etc. Oral administration of these substances produces no harmful effects.

Botulism. This dread disease, having a mortality of 60–85%, is caused by the soluble exotoxin of *Clostridium botulinum*. Growth of this organism and production of its toxin require anaerobiosis (although it has been known to grow under aerobic conditions in the presence of some aerobes) and a period of incubation. These are furnished most readily by improperly prepared sausages and canned foods eaten cold or after warming. Fresh foods are not involved, nor freshly cooked foods, as the toxin is destroyed by heat. This most potent bacterial toxin is the only such toxin known that is not destroyed by gastrointestinal secretions. If kept in the dark under anaerobic conditions, as is the case in preserved food, potency is maintained for six months.

In the United States and Canada in the period 1899–1943, 404 outbreaks of botulism involving 1125 cases and 732 deaths were reported. Commercially packed foods were involved in 50 of these outbreaks, with olives and spinach or chard the worst offenders. Home-canned foods were responsible for 315, with string beans and corn the most frequent culprits.

The symptoms of the disease are not necessarily gastrointestinal. Rather they reflect the action of the toxin on the peripheral nerves with a resulting flaccid paralysis, particularly of the face and throat.

Antitoxin has been found effective in protecting experimental animals. In human beings, the results have been encouraging, but the number of cases in which antitoxin has been used is small. It is effective only before the nerve tissue has been damaged, as is evidenced by the onset of symptoms.

For further details on the organism and the disease, see pp. 256–258, Chapter XXVI.

Staphylococcus Food Poisoning. Certain Micrococci, generally hemolytic strains of the *Micrococcus pyogenes* var. *aureus* group, cause food poisoning with characteristic gastrointestinal disturbances. The symptoms are often associated with consumption of contaminated foods containing starch thickening, such as éclairs, cream puffs, some cake fillings and salad dressings, etc.

The organisms produce a soluble toxic substance or enterotoxin. The incubation period is usually 2–6 hours, with complete recovery in 24–48 hours, and extremely rare fatalities.

The Micrococci may contaminate food via boils or abscesses on the skin, nose, or mouth of food handlers. Inadequate refrigeration allows growth and toxin formation. Subsequent inadequate heating may kill the organisms but not the toxin, as the latter is quite heat-resistant.

This type of food poisoning can be prevented by refrigeration of baked goods containing starchy creams, and similar precautions with other foods.

Other organisms that have been incriminated in similar cases of food poisoning include members of the colon-aerogenes group, Streptococci, and Proteus. Whether or not these organisms produce an enterotoxin is still open to question.

Salmonella Infection. This is a quite common type of food infection, caused mainly by *Salmonella typhimurium* (*aertrycke*), *Salmonella schottmuelleri*, *Salmonella enteritidis*, and *Salmonella choleraesuis*. Indications that actual infection takes place are: (1) the somewhat longer incubation period—usually over 12 hours—than with Micrococci and (2) the presence of the organisms in the feces. The illness is characterized by gastrointestinal disturbances of short duration and a low mortality of 0–10%. The disease is spread by human carriers, by the excreta of mice and rats, and by ingestion of meat of infected animals. Almost any kind of food may be involved, but high protein foods predominate. The disease is also spread by raw or improperly cooked foods, or cooked foods that are left uncovered, exposing them to excreta of rodents, or that are left unrefrigerated, giving the organisms a chance to grow.

Preventive measures include cleanliness in the kitchen and on the part of food handlers, proper cooking and refrigeration, and avoidance of uncooked foods at club suppers or picnics, where the disease is often spread because of a long incubation time in the kitchen or in transit.

Other Infections. *Enteric diseases*, such as typhoid, paratyphoid, dysentery, etc., are often spread by contaminated oysters and other shellfish and vegetables like celery and watercress, grown in association with contaminated water.

Tuberculosis may be spread to human beings by the ingestion of meat from infected cattle.

BIBLIOGRAPHY

American Can Company. *The Canned Food Reference Manual.* New York: 1947.

Canning Trade. *A Complete Course in Canning.* Baltimore: 1936.

Clark, C. C., and R. H. Hall. *This Living World.* New York: McGraw-Hill Book Company, 1940. Pp. 450–477.

Cruess, W. V. *Commercial Fruit and Vegetable Products.* Third Edition. New York: McGraw-Hill Book Company, 1948.

Dack, G. M. *Food Poisoning.* Revised Edition. Chicago: University of Chicago Press, 1949.

Frazier, W. C. *Food Microbiology.* New York: McGraw-Hill Book Company, 1958.

Gale, E. F. *The Chemical Activities of Bacteria.* New York: Academic Press, 1951.

Jacob, Morris B. *The Chemistry and Technology of Food and Food Products,* Vols. I and II. New York: Interscience Publishers, 1944.

Jensen, L. B. *Microbiology of Meats.* Third Edition. Champaign, Ill.: Garrard Press, 1954.

Maxcy, K. F., ed. *Rosenau Preventive Medicine and Public Health.* Eighth Edition. New York: Appleton-Century-Crofts, 1956.

Prescott, Samuel Cate, and B. E. Proctor. *Food Technology.* New York: McGraw-Hill Book Company, 1937.

Rosebury, T. *Experimental Air-borne Infection.* Baltimore: Williams & Wilkins Company, 1947. Pp. 1–137.

Sherman, Henry Clark. *Chemistry of Food and Nutrition.* New York: Macmillan Company, 1941.

Standard Methods for the Microbiological Examination of Foods. New York: American Public Health Association, 1958.

Swingle, D. B. *General Bacteriology,* rev. by G. W. Walter. New York: D. Van Nostrand Company, 1947. Pp. 243–249.

Tanner, F. W. *The Microbiology of Foods.* Second Edition. Champaign, Ill.: Garrard Press, 1944.

Chapter XIV

INDUSTRIAL MICROBIOLOGY

We have already considered several fields in which microorganisms perform positive functions in our daily lives. These include the activities of the organisms of the soil, the manufacture of dairy products, and the production of the antibiotic penicillin. We have also indicated the negative role, so important to public health and industry, of spoilage and contamination by organisms of food and water. Now we shall investigate other industries which are concerned with microorganisms—those that put bacteria, yeasts, and molds to work making a wide variety of products, as well as those in which bacterial destruction may cause great losses.

VINEGAR-MAKING

When an alcoholic liquid is exposed to air for a time, a surface film appears, and the liquid becomes sour due to oxidation of the alcohol to acetic acid by members of the genus Acetobacter. The product is vinegar. The film is composed of a viscous gelatinous zooglea, in which are found a large number of the organisms. This scum has been termed "mother of vinegar" as a small portion of it acts as a starter.

Apple cider may be exposed to the action of yeasts to form alcoholic hard cider, which, when oxidized by Acetobacter, yields cider vinegar. Oxidation of wines results in wine vinegar. Any substance containing 12–16% sugar, such as fruit juice, malt extract, diluted honey, or molasses, is a suitable raw material for production of the alcoholic liquid.

Commercial vinegars contain about 4% acetic acid. Traces of esters, alcohol, sugar, and glycerin formed during the fermentations are largely responsible for the pleasant odor and flavor of the product.

Vinegar may lose strength on standing, due to the oxidation of acetic acid to carbon dioxide and water by some Acetobacter species. This takes place only in the presence of considerable oxygen, and may be prevented by storage in well-filled, tightly stoppered bottles or destruction of the organisms by pasteurization.

Acetobacter Aceti (Mycoderma Aceti). 1. Family—Pseudomonadaceae. 2. Habitat—vinegar. 3. Discoverer—Kützing, in 1837.

Morphological Characteristics. 1. Form—rods. 2. Cell groupings—occurs singly and in long chains, with large club-shaped forms. 3. Stains yellow with iodine solution. 4. Nonmotile.

Cultural Characteristics. 1. Beer gelatin with 10% sucrose—large shiny colonies. 2. Fluid media—slimy pellicle formed.

Physiological Characteristics. 1. Optimum temperature, 30° C. 2. Aerobic. 3. Acid is formed from dextrose, ethyl alcohol, propyl alcohol, and glycol.

Acetobacter Pasteurianum (Bacterium Pasteurianum). 1. Family—Pseudomonadaceae. 2. Habitat—vinegar, beer, and beer wort. 3. Discoverer—Hansen, in 1879.

Morphological Characteristics. 1. Form—rods. 2. Cell groupings—occurs singly and in chains, with club-shaped forms. 3. Stains blue with iodine solution. 4. Variable motility.

Cultural Characteristics. 1. Wort gelatin—small circular colonies with entire edge. 2. Meat infusion gelatin—widespread growth. 3. Beer with 1% alcohol—dry, wrinkled, folded pellicle.

Physiological Characteristics. 1. Optimum temperature, 30° C., but growth occurs between 5° and 42° C. 2. Aerobic. 3. Acid is formed from dextrose, ethyl alcohol, propyl alcohol, and glycol.

Related Organisms. 1. *Acetobacter acetosum*—beer. 2. *Acetobacter xylinum*—vinegar. 3. *Acetobacter industrium*—beer wort. 4. *Acetobacter suboxydans*—beer.

SAUERKRAUT-MAKING

Members of the Lactobacillus group are responsible for the fermentative process which occurs in shredded cabbage leaves. These organisms produce lactic acid, other organic acids, alcohol, and carbon dioxide.

Usually the organisms present on the cabbage are allowed to process it naturally, but sometimes a pure culture is added.

The sauerkraut fermentation can be carried on successfully only when air is at least partially excluded from the fermenting mass, since lactic acid formation is inhibited by the presence of oxygen. The acid produced prevents putrefaction and decay.

ALCOHOLIC BEVERAGE PRODUCTION

The beer and wine industries utilize sugar fermentation by yeasts with the formation of alcohol and carbon dioxide. This process is best carried out under anaerobic conditions at about 30° C. in the presence of an optimum concentration of about 25% sugar. When the alcohol concentration reaches about 13%, the fermentation stops.

Saccharomyces cerevisiae is used for beer-making; *Saccharomyces ellipsoideus* for wine. When fruit juices are used as the starting material, the end product is wine. Specially selected strains of organisms are inoculated in pure cultures to control the taste, odor, and flavor of the product. Occasionally, contaminants interfere with proper fermentation and spoil the beer or wine by altering its flavor or acidity.

TEXTILE BACTERIOLOGY

Retting of Flax and Hemp. The "Lushi," a Chinese work of the Sung dynasty, about 500 A.D., contains a statement that the Emperor Shen Mung in the twenty-eighth century B.C. first taught the people of China to cultivate "Ma" (hemp) for making hempen cloth. Both flax and hemp are now important sources of fabrics. The bast fibers of the plants are removed by retting or rotting. The process involves the bacterial hydrolysis and fermentation of pectin, which, with cellulose and hemicellulose, comprises the polysaccharide framework of plants.

Water Retting. Stalks are immersed in water, either in a slow-flowing warm stream or in tanks or vats, and weighted down. Water and soluble substances are extracted, and this solution serves as a culture medium. Aerobic organisms grow first, using up the dissolved oxygen. Then the anaerobes grow, and ferment and dissolve the pectin, leaving the fibers intact. The hydrolysis and fermentation are accomplished chiefly by *Clostridium butyricum* or *Clostridium felsineum*, the main products being acetic acid, butyric acid, carbon dioxide, and hydrogen. The process takes about 10 days.

Dew Retting. Plants are spread on the ground in thin layers, and kept there for several months. Dew keeps them moist, enabling the biological activity to proceed. *Mucor stolonifer* is chiefly responsible.

Textile Spoilage. Huge losses of textile goods in process or in the finished state are caused by bacteria and molds. The bacteria involved appear to be largely the more common aerobic soil spore-formers. *Bacillus mesentericus, Bacillus subtilis*, certain of the Actinomyces, and molds of the genera Aspergillus and Penicillium are capable of attacking woolen fabrics previously sterilized by steaming. *Bacillus mycoides* and *Proteus vulgaris* can produce tendering at ordinary temperatures. Discoloration in silk may be caused by molds.

The types of activity of microorganisms in damaging textiles may be classified as follows:

1. Staining or discoloration of the material because of:
(a) Pigment formation by chromogenic bacteria.
(b) Development of colored spores of microorganisms.
2. Direct attack on the fibers with consequent loss of strength.
(a) Chemical action by bacterial and mold enzymes.
(b) Possible physical effects such as malformations due to the growth of microorganisms.
3. Action on the surface of the fiber which may result in:
(a) Uneven dyeing.
(b) Difficulties in finishing.
(c) Irregular surface appearance, luster, etc.
4. Action on the dyes by:
(a) Changing the pH (hydrogen ion concentration) of the fibers so that the shades of the dyes vary. This action is due to members of the genus Acetobacter.
(b) Possible direct chemical action on the dyes by the products resulting from microbial growth and metabolism.

Prevention of Damage. Certain chemicals and antiseptics have been used to check damage to textiles caused by microbes.

1. *Chemical treatment* is rather widely used to check growth of bacteria and molds, being applied either directly or by mixing with the sizing which is necessary in many textile products.
2. *Gaseous treatment* has been found to suppress the growth of certain mildews in cotton subjected to a storage atmosphere containing an equal mixture of air and carbon dioxide gas.

BREAD

Through the action of the yeast *Saccharomyces cerevisiae*, carbon dioxide is produced in bread. This gives porosity and is responsible for "rising." Starches and proteins are also split and made more digestible. Characteristic flavors and aromas are produced, and the gluten is softened.

Ropy or *slimy bread* may result from contamination with spore-forming bacilli of the subtilis-mesentericus group. The spores, introduced with the flour or the yeast, are highly resistant and survive the high temperature of baking. Ropiness in bread is not harmful, but it is hardly appetizing. It may be retarded by maintaining a cool temperature.

Sour bread results from the overgrowing of the yeast by undesirable organisms, largely lactic acid and butyric acid bacteria. They are introduced with the flour, water, or yeast.

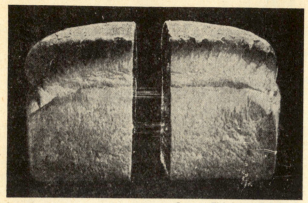

Courtesy, Fleischmann Laboratories.

Fig. 36.   Ropy bread.

Moldy bread is caused by contamination with mold spores during cooling, slicing, or wrapping, and their subsequent multiplication if optimum conditions of moisture and temperature prevail.

In addition to proper sanitation and cooling, propionic acid has been used to reduce rope and mold in bread.

TANNING OF LEATHER

The object of processing hides is to prevent their decomposition without interfering with the properties which make leather adaptable to its many uses.

Making hides into leather usually involves the following steps:

1. Preservation by salting or drying.

2. Disinfection by soaking in 2% mercuric chloride solution containing 1% formic acid, or in a 10% solution of salt in 2% hydrochloric acid. The main object of this process is to destroy *Bacillus anthracis.*

3. Soaking and fleshing to remove excess tissue.

4. Depilation or removal of hair by liming, sweating, use of acids, alkalis, or enzymes, or by carefully controlled bacterial decomposition, usually with *Proteus vulgaris.*

5. Bating—placing in a weak fermenting infusion of bird or dog dung or a bran mash. Attempts are now being made to substitute specific bacterial cultures.

6. Final tanning in the tan liquors, in which the action of microorganisms plays a role.

CURING OF TOBACCO

Tobacco is cured to improve its aroma and texture. The process is one of fermentation, but the exact nature and mechanism of the changes are not clearly understood. Certain plant enzymes and the organisms *Bacillus subtilis*, *Bacillus mycoides*, and some species of Proteus may be involved. Some workers, however, believe the changes to be purely oxidative and not catalyzed by enzymes. During the curing, about 28% of the nicotine content is lost, and the citric acid is increased.

The process involves stacking the leaves, which are then allowed to ferment. Heat is generated, oxygen consumed, and carbon dioxide, ammonia, and other gases released. The flavor may be improved by moistening the leaves with sugar, sirups, or molasses, which results in the production of desirable aromatic esters and other compounds.

SILAGE

Production of silage is a simple method for the preservation of foods for stock animals. Any farm product can be siloed if it contains sufficient sugar for the production of the acid needed for its preservation. Corn makes particularly good silage. Oats, peas, alfalfa, clover, straw, and even weeds may be used.

The material is cut into small pieces, packed tightly in the silo, and moistened if necessary. Sap is exuded, forming an ideal pabulum for microbial growth. Because of the carbohydrates present, fermentative rather than putrefactive changes occur. The oxygen is soon consumed and the CO_2 content increases, until anaerobic conditions prevail. Then the anaerobes grow and produce most of the changes, generating large quantities of heat and carbon dioxide.

In the silage juice are billions of microorganisms, mostly *Lactobacillus vulgaricus*, *Lactobacillus acidophilus*, and certain Streptococci, together with acids and alcohols produced by the sugar decomposition. The acids include lactic, which should predominate, acetic, butyric, and propionic. The temperature of the silo is important, for butyric acid organisms, imparting an objectionable odor and taste, predominate at temperatures above 60° C., and lactic acid organisms at temperatures around 50° C.

The acceleration growth phase continues for three weeks when the lag phase sets in, with few organisms remaining. Fermentation ceases by the end of the first month.   Complete loss of silage may result when putrefactive changes and acid accumulations occur in heavily soiled silage.   If the spoiled silage is consumed by dairy cattle, an unpleasant taste is imparted to the milk.

During the process acids, alcohols, and carbon dioxide are produced from the sugars.   The plant tissues are broken down and the entire mass is changed in appearance, flavor, and nutritive value. The material is preserved and will keep for months or even years.

LACTIC ACID PRODUCTION

Lactic acid is made industrially either by synthesis or by the action of Lactobacilli or *Streptococcus lactis*.   Starting materials for bacterial action may be sugars such as glucose, sucrose, or lactose, or starches of various kinds which are first hydrolyzed to sugars by acids or enzymes.   Corn, molasses, and whey are excellent carbohydrate sources for lactic acid production.

The product finds many industrial uses: in the leather industry, in the dyeing and finishing of textiles, in the production of drugs and chemicals, in the preparation of foods and beverages, and in the manufacture of baking powder, plastics, and lacquers.

PRODUCTION OF ALCOHOL AND ACETONE

Ethyl alcohol is prepared commercially by the action of microorganisms, chiefly yeasts, on various carbohydrates—crude molasses, potatoes, sugar beets, hydrolyzable starches, and cellulose.   Certain aerobic bacteria, *Bacillus acetoethylicus* and *Bacillus macerans*, are also used to produce alcohol as well as acetone from carbohydrates.

Acetone may be produced by the fermentation of a saccharine fluid, usually diluted molasses, allowing the alcoholic fermentation to continue until acetic acid is produced.   To this is added calcium hydroxide solution, and the resulting calcium acetate is distilled to yield acetone.

Acetone and normal butyl alcohol can be obtained by the action of organisms, especially anaerobes like *Clostridium acetobutylicum*, on starch or corn-meal mash.   The bacterial enzymes are responsible for both the hydrolysis of the starch to sugar and the oxidation of this to butyric and acetic acids, which are converted to butyl alcohol and acetone by reduction.

Numerous species of Clostridium have been employed for the fermentation of carbohydrates to butyric acid, butyl alcohol, and other compounds.

COCOA

Before cocoa beans are roasted, they undergo a fermentation which removes a large amount of the pulp, destroys the seed vitality, and imparts desirable flavor, aroma, and color. Some of the changes include the fermentation of the sugar present by yeasts to form carbon dioxide and alcohol and the oxidation of the latter to acetic acid by Acetobacter.

VITAMIN ASSAYS

Taking advantage of the selective nutritive requirements of microorganisms, procedures have been devised to determine vitamins in food or other material. *Lactobacillus casei* is used for the determination of riboflavin (B_2 or G), pyridoxine (B_6), and pantothenic acid; *Lactobacillus arabinosus* for niacin. *Saccharomyces cerevisiae* is employed to determine amounts of thiamin (B_1) by measuring the rate of fermentation of glucose to carbon dioxide and alcohol.

GLYCERIN

Glycerin may be prepared commercially by the action of certain yeasts, e.g., *Saccharomyces ellipsoideus*, on sugar.

ORGANIC ACIDS

Citric Acid. Yields of 87–100% of citric acid are obtained commercially by the action of molds, especially *Aspergillus niger*, on a mixture of cane sugar and technical glucose. The product is widely used in medicines, citrate production, baking, the making of candy, flavoring extracts, and soft drinks, textile printing and dyeing, engraving, silvering, and the manufacture of ink.

Propionic Acid. This may be produced by the action of Propionibacter on a large number of carbohydrates. The organisms are nonmotile, Gram-positive, non-spore-formers. Uses of propionic acid include the manufacture of perfumes, solvent for the explosive pyroxylin, and the production of propionates.

Gluconic Acid. Gluconic acid, salts of which are widely used in medicines, is produced by the action of Acetobacter on glucose. Penicillium or Aspergillus may also be used.

Butyric Acid. See discussion of butyl alcohol on p. 156.

ENZYMES

Bacillus mesentericus is used for the commercial production of amylase and protease; *Bacillus macerans* and *Bacillus polymyxa* may also be used for amylase.   Commercial enzymes take part in brewing by converting the raw starches to materials fermentable by yeasts. Bacterial amylase is used for partial starch digestion for sizing cotton fibers and paper.   Bacterial protease is employed to desize cotton and degum silk.

MISCELLANEOUS APPLICATIONS

Synthetic Rubber.   Liebemann in 1943 developed a variety of organism that produced commercially practicable amounts of 2.3-butylene glycol (butanediol) used in the manufacture of synthetic rubber. The compound is formed directly during the fermentation of corn-starch mash.

Petroleum.   Certain bacteria, e.g., Pseudomonas, Achromobacter, and other soil organisms, and molds attack hydrocarbons and mineral oils.

Other Uses.   The use of bacterial preparations as antigens will be discussed in the section on medical bacteriology.   Bacteria are used in research to help identify sugars of unknown origin.   They probably play a role in the pulp and paper industries, but their exact status is not known.   They have a marked effect on the keeping qualities and behavior of rubber latex and wood.

BIBLIOGRAPHY

Barker, H. A. *Bacterial Fermentations*. New York: John Wiley and Sons, 1957.

Cruess, W. V. *Commercial Fruit and Vegetable Products*. Third Edition. New York: McGraw-Hill Book Company, 1948.

————. *The Principles and Practice of Wine Making*. Westport, Conn.: Avi Publishing Company, 1947.

Jacobs, Morris B. *The Chemistry and Technology of Food and Food Products*, Vols. I and II.   New York: Interscience Publishers, 1944.

Maxcy, K. F., ed. *Rosenau Preventive Medicine and Public Health*. Eighth Edition. New York: Appleton-Century-Crofts, 1956.

Porter, J. R. *Bacterial Chemistry and Physiology*. New York: John Wiley and Sons, 1946. Pp. 896–1011.

Prescott, S. C., and C. G. Dunn. *Industrial Microbiology*. Third Edition. New York: McGraw-Hill Book Company, 1959.

Rahn, O. *Microbes of Merit*. New York: Ronald Press, 1945.

Stephenson, Marjory. *Bacterial Metabolism*. Third Edition. New York: Longmans, Green and Company, 1949.

Wadsworth, A. B. *Standard Methods*. Baltimore: Williams & Wilkins Company, 1947. Pp. 257–538.

Woodruff, H. B., and L. E. McDaniel. *Commercial Fermentations*. New York: Chemical Publishing Company, 1949.

Chapter XV

VETERINARY BACTERIOLOGY

There are many diseases of animals that may be transmitted directly or indirectly to man. These include a wide variety of bacterial infections as well as diseases caused by viruses, pathogenic protozoa, and Rickettsia. Bacterial infections of animals may be contracted by man in different ways: 1. Through an insect vector, such as lice, fleas, or ticks. 2. Through eating food or drinking milk coming from diseased animals. 3. Through direct contact with animals that are infected. Veterinary medicine and bacteriology, then, are closely interrelated subjects and an integral part of medical bacteriology. The veterinarian fills a critical public health role as inspector of all foods of animal origin, including meat, poultry, milk, and sea foods, and as inspector of dairy activities, insuring conformity with federal, state, and municipal standards. The U. S. Armed Forces still use veterinary officers to inspect all classes of foodstuffs.

MYCOBACTERIUM PARATUBERCULOSIS

Mycobacterium Paratuberculosis (Bacillus of Johne's Disease).
1. Group—acid-fast. **2.** Family—Mycobacteriaceae. **3.** Habitat— intestinal mucosa; cause of Johne's disease, a chronic diarrhea of cattle. **4.** Discoverers —Johne and Frothingham, in 1895.

Morphological Characteristics.
1. Form—more or less pleomorphic rods.
2. Cell groupings—occurs singly, in pairs, or in clumps. **3.** Size—0.5μ broad and 1 to 2μ long. **4.** Staining properties— usually stains uniformly, but the longer forms show alternate stained and unstained segments. With Ziehl-Neelsen stain, it is

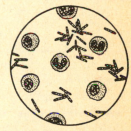

Fig. 37. *Mycobacterium paratuberculosis* in lymph.

acid-fast. It is not stained by ordinary aniline dyes. **5.** Gram-positive. **6.** Nonmotile. **7.** Nonsporing.

Cultural Characteristics. The organism has been difficult to cultivate and grows poorly on agar, potato, and gelatin stabs. Growth is more profuse if killed cultures of other acid-fast bacteria are added. **1.** Glycerol broth—grows in clumps, with sediment in bottom of tube and clear liquid above; adapted cultures may produce a thin surface pellicle which becomes wrinkled and thickened. **2.** Dorset's glycerol egg medium—after several days of incubation, very minute, slow-growing, dry, grayish-white colonies appear; as colony grows older and larger, it becomes wrinkled.

Physiological Characteristics. 1. Optimum temperature, 39° C.; minimum, 27° C.; maximum, 43° C. 2. Grows best aerobically; slight growth under anaerobic conditions. 3. Resistance—the organisms will resist desiccation. Direct sunlight kills them in a few minutes. Disinfectants are effective in the usual dilutions, if the organism is free from albuminous matter.

Pathogenicity. The disease affects cattle and, less frequently, sheep and deer, following ingestion of fodder that has been soiled with feces of infected animals. The disease is a chronic enteritis characterized by gradual emaciation, running a long course extending from a period of several months, finally ending in death in practically all cases. The incubation period may run up to 18 months. Aside from evidence of emaciation, the post-mortem examination reveals no specific lesions apart from those in the intestinal tract. The intestinal mucosa, particularly that of the small intestine, is greatly thickened and presents a marked, wrinkled, or corrugated appearance. On the surface of the corrugations there are hemorrhages, and occasionally small nodules are observed. The bacilli are usually found in the mesenteric glands which are enlarged, edematous, and pigmented.

Diagnosis. 1. A preparation, *johnin*, which is similar to tuberculin, has been made by growing Johne's bacillus in a special broth medium containing dead *Mycobacterium phlei* bacilli. When 0.2 ml. of johnin is injected intradermally into an animal and 0.2 ml. is injected 48 hours later into the site of the previous injection, a positive reaction of diffuse swelling occurs in infected animals. The injection of avian tuberculin in large amounts may produce a reaction in animals with Johne's disease. The test with johnin is believed to be specific and of diagnostic value.

2. The organism may be cultivated on glycerin egg medium containing dead tubercle bacilli, by inoculating the medium with intestinal mucosa washed in 20% antiformin.

3. A smear may be prepared using a piece of the intestinal mucosa and staining by the Ziehl-Neelsen method.

Production of Disease. Although attempts have been made to infect guinea pigs, rats, and mice, it is considered that the typical disease has not yet been transferred to animals. No case of human infection has been recorded.

Prophylaxis. No method of prophylaxis has been devised. Control of the disease is based on the destruction of infected animals.

Related Organisms. 1. *Mycobacterium avium*, 2. *Mycobacterium tuberculosis*, 3. *Mycobacterium leprae*, 4. *Mycobacterium piscium*, 5. *Mycobacterium marinum*, 6. *Mycobacterium ranae*, and 7. *Mycobacterium friedmannii*.

BACILLUS ANTHRACIS

Bacillus Anthracis (Anthrax Bacillus). 1. Group—aerobic, spore-bearing. 2. Family—Bacillaceae. 3. Habitat—found in soil and parasitic in cattle, sheep, man, and other animals. 4. Discoverers—Cohn, in 1872, and Koch, in 1876.

Morphological Characteristics. 1. Form—rods with square or concave ends. 2. Cell groupings—occurs singly with diplobacillary forms, and also in chains. 3. Size—1 to 2μ by 5 to 10μ. 4. Staining properties—stains well with carbol fuchsin, methylene blue, and other aniline dyes. 5. Gram-positive and non-acid-fast. 6. Spores are equatorial, ellipsoidal, and do not bulge. Their germination is polar and they are not produced in the animal body; they are formed only in the presence of oxygen. 7. Capsule is found in the animal body and is present in cultures grown on serum media, but not on agar. 8. Nonmotile.

Cultural Characteristics. *Bacillus anthracis* is easily cultivated on culture media generally used in the laboratory. 1. Agar cultures—the colonies are raised, grayish-white, and irregular, with an uneven surface. They give a Medusa head appearance which is characteristic of the organism; the colony is made up of characteristic parallel interlacing chains of bacilli. 2. Agar slant—the growth is thick, grayish-white or cream color, with an uneven surface, resembling matted hair. 3. Gelatin stab—slight filiform growth which gradually resembles an inverted fir tree. Liquefaction is crateriform, becoming stratiform. 4. Litmus milk—growth is rapid, with decolorization in 2 to 3 days and coagulation with subsequent peptonization. 5. Potato—growth is dry, elevated, tending to expand, and white to grayish-white in color. 6. Broth—growth in flocculent masses, with a thick pellicle forming on the surface. 7. Blood serum—growth is abundant and creamy-yellow, with slight liquefaction of the medium.

Physiological Characteristics. 1. Optimum temperature, 37° C., with limits between 12° and 44° C. 2. Aerobic, and facultative anaerobe. 3. Acid, but no gas, is formed in glucose, maltose, sucrose, and salicin. 4. Nitrates are reduced to nitrites. 5. The organism does not produce indol, nor is starch hydrolyzed. 6. Ammonia production is slight. 7. Resistance—spores are killed by boiling for 10 minutes, but in dry state may remain alive for years. 8. Pigment production—some strains form a brownish-yellow and, occasionally, a pink pigment.

Pathogenicity. The natural disease affects herbivora, chiefly sheep and cattle, in which the mortality may be as high as 80%. The disease generally involves septicemia. In acute cases, the animal may appear in customary health until a short time before death, when convulsions, rigors, and elevation of temperature occur, with blood extravasations from the nose, mouth, and rectum. The most marked post-mortem finding is enlargement of the spleen. This organ becomes pliable and soft and contains enormous numbers of bacilli.

Man may be infected by the cutaneous, respiratory, or intestinal routes.

1. *Cutaneous Anthrax.* This form occurs most frequently upon the hands and forearms, especially of persons working with livestock. It is referred to as *malignant pustule* and is characterized by the appearance of a small furuncle within 12 to 24 hours after infection. The lesion becomes vesicular, with a seropurulent exudate, and then undergoes necrosis. This may be excised and the patient will, in general, recover, or local gangrene may set in followed by a systemic infection and death in 5 to 6 days.

2. *Pulmonary Anthrax (Woolsorter's Disease).* This form is contracted by inhalation or by swallowing spores of *Bacillus anthracis*. The disease is characterized by many of the symptoms of pneumonia and often passes into a fatal septicemia. Before death the organism may be isolated in the sputum.

3. *Intestinal Anthrax.* This form is rare and is contracted by ingestion of uncooked meat of infected animals. There is a severe enteritis with bloody stools and prostration, with death generally. The presence of bacilli in the feces is indicative of this type of anthrax infection.

Serological Reactions. Agglutination and precipitation tests have been used to differentiate *Bacillus anthracis* from other members of the aerobic spore-bearing group. The agglutinins produced in response to injections of the organism have given cross reactions with other members of the group.   The performance of the agglutination test with *Bacillus anthracis* is hampered by the difficulty in obtaining homogeneous suspensions.

The formation of precipitins by injection of *Bacillus anthracis* into animals has been utilized by Ascoli in a precipitin test for the diagnosis of anthrax in animals dead of the disease or to detect contamination of hides.   The test is carried out as follows:

1. An extract of a small piece of the spleen or other tissue of a dead animal is made by boiling it in saline, cooling, and filtering.

2. The extract is layered on high-titer antiserum.

3. The formation of a zone of precipitate at the junction of the extract and antiserum indicates a positive reaction.

4. The serum for the test should be prepared by immunizing rabbits with a suitable strain of encapsulated anthrax bacilli.

Diagnosis.   1. *Collection of Material.*   This will depend upon the type of infection—pus or fluid from local skin lesions in malignant pustule, blood in the systemic stage of the disease, sputum in the pulmonary infection, feces in the intestinal infection, or the spinal fluid in the occasional meningeal infections.

2. *Smears* are made and examined for characteristic large, Grampositive, encapsulated organisms.   Blood examinations may give negative results, for the bacilli are not present in the blood in large numbers until shortly before death.

3. *Cultures* are made by inoculating some of the material in broth and on agar plates and incubating.

4. *Animal inoculations* are made as confirmatory tests.   The subcutaneous injection of white mice, guinea pigs, or rabbits with a small portion of broth culture, or a suspension made of the growth on nutrient agar, results in the death of the animal in 12 hours to 2 or 3 days with a fatal septicemia, if anthrax bacilli are present.   The material isolated from the lesion may be directly inoculated into animals.

Prophylaxis and Therapy.   General prophylaxis consists chiefly in destruction of infected animals, burying of cadavers, and the disinfection of stables.

1. *Active Immunization of Animals.*   The method of Pasteur is still used today.   It consists of the subcutaneous inoculation of attenuated cultures of *Bacillus anthracis*.   Two vaccines are injected; the first is a culture which has lost its virulence for guinea pigs and rabbits, and is potent only for mice; the second is a culture which is still potent for mice and guinea pigs, but is harmless for rabbits.   The second

vaccine is given 12 days after the first.   Various modifications of this method are in use, but it is still essentially the same.

2. *Serum Therapy*.   Inoculation with anti-anthrax serum confers some degree of immunity and has been used for prophylaxis and therapy.

3. *Arsenicals* are frequently used in conjunction with antiserum therapy.

Related Organisms. 1. *Bacillus subtilis*, 2. *Bacillus mesentericus*, and 3. *Bacillus mycoides*.   These are all aerobic, spore-bearing organisms and may be differentiated from *Bacillus anthracis* in that they are motile and nonpathogenic, usually found in soil and feces.

PASTEURELLA TULARENSIS

Pasteurella Tularensis (Bacterium Tularense). 1. Group—hemorrhagic septicemia. 2. Family—Parvobacteriaceae. 3. Habitat—in ground squirrels, rabbits, hares, and other rodents and may be transmitted to man by bites of insects or by contact with infected animals. 4. Discoverers—McCoy and Chapin, in 1910.

Morphological Characteristics. 1. Form—rods which are pleomorphic, with bacillary, coccoid, and bipolar forms. 2. Cell groupings—occurs singly or in pairs, or in short chains. 3. Size—0.2μ in thickness and from 0.3 to 0.7μ in length. 4. Capsule—organism is enclosed in a capsule which can be demonstrated in films made directly from tissues or when the bacteria are mixed with serum. 5. Staining properties—the organism is stained with carbol fuchsin and gentian violet. 6. Gram-negative and non-acid-fast. 7. Nonmotile. 8. Nonsporing.

Cultural Characteristics.   *Pasteurella tularensis* does not grow on ordinary culture media.   The addition of egg, blood, fresh sterile rabbit spleen, or serum to media is favorable to the growth of the organism. 1. Dorset's egg medium—transparent droplike colonies appear in 3 to 5 days. 2. Egg yolk medium—mucoid minute colonies on the second or third day, becoming larger later. 3. Serum glucose agar—colonies are at first small, but then they appear as droplets. 4. Blood glucose cystine agar—on this medium, which is generally used to cultivate *Pasteurella tularensis*, minute, viscous, easily emulsified, grayish-white colonies develop.

Physiological Characteristics. 1. Optimum temperature, 37° C. 2. Aerobic, and facultative anaerobe. 3. Resistance—the organism is destroyed at a temperature of 56° C. for 10 minutes.   It is killed by 2% tricresol in 2 minutes, and by desiccation and sunlight in 4 to 5 hours. 4. Slight acid is formed in dextrose, maltose, and mannose.

Pathogenicity. The organism is responsible for a disease occurring naturally in rodents, especially the rabbit. The disease in human beings is known as tularemia. It is transmitted by the handling of infected animals. Infection may take place by inoculation, through the unbroken skin, through the conjunctiva, and possibly by inhalation, and also by intermediary infected blood-sucking insects—flies (*Chrysops discalis*), ticks (*Dermacentor andersoni*), lice, fleas, and others. In animals the disease resembles plague and is fatal, but in man it is milder and rarely fatal, although of long duration. In man it is of four clinical types:

1. *Ulceroglandular type* in which the primary lesion is a papule of the skin that later becomes an ulcer and is accompanied by enlargement of the regional lymph glands.

2. *Oculoglandular type* in which the primary lesion is a conjunctivitis with an accompanying enlargement of the regional lymph glands.

3. *Glandular type* in which there is no primary lesion at the site of infection, but there is enlargement of the regional lymph glands.

4. *Typhoid type* in which there is neither a primary lesion nor glandular enlargement.

Serological Reactions. In response to infection with tularemia, agglutinins and complement-fixing antibodies are developed. The agglutination test is a valuable diagnostic procedure in tularemia. Since there is some cross agglutination between antiserum prepared against *Pasteurella tularensis* and *Brucella melitensis* and *Brucella abortus*, it is necessary to differentiate the causative agent of tularemia from these organisms. This is determined from the fact that *Brucella melitensis* and *Brucella abortus* agglutinate *Pasteurella tularensis* antiserum to about one-fourth to one-sixth its titer, which may run up to as high as 1:2560 or 1:5120 against *Pasteurella tularensis*. The members of the Brucella group, however, do not absorb agglutinins for *Pasteurella tularensis*. Therefore, a reciprocal absorption test is used to differentiate between them.

Diagnosis. 1. *Cultures* are made by inoculating the infected material into blood glucose cystine agar, incubating at 37° C. for from 3 days to 3 weeks, and examining for characteristic colonies.

2. *Animal Inoculations.* The suspected material is inoculated into guinea pigs by rubbing some of the material into the shaven abraded skin or by inoculating it subcutaneously. If *Pasteurella tularensis* is present, the animals will die in 5 to 10 days with charac-

teristic lesions of cervical, axillary, or inguinal buboes, enlarged glands, enlarged spleen, and a granulated liver.

3. *Agglutination Test.* Agglutinins appear in the blood of cases with tularemia after the first week of the disease. The blood is collected at this period and serum dilutions of 1:10 to 1:320 or higher are prepared and tested in 0.5 ml. amounts against 0.5 ml. of a saline suspension of *Pasteurella tularensis.* When agglutination occurs in serum dilution 1:80 or higher, this is considered diagnostic, if there is no cross agglutination with *Brucella abortus* or *Brucella melitensis.*

Prophylaxis and Immunity. The disease is preventable in human beings by washing the hands with a strong antiseptic following the handling of wild rabbits, by using rubber gloves when skinning or handling rabbits, and by thorough cooking of rabbits before eating them. One attack confers immunity in man. A toxoid vaccine is reported to have given successful protection. Streptomycin is the best treatment for human tularemia.

Related Organisms. 1. *Pasteurella pestis* (plague bacillus), 2. *Pasteurella avicida* (fowl cholera bacillus), 3. *Pasteurella muricida* (septicemia of wild rats), 4. *Pasteurella suilla* (swine plague bacillus), and 5. *Pasteurella bollingeri* (hemorrhagic septicemia in domestic cattle, hogs, and horses).

PASTEURELLA AVICIDA

Pasteurella Avicida (Bacillus Avisepticus, Bacterium Avicidum). 1. Group—hemorrhagic septicemia. 2. Family—Parvobacteriaceae. 3. Habitat—cause of fowl cholera and septicemia in other domestic animals. 4. Discoverer—Pasteur, in 1880.

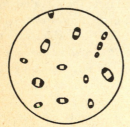

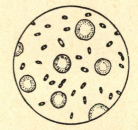

FIG. 38. *Pasteurella avicida.* Note bipolar staining and involution forms.

FIG. 39. *Pasteurella bollingeri* in bloodstream of cow with hemorrhagic septicemia.

Morphological Characteristics. 1. Form—short oval rods. 2. Cell groupings—occurs singly, in pairs, or in short chains. 3. Size—0.3

to 1.25μ in length. 4. Staining properties—stains easily with ordinary aniline dyes, and also exhibits marked polar staining. 5. Gramnegative and non-acid-fast. 6. Nonmotile. 7. Nonsporing. 8. Occasionally exhibits vacuolated forms.

Cultural Characteristics. The organism is easily cultivated from blood and organs of infected animals on the simplest media. 1. Agar cultures—within 24 to 48 hours, minute white or yellowish colonies are observed. 2. Agar slant—white or yellowish growth which is at first transparent, but later opaque. 3. Gelatin stab—growth without liquefaction. 4. Litmus milk—growth is slow, with no visible change in the medium. 5. Potato—no growth. 6. Broth—uniform turbid growth with pellicle.

Physiological Characteristics. 1. Optimum temperature, 37° C., with limits of growth between 12° and 43° C. 2. Aerobic, and facultative anaerobe. 3. Nitrates are reduced to nitrites. 4. Indol is produced. 5. Ammonia is produced. 6. Hydrogen sulfide is formed. 7. Acid, but no gas, is produced in dextrose, mannitol, sucrose, levulose, galactose, mannose, and xylose. 8. Resistance—the organism is killed by 0.5% phenol in 15 minutes, and by temperatures above 45° C. within a few minutes.

Pathogenicity. *Pasteurella avicida* produces a disease known as fowl cholera. This is widely prevalent among barnyard fowl, attacking chickens, ducks, geese, and a large variety of smaller birds. The infection is extremely acute, ending fatally in a few days. It is accompanied by diarrhea, often with bloody stools, great exhaustion, and a drowsiness bordering on coma. Autopsy reveals hemorrhagic inflammation of the intestinal mucosa, enlargement of the liver and spleen, and often a bronchopneumonia. The specific bacilli may be found in the blood, in the organs, and in the dejecta. Infection takes place probably through the ingestion of food and water contaminated by the discharges of diseased birds. The subcutaneous injection of a 24 hour broth culture of the bacillus brings about the death of mice in 18 to 72 hours, with evidences of local edema and congestion, and the presence of large numbers of organisms in the blood and organs.

Serological Reactions. By means of agglutination reactions two or more antigenic types of *Pasteurella avicida* may be identified. Antiserum prepared against this organism will also agglutinate *Pasteurella pestis* and *Pasteurella pseudotuberculosis* to a certain degree. *Pasteurella avicida* does not produce a true exotoxin. It has been possible to produce immune sera of prophylactic and therapeutic value by the inoculation of horses with living or dead bacilli.

PASTEURELLA SUILLA

Pasteurella Suilla (**Pasteurella Suiseptica, Bacterium Suicida**).
1. Group—hemorrhagic septicemia.   2. Family—Parvobacteriaceae.
3. Habitat—causes swine plague.   4. Discoverer—Löffler, in 1886.

Morphological Characteristics. 1. Form—short, ovoid, bipolar rods with rounded ends.   2. Cell groupings—occurs singly, in pairs, or occasionally in short chains.   3. Size—0.5 to 1μ.   4. Staining properties—bipolar staining brought out by Wright stain.   5. Gram-negative.   6. Nonmotile.   7. Nonsporing.   8. Occasionally involution forms are noted.   9. Generally no capsules are formed.

Cultural Characteristics. 1. Agar cultures—small, transparent, slightly raised colonies appear after 24 hours.   2. Agar slant—grayish-white, slightly raised, glistening growth.   3. Gelatin stab—white growth along the line of stab, with no liquefaction.   4. Litmus milk—the medium is rendered slightly acid, with no coagulation. 5. Broth—uniform turbidity, with subsequent sediment.

Physiological Characteristics. 1. Optimum temperature, 37.5° C. 2. Aerobic, and facultative anaerobe.   3. Resistance—the organism is not very resistant.   A temperature of 58° C. will destroy it in 20 minutes, while phenol and bichloride of mercury are effective in shorter periods of time.

Pathogenicity. *Pasteurella suilla* produces in hogs an epidemic disease which is characterized by a bronchopneumonia followed by a general septicemia.   Swine plague manifests itself either as a peracute, hemorrhagic septicemia in which extensive pulmonary changes have not had time to develop, or as a somewhat less acute infection in which marked pulmonary involvement with hepatization and necrotic areas are present.   A chronic form of the disease is recognized in which the animal becomes greatly emaciated.   In such cases, autopsy reveals old necrotic foci in the lungs, caseous foci in various lymph glands, and joint involvement.

There is often a serosanguinous pleural exudate, and a swelling of bronchial lymph glands and of the liver and spleen.   The gastro-intestinal tract is rarely affected.   The bacilli are numerous in the lungs, in the exudates, in the liver and spleen, and in the blood.   It is probable that spontaneous infection usually occurs by inhalation. The disease may be transmitted experimentally to mice, guinea pigs, and rabbits by subcutaneous inoculations of the organism, resulting in death in 3 to 4 days.   Pigs have also been experimentally infected by inhalation of the organism.

Prophylaxis and Therapy. Polyvalent strains for immunization purposes and also for the production of antisera are advisable for members of the Pasteurella group because of their interrelationship. Killed or attenuated cultures of *Pasteurella suilla* have failed to give uniform results because hogs may be immune to one strain and susceptible to another.

MALLEOMYCES MALLEI

Malleomyces Mallei (Bacillus Mallei, Pfeifferella Mallei, Actinobacillus Mallei). 1. Family—Parvobacteriaceae. 2. Habitat—cause of glanders in horses, sheep, and goats, and occasionally infecting man. 3. Discoverers—Löffler and Schütz, in 1882.

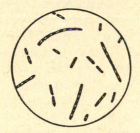 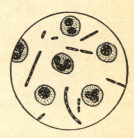

Fig. 40. *Malleomyces mallei.* Left: Variants in pure culture from short filamentous forms to forms up to 12μ in length. Right: Extracellular bacilli with leucocytes.

Morphological Characteristics. 1. Form—slender rods with rounded ends. 2. Cell groupings—occurs singly or in pairs, and sometimes branched. 3. Size—0.5 to 1μ by 2 to 5μ. 4. Staining properties—bipolar staining is common; stains poorly with aqueous solutions of aniline dyes, but satisfactorily with carbol fuchsin and Löffler's alkaline methylene blue. 5. Gram-negative and non-acid-fast. 6. Nonmotile. 7. Nonsporing. 8. Pleomorphism is exhibited in cultures, with bacilli varying in length and width, often showing irregularly shaped forms, and long filaments with true or false branching.

Cultural Characteristics. 1. Agar cultures—moist, grayish-white, translucent, ropy colonies with regular borders appear within 24 hours, and later become yellowish and yellowish-brown in color. 2. Agar slant—moderate, moist, glistening, ropy growth which is grayish-white and later becomes yellowish in color. 3. Gelatin stab—slight pin-point growth with no liquefaction. 4. Litmus milk—slow coagulation within a week with slight acid production. 5. Potato—pale yellow, honey-drop-like colonies which appear in 36 to 48 hours and become darker, the medium having a faint greenish tinge around the growth.   6.   Broth—slowly clouds with subsequent slimy or ropy sediment and a thin surface pellicle.   7.   Löffler's serum medium—moist, viscid, yellowish colonies develop after 36 to 48 hours.

Physiological Characteristics.   1.   Optimum temperature, 37° C., with limits of 22° to 44° C.   2.   Aerobic, and facultative anaerobe. 3.   Sugars are generally not fermented, although some strains ferment dextrose with slight acid formation.   4.   Resistance—the organism is killed at a temperature of 55° C. for 10 minutes or 75° C. for 5 minutes; it resists drying for only 2 or 3 weeks; it is killed by direct sunlight in a few hours, and by ordinary antiseptics in about 10 or 15 minutes.

Toxic Products.   *Malleomyces mallei* produces an endotoxin, *mallein,* which is thermostable at 120° C. and withstands long periods of storage with retention of potency.   This substance, when injected into a horse suffering from glanders, produces within 6 to 8 hours a sharp rise in temperature which lasts for several hours and then gradually begins to fall.   At the site of injection there appears, within a few hours, an area of erythema and swelling 20 to 30 cm. in diameter, which lasts for 3 to 9 days.   Injections of mallein in normal animals are attended by only a slight rise in temperature and a local reaction which is much smaller and less enduring than that encountered in infected animals.   For the test, 1 ml. of mallein is injected beneath the skin of the neck or into the conjunctival sac.   The temperature should be taken every 2 hours for a least 18 hours after the injection, and the size, appearance, and duration of the local reaction recorded.

Pathogenicity.   *Malleomyces mallei* causes the disease known as glanders in the horse, and to which nearly all equines are susceptible. The organism may be found in lesions in the liver, spleen, pancreas, lungs, kidneys, ovaries, testes, nasal mucosa, and on all mucous surfaces.   Large nodules of pus are found in the lesions.   Infection usually takes place by droplet infection or through breaks in the skin. Glanders in man is a rather rare disease, transmissible from horse to man, but laboratory infections have occurred.

There are two forms of the disease—the *acute form* which is limited to the nasal mucosa and the upper respiratory tract, and the *chronic type* which is characterized by farcy buds, ulcers, and suppurations of lesions in the joints and muscles.

1.   *Acute glanders*—this usually has a fatal outcome within 4 to 6 weeks.   The animal becomes ill with fever and prostration.   After 2 or 3 days there is nasal discharge, which is at first serous, but later seropurulent.   There are ulceration of the nasal mucosa and acute

swelling of the lymph nodes, which may break down and form deep suppurating sinuses and ulcers.

2. *Chronic glanders*—this has a more gradual onset and is attended by persistent abscess formation, purulent arthritis, nasal ulcers, multiple swellings of the skin, and general lymphatic enlargement.

Straus Test. A test used to diagnose glanders is the intraperitoneal injection of male guinea pigs with suspected material. Within 2 or 3 days there is an orchitis, with tumefaction and purulent inflammation of the testicles. Material taken from lesions in the liver, spleen, pancreas, and lungs of the inoculated animal is then cultured and examined for typical organisms.

Serological Reactions. The development of complement-fixing antibodies has been employed practically in the diagnosis of glanders. Agglutinins have also been utilized in diagnostic procedures. Animals who have recovered from the disease have in their sera specific agglutinins against the organism, but normal serum also agglutinates *Malleomyces mallei*. Therefore, to be of significance, tests should be done with sera of a titer of at least 1:500.

Immunity and Prophylaxis. The use of attenuated cultures of *Malleomyces mallei* and mallein to confer immunity against glanders has not been very successful. The disease itself does not confer immunity against a second attack.

Prophylaxis consists of slaughtering animals showing clinical signs of glanders or giving a positive mallein test, and properly disposing of the carcass.

Related Organisms. *Malleomyces pseudomallei*, isolated by Whitmore in 1913, resembles *Malleomyces mallei*, but it is motile and liquefies gelatin. The organism gives a positive Straus test. It produces a disease which is characterized by septicemia, pyemia, and the formation of characteristic granulomatous nodules in nearly all parts of the body. The disease is present in wild rats, guinea pigs, and rabbits, and also in man.

SALMONELLA CHOLERAESUIS

Salmonella Choleraesuis (Bacillus Suipestifer, Bacillus Cholerae Suum, Salmonella Suipestifer). 1. Family—Enterobacteriaceae. 2. Habitat—occurs naturally in pig, and is an important secondary invader in hog cholera. It may occasionally cause an acute gastroenteritis in man. 3. Discoverers—Salmon and Smith, in 1885.

Morphological Characteristics. 1. Form—short rods. 2. Cell groupings—occurs singly. 3. Size—0.6 to 0.7μ by 2 to 3μ. 4. Staining properties—stains well with ordinary aniline dyes. 5. Gram-negative and non-acid-fast. 6. Motile with 4 to 5 peritrichous flagella. 7. Nonsporing.

Cultural Characteristics. 1. Agar cultures—round, moist, translucent, grayish colonies with smooth surface. 2. Agar slant—round, moist, grayish-blue colonies. 3. Gelatin stab—growth on surface and along stab, with no liquefaction. 4. Litmus milk—the medium is at first slightly acid and becomes alkaline with the continued growth of the organism. 5. Potato—grayish-white growth which becomes brownish. 6. Broth—uniform white turbidity with a slight sediment and a thin pellicle.

Physiological Characteristics. 1. Optimum temperature, 37°C. 2. Aerobic, and facultative anaerobe. 3. Nitrates are reduced to nitrites. 4. Hydrogen sulfide is produced. 5. Acid and gas are formed in dextrose, levulose, galactose, mannose, arabinose, xylose, maltose, glycerol, mannitol, dulcitol, and isodulcitol. 6. Resistance—the organism is killed in 1 hour at 56° C. and, in 20 minutes at 58° C. The usual dilutions of various antiseptics, such as corrosive sublimate, formalin, phenol, and chloramine, kill it in 15 to 20 minutes.

Serological Reactions. Salmonella choleraesuis produces cross reactions with sera of a number of other species of Salmonella, from which it is differentiated by a combination of serological and cultural methods.

Pathogenicity. This organism is responsible for food infection in man (see p. 148) and is a secondary invader in hog cholera, the primary agent being a filterable virus. *Salmonella choleraesuis* is commonly found in the hog intestine, but only in cases of weakened resistance is it pathogenic. In hog cholera it is associated with other bacteria in ulcers found in various organs, such as the spleen, liver, and intestine. Inoculated into rabbits, mice, or guinea pigs, it produces an infection which is fatal. (See also p. 308.)

SALMONELLA PULLORUM

Salmonella Pullorum (Bacillus Pullorum). 1. Family—Enterobacteriaceae. 2. Habitat—causes white diarrhea in chicks, and infects the ovaries and eggs of adult birds. 3. Discoverers—Rettger and Harvey, in 1908.

Morphological Characteristics. 1. Form—short rods. 2. Cell groupings—occurs singly. 3. Size—0.3 to 0.5μ by 1 to 2.5μ. 4. Staining properties—stains well with ordinary aniline dyes. 5. Gram-negative. 6. Nonmotile. 7. Nonsporing.

Cultural Characteristics. 1. Agar cultures—smooth, glistening, grayish-white colonies. 2. Agar slant—circular colonies which are discrete and translucent. 3. Gelatin stab—slight surface growth, grayish in color, with no liquefaction. 4. Litmus milk—the medium is first acid and then turns alkaline. 5. Potato—slight grayish growth. 6. Broth—uniform white turbidity.

Physiological Characteristics. 1. Optimum temperature, 37°C. 2. Aerobic, and facultative anaerobe. 3. Nitrates are reduced to nitrites. 4. Hydrogen sulfide is formed. 5. Acid and gas are formed from dextrose, levulose, galactose, mannose, arabinose, xylose, mannitol, and isodulcitol. 6. Resistance—a temperature of 56° C. for 1 hour kills the organism. Exposure to 58° C. for 20 minutes, and for the same period of time to antiseptics, such as phenol, mercuric chloride, and formalin, is also destructive.

Serological Reactions. Following infection with *Salmonella pullorum,* agglutinins are present in animals that recover, which fact is utilized in a microscopic agglutination test employed in diagnosis. A titer of 1:100 is considered significant.

Pathogenicity. *Salmonella pullorum* is the cause of white diarrhea in young chicks and is transmitted to the chick through the egg. Other members of the brood may contract the disease from infected chicks. The point of predilection is the ovaries, from which the organism travels to the intestines to produce diarrhea. The control of the disease lies in the eradication of infected and recovered cases because they are sources of infection; the latter may remain carriers of the disease, although apparently well.

Related Organisms. Most of the other members of the Salmonella group are important because of their role in food infection. These will be discussed in greater detail in the chapter on the Coliform Group. (See page 244.) They include: 1. *Salmonella schottmuelleri* —cause of food infection and enteric fever in man. 2. *Salmonella enteritidis*—which has also been isolated in epidemics of enteric fever due to ingestion of foods contaminated with it. 3. *Salmonella paratyphi*—cause of paratyphoid fever. 4. *Salmonella gallinarum*— cause of fowl typhoid. 5. *Salmonella abortivoequina*—produces abortion in mares. 6. *Salmonella abortusovis*—cause of abortion in sheep.

CLOSTRIDIUM CHAUVOEI

Clostridium Chauvoei (Bacillus of Symptomatic Anthrax).
1. Group—anaerobic spore-bearers.   2. Family—Bacillaceae.

3. Habitat—soil. 4. Discoverers—Arloing, Cornevin, and Thomas, in 1887.

Morphological Characteristics. 1. Form—rods with rounded ends. 2. Cell groupings—generally occurs singly, also in pairs joined end to end, but never growing out into chains or long filaments. 3. Size—0.5 to 1μ broad and 3 to 8μ long. 4. Staining properties—stains readily with aniline dyes. 5. Gram-positive. 6. Motility—actively motile with peritrichous flagella. 7. Spores—oval, elongated, excentric to subterminal, wider than the bacillus.

Cultural Characteristics. 1. Agar cultures—filamentous, transparent colonies on surface. They are finely granular with a fernlike edge and appear grayish by reflected light and bluish-gray by transmitted light. 2. Agar slant—poor spreading surface growth, which is barely visible, with a smooth surface or a finely granular one. The growth, however, is more compact than that of *Clostridium septicum*, and filamentous projections are sent out. 3. Deep glucose agar shake tube—the growth is abundant throughout the medium except 3 mm. below the surface. Colonies are irregularly round with an opaque grayish center. There is moderate gas formation. 4. Gelatin stab—liquefaction with gas bubbles. 5. Litmus milk—coagulation of the medium with gas production, and formation of acid. 6. Broth—growth is poor, with formation of a slight powdery deposit. 7. Cooked meat medium—growth is slight with production of gas bubbles in about 8 hours, and turbidity of the medium. The meat becomes pinkish, but there is no blackening or digestion.

Physiological Characteristics. 1. Optimum temperature, 37° C. 2. Anaerobic. 3. Hydrogen sulfide is produced. 4. Acid and gas are formed in dextrose, maltose, lactose, and sucrose, but inulin, salicin, mannitol, glycerol, and dextrin are not fermented.

Serological Reactions. Agglutinins are formed in response to injection or infection with *Clostridium chauvoei*. These are specific for this organism since agglutinating serum against *Clostridium chauvoei* does not agglutinate *Clostridium septicum*, which is closely related to it, except to a very slight degree. The antitoxin produced against *Clostridium chauvoei* is also specific, in that it has been reported to protect against infection with this organism but not against infection with *Clostridium septicum*.

Pathogenicity. *Clostridium chauvoei* is the cause of a disease in cattle and sheep which is known as blackleg, quarter evil, or symptomatic anthrax. It is characterized by a peculiar emphysematous swelling of the subcutaneous tissues and muscles, especially over the quarters. The mechanism seems to be that of a progressive bacteremia. Sick animals usually die in 1 to 2 days. The disease occurs chiefly in cattle, more rarely in sheep and goats; horses are not attacked spontaneously. In man, infection has not been reported, and the organism is believed to be nonpathogenic for human beings. The portal of entry is unknown. Experimentally, the organism produces a fatal infection in guinea pigs, and less often in mice. Rabbits and pigeons are fairly resistant. The intramuscular injection of 0.25 ml. of a 24 hour culture of the organism kills a guinea pig in 24 to 48 hours, the abdominal muscles being deep red and containing numerous small gas bubbles. At autopsy the bacillus is found in the local lesion and in the bloodstream. *Clostridium chauvoei* produced a specific weak exotoxin which is thermolabile and is destroyed at a temperature of 52° C. in 30 minutes.

Immunity and Prophylaxis. Recovery from one attack of symptomatic anthrax protects an animal against a second infection. Active immunity can be produced by vaccines of attenuated organisms. A dried powder of the muscles of animals which have succumbed to the disease is used as a vaccine and is subjected to a suitable temperature to insure attenuation of the virulence of the spores contained therein. Two vaccines are prepared by subjecting the material to heating for various periods. Good results have been obtained by inoculating animals into the end of the tail, first with the weaker vaccine and then with the stronger preparation. Passive immunization with serum of actively immunized sheep and goats has been practiced simultaneously with active immunization.

CLOSTRIDIUM SEPTICUM

Clostridium Septicum (Vibrion Septique, Bacillus Septicus).
1. Group—anaerobic spore-bearers. 2. Family—Bacillaceae.
3. Habitat—soil, animal intestine, dust, and polluted water. 4. Discoverers—Pasteur and Joubert, in 1877.

Morphological Characteristics. 1. Form—rods which are straight or slightly curved. 2. Cell groupings—occurs singly, in pairs, or in chains, and occasionally in long filaments. 3. Size—0.8 to 1μ by 3 to 8μ. 4. Staining properties—easily stained by aniline dyes. 5. Gram stain—young cultures are Gram-positive, but older ones may be Gram-negative. 6. Motility—motile by four to sixteen peritrichous flagella. 7. Spores—oval, excentric to subterminal, slightly wider than the bacilli, and readily formed. 8. Pleomorphism—the organism has a tendency toward marked pleomorphism in agar cultures where it may vary in size, shape, and depth of staining; shadow forms are seen.

Cultural Characteristics.   1. Agar cultures—spreading, filamentous, translucent colonies which are irregularly round and are grayish in color.   2. Agar slant—slight, spreading, translucent, glistening, grayish-yellow growth, with smooth surface.   3. Deep glucose agar shake tube—delicate, arborescent, and flocculent colonies throughout the medium, with abundant gas formation.   4. Gelatin stab—liquefaction with gas bubbles.   5. Litmus milk—coagulation with formation of acid and gas.   6. Broth—slight turbidity with a powdery deposit.   7. Cooked meat medium—growth is moderate with slight turbidity and gas formation, the meat turning pinkish in color without blackening or digestion.

Physiological Characteristics.   1. Optimum temperature, 37° C.   2. Anaerobe.   3. Hydrogen sulfide is formed.   4. Acid and gas are formed in glucose, maltose, lactose, and salicin, but not in mannite or sucrose.   5. Nitrates are reduced to nitrites.   6. Resistance—the spores resist boiling for hours.

Toxic Products.   A powerful exotoxin is produced by *Clostridium septicum*. The injection of the toxin subcutaneously or intramuscularly does not always produce a fatal reaction but produces local necrosis. The toxin is prepared in a 0.2% glucose broth containing 10% horse serum, the broth being incubated for 24 to 48 hours after inoculation. *Clostridium septicum* also produces a hemolysin which causes hemolysis of human and sheep red blood cells.

Serological Reactions.   Antitoxins against *Clostridium septicum* are prepared by injecting horses or sheep with the toxin. They are specific against the organism and do not protect against *Clostridium novyi*. By means of agglutination tests, strains of *Clostridium septicum* have been divided into four antigenic groups. There is no cross agglutination between *Clostridium septicum* antiserum and *Clostridium chauvoei*.

Pathogenicity.   *Clostridium septicum* is the cause of malignant edema, a fatal infection of horses following surgical or traumatic wounds, and occasionally in cattle and sheep. Pathologically, the disease is characterized by edematous, emphysematous swellings, and infiltration of cellular tissues, in which lesions the bacillus is found. In man the virulence of the organism varies greatly with the strain. Some types are almost or completely nonpathogenic.   In infections of man it is important as a cause of gas gangrene, although it is much less commonly encountered than is *Clostridium perfringens*. The form of gas gangrene due to this organism differs somewhat from that due to *Clostridium perfringens* in that edema with blood-stained fluid is more marked, and necrosis and gas production are less violent. *Clostridium septicum* is rarely found alone, but pure infections have been recorded.

Experimentally, the organism is pathogenic for guinea pigs, mice, rabbits, and pigeons. An intramuscular injection of 0.01 to 0.5 ml. of a 24 hour culture into guinea pigs produces death in 12 to 24 hours, with edema and gas production.

Prophylaxis. Vaccination, using attenuated spores, has proved effective in the hands of some workers. Various investigators have produced antitoxic and antibacterial sera for which both prophylactic and therapeutic value are claimed in the early stages of gas gangrene.

Related Organisms. The other members of the anaerobic spore-bearing group of organisms, genus Clostridium, are discussed in detail in the chapter dealing with this group. These include: 1. *Clostridium tetani*, 2. *Clostridium novyi*, 3. *Clostridium perfringens*, 4. *Clostridium botulinum*, 5. *Clostridium fallax*, 6. *Clostridium histolyticum*, and 7. *Clostridium lentoputrescens*.

Differentiation of Clostridium Septicum from Clostridium Chauvoei.

1. *Clostridium septicum* frequently infects wounds in animals and man, whereas *Clostridium chauvoei* has not been isolated from wound cultures and has never been found in human infections.

2. *Clostridium chauvoei* ferments sucrose and not salicin, and *Clostridium septicum* ferments salicin and not sucrose.

3. *Clostridium septicum* is more pathogenic for laboratory animals and produces more gas in tissues than does *Clostridium chauvoei*.

4. *Clostridium chauvoei* grows more slowly than *Clostridium septicum*.

BRUCELLA GROUP

In the Brucella group are included four organisms, *Brucella melitensis*, *Brucella abortus*, *Brucella suis*, and *Brucella bronchiseptica*. All are nonmotile, nonsporing, Gram-negative rods which do not liquefy gelatin, nor ferment carbohydrates. They are strict parasites, producing infection of the genital tract, the mammary gland or the lymphatic tissues, and the intestinal tract.

A. Brucella Abortus (**Bacillus Abortus, Alcaligenes Abortus**). 1. Group—Brucella. 2. Family—Parvobacteriaceae. 3. Habitat—present in milch cow and the cause of contagious abortion in cattle and undulant fever in man. It also invades mares, sheep, rabbits, and guinea pigs, producing contagious abortion. 4. Discoverer—Bang, in 1897.

Morphological Characteristics. 1. Form—short, slender, pleomorphic rods with rounded ends. 2. Cell groupings—occurs singly, in pairs, or in small chains. 3. Size—0.4 by 1μ. 4. Staining properties—stains easily with ordinary aniline dyes. 5. Gram-negative. 6. Nonmotile. 7. Nonsporing.

Cultural Characteristics. *Brucella abortus* is microaerophilic. It requires 25% CO_2 on primary isolation and 10% CO_2 on subsequent transfers. After prolonged culture, however, the organism grows freely under aerobic conditions. 1. Agar cultures—small, round colonies with smooth, glistening surface and entire edge. 2. Agar slant—slow grayish growth. 3. Gelatin stab—poor growth with no liquefaction. 4. Potato—slight grayish-brown growth. 5. Litmus milk—the medium is rendered slightly alkaline, with no coagulation. 6. Broth—slight turbidity.

Physiological Characteristics. 1. Optimum temperature, 37° C. 2. Aerobic, and facultative anaerobe. 3. Hydrogen sulfide is produced. 4. Sugars are not fermented. 5. Resistance—resists cold and drying. Killed at pasteurizing temperatures.

Pathogenicity. There are two organisms which have been designated in the literature as *Brucella abortus*. The porcine variety is now called *Brucella suis*. The bovine variety has retained the name *Brucella abortus*. The two organisms are similar in most respects, except that *Brucella suis* requires no added CO_2 for its growth. Both the porcine and bovine types can cause undulant fever in man after an incubation period of 1 week to several months. There is, typically, a long continued pyrexia, often with remissions, which may be accompanied by joint pains, skin rashes, sweating at night, and other symptoms. Whether or not these organisms are related to cases of miscarriage in human beings has not been determined. *Brucella abortus* has been isolated in several instances where the diagnosis of endocarditis was made.

Epidemiology. *Brucella abortus* may appear in the milk of cows that have aborted and may also appear in the milk of cows that are carriers but have suffered no abortion themselves. Human infection may result from attending to infected animals or, at other times, from handling the flesh or drinking raw milk of infected cows.

Serological Reactions. Agglutinins and complement-fixing antibodies are produced in response to infection with *Brucella abortus*, and a diagnosis of the disease is often made by ascertaining the agglutination titer of a patient's serum. The finding of an agglutinating titer of 1:80 or 1:100 is usually considered significant. This is not pathognomonic, however, since titers of this magnitude or higher are often encountered in healthy human beings. Therefore, when blood cultures are negative (the disease is one in which septicemia is often present), a positive diagnosis should not be made unless the serum agglutinin titer is at least 1:500.

Immunity and Prophylaxis. Calves, nonpregnant cows, and human beings seem to have a high natural resistance. One attack of undulant fever protects against a second attack. It is difficult to ascertain a cure, since remissions occur. Vaccines have been used in cattle, and the results thus far have been inconclusive. Experiments on active immunization of human beings have also given questionable results. The use of immune serum and brucellin has been advocated by many workers, but here again most of the reports have been unconvincing or negative. Recently, cures have been reported in cases treated with sulfanilamide, aureomycin, and chloromycetin.

B. Brucella Melitensis (Bacillus Melitensis, Alcaligenes Melitensis). 1. Group—Brucella. 2. Family—Parvobacteriaceae. 3. Habitat—strict parasite of goats and cause of Malta fever in man and contagious abortion in goats. 4. Discoverer—Bruce, in 1887.

Morphological Characteristics. 1. Form—short rods. 2. Cell groupings—occurs singly, in pairs, or in short chains. 3. Size—0.5 to 0.7μ wide by 0.6 to 1.2μ long. 4. Staining properties—stains well with ordinary aniline dyes. 5. Gram-negative and non-acid-fast. 6. Nonmotile. 7. Nonsporing. 8. Nonencapsulated.

Cultural Characteristics. 1. Agar cultures—grayish-white, round colonies about 0.5 mm. in diameter. 2. Gelatin stab—slow growth, with no liquefaction. 3. Litmus milk—medium becomes alkaline. 4. Potato—slight grayish-brown to chocolate-colored growth. 5. Broth—slight turbidity, with no pellicle or deposit.

Physiological Characteristics. 1. Optimum temperature, 37° C., with limits of 20° to 40° C. 2. Aerobic. 3. Pigment production—brown pigment on potato and in old agar cultures. 4. No fermentation of carbohydrates. 5. Nitrates are occasionally reduced, with disappearance of the nitrites formed. 6. Resistance—the organism is killed by moist heat at 60° C. in 10 minutes, at 65° C. in 5 minutes, and by 0.5% phenol in 15 minutes. It has good keeping qualities. In the dry state it may survive for 3 months, and hermetically sealed it may remain potent for 6 months at room temperature.

Serological Reactions. By agglutination two antigenic groups are distinguishable, the melitensis (smooth) type and the paramelitensis (rough) type. Carriers of the disease and those who have recovered have agglutinins and complement-fixing antibodies in their sera. Agglutinin absorption tests are necessary for diagnosis since such antisera will cross agglutinate with *Brucella abortus* and *Brucella suis*. Agglutinins occur in serum about the tenth or twelfth day of the fever. The titer rises to 1:100 and 1:300. Complement fixation is positive in nearly all cases during the fever.

Pathogenicity. Malta fever is spread to man by means of goat's milk infected with *Brucella melitensis*. The disease in both goats and man is a bacteremia, and the etiologic agent may be recovered from the blood. In goats, the most obvious clinical symptom is abortion, although this need not occur. Goats may have the disease with the organism circulating in the blood and being excreted in the urine, without the animal's showing signs of the infection. Many goats, however, show evidences of infection by losing weight, developing a cough, and in some instances developing mastitis and arthritis. In man a blood culture is positive in about 80% of the cases after the second day of the disease, which may prevail in the bacteremia form for several months.

Control of Undulant Fever and Malta Fever. The prevention of the spread of these infections depends upon the following measures:

1. Eradication of the diseases in herds of cattle, goats, and swine.

2. Proper precautions in handling infected animals, and in so far as possible avoiding contact with them.

3. Pasteurization of milk and milk products.

4. Disinfection and proper disposal of urine and feces of patients.

5. Treatment with the antibiotics aureomycin and chloromycetin in human beings.

Differentiation of Members of the Brucella Group. The following table presents a scheme for differentiating *Brucella abortus*, *Brucella suis*, and *Brucella melitensis* from each other.

Species	Infectivity for guinea pigs	CO_2 required for growth	H_2S formation	Growth on media containing:		
				Thionin 1:50,000	Basic Fuchsin 1:25,000	Pyronin 1:100,000
Brucella melitensis	++	−	±	+++	+++	+
Brucella abortus	+	+	++	−	+++	+++
Brucella suis	++	−	++++	+++	−	−

1. *Brucella melitensis* and *Brucella suis* are more highly infective for guinea pigs than is *Brucella abortus*.

2. *Brucella abortus* when first isolated requires CO_2, but *Brucella melitensis* and *Brucella suis* may be cultivated under aerobic conditions.

3. *Brucella suis* is the most active producer of hydrogen sulfide, and *Brucella melitensis* may form this gas only slightly.

4. *Brucella melitensis* is not inhibited by either thionin, basic fuchsin, or pyronin. *Brucella abortus* is inhibited by thionin and grows well in media containing basic fuchsin and pyronin. *Brucella suis* grows well in the presence of thionin, but its growth is deterred by basic fuchsin and pyronin.

BIBLIOGRAPHY

American Can Company. *The Canned Food Reference Manual*. Third Edition. New York: 1947.

Baumgartner, J. G. *Canned Foods—An Introduction to Their Microbiology*. Second Edition. London: Churchill, 1946.

Bergey, David H. *Bergey's Manual of Determinative Bacteriology*. Seventh Edition. Baltimore: Williams & Wilkins Company, 1957. Pp. 394–406.

Edelmann, R. *Textbook of Meat Hygiene*. Eighth Edition. Philadelphia: Lea and Febiger, 1943.

Elliker, P. R. *Practical Dairy Bacteriology*. New York: McGraw-Hill Book Company, 1949.

Fabian, F. W. *Home Food Preservation: Salting, Canning, Drying, and Freezing*. New York: Avi Publishing Company, 1943.

Frazier, W. C. *Food Microbiology*. New York: McGraw-Hill Book Company, 1958.

Hagan, W. A., and D. W. Bruner. *The Infectious Diseases of Domestic Animals*. Third Edition. Ithaca, N. Y.: Cornell University Press, 1957.

Hopkins, E. S., and W. H. Schulze. *The Practice of Sanitation*. Third Edition. Baltimore: Williams & Wilkins Company, 1958.

Hull, T. G. *Diseases Transmitted from Animals to Man*. Fourth Edition. Springfield, Ill.: Charles C. Thomas, 1955.

Jensen, L. B. *Microbiology of Meats*. Third Edition. Champaign, Ill.: Garrard Press, 1954.

Schaub, I. G., *et al*. *Diagnostic Bacteriology*. Fifth Edition. St. Louis: C. V. Mosby Company, 1958.

Spink, W. W. *The Nature of Brucellosis*. Minneapolis: University of Minnesota Press, 1956.

Standard Methods for the Microbiological Examination of Foods. New York: American Public Health Association, 1958.

Stefferud, A., ed. *Animal Disease: The Yearbook of Agriculture*. Washington, D. C.: Government Printing Office, 1956.

Todd, J. C., *et al*. *Clinical Diagnosis by Laboratory Methods*. Twelfth Edition. Philadelphia: W. B. Saunders Company, 1953.

U. S. Department of Agriculture. *Facts about the Pasteurization of Milk*. Leaflet No. 408. Washington, D. C.: Superintendent of Documents, 1956.

CHAPTER XVI

PLANT PATHOGENS

Bacteria* pathogenic for plants do not vary greatly in morphology, and resemble many of the common soil and water forms. The organisms are small, straight, Gram-negative, non-spore-forming rods, aerobes or facultative anaerobes, which grow well on ordinary laboratory media and are generally metabolically active. They are found in three genera: Pseudomonas, Xanthomonas, Erwinia. Formerly they were placed together in one tribe. Now, in a still unsettled classification, the Erwinia have been placed among the intestinal Enterobacteriaceae and the Pseudomonas and Xanthomonas among the Pseudomonadaceae.

Motility is exhibited by all species of Erwinia (peritrichous flagella) and many species of Pseudomonas and Xanthomonas. All Xanthomonas and many Erwinia form yellow pigmented, moist colonies. However, the Pseudomonas produce a greenish, water-soluble pigment. Some of the plant pathogens in the Pseudomondaceae are more like Rhizobium in being metabolically rather inert.

The phytopathogens show selective action in infecting only certain species or even specific parts of plants. They may also have their own particular portal of entry. They are virulent and have invasive powers which bring about pathological alterations in plant tissues and structures. Individual plants or species have powers of resistance or immunity to certain infections which enable them to keep out invading bacteria. Natural plant immunity comes from (a) mechanical resistance of the cuticle, (b) cell wall, (c) chemicals dissolved in sap, such as essential oils, glucosides, and, in addition, quinine, turpentine, pine tar, etc. No immunity comparable with that of animals has been demonstrated, such as acquired immunity, phagocytosis, or antibodies.

Portals of Entry. Because the waxy epidermis and bark are impermeable to bacteria, these must enter the plant through lenticels, stomata, nectaries of flowers, or, most often, through wounds.

* Phytopathogenic viruses will be considered in Chapter XXXII.

Wounds in plants may be caused by cultivating or pruning, insects, wind, hail, or animal depredations.

Types of Diseases. Bacterial plant diseases are generally classified in five groups: (1) rots, (2) wilts, (3) blights, (4) spots, and (5) galls or tumors.

1. *Rots.* By means of an extracellular enzyme which dissolves pectin, certain plant pathogens, generally of the genus Erwinia, can reduce plant tissue to a soft, very moist, pulpy mass. The action is probably hydrolytic, resulting in the liberation of soluble sugars used for bacterial food. The causative organism is usually accompanied or closely followed by secondary invaders, soil bacteria and fungi. These grow in the exposed cells and deaminize the proteins of the dead plant, and the ammonia formed destroys neighboring plant cells. The foul odor of plants thus attacked is due to these secondary invaders.

2. *Wilts.* Wilt is caused by interruption of the flow of sap due to multiplication and accumulation of large numbers of organisms in the vascular system of the plant. Depending on the point and extent of the interruption, wilting may occur rapidly or produce its effects slowly, resulting in a sickly plant. Many times death is due to a secondary invader.

3. *Blights.* The bacteria grow in the plant juices and may penetrate considerable distances between the cells. There is no digestion of tissue. The discoloration or death of whole leaves or branches, characteristic of the disease, is probably due to interference with the flow of cell sap.

4. *Spots.* These are local infections restricted to a small area around the portal of entry, which is usually the stomata on the leaves. Following a vigorous attack by the bacteria on the plant tissue, the cells become heavily infected and discolored. These discolored areas frequently dry up and fall out, leaving "shot holes" in the leaves.

5. *Galls or Tumors.* Overgrowths may result from the rapid division of plant tissue due to bacteria-produced irritation of plant cells. Some have proposed that these growths are similar to cancer of animals, but this theory has not received much backing. The tumors vary in size, sometimes being quite small, and sometimes growing larger than the original plant.

Control of Plant Diseases. The general precautions to be taken in controlling plant diseases caused by bacteria include the following:

1. Burn plants or portions of them that are diseased.

2. In handling diseased plants, take care that they do not come in contact with other susceptible plants.

TABLE OF IMPORTANT BACTERIAL PLANT DISEASES

Causative Organism	Name of Disease	Plants Affected	Remarks
	Rots		
Erwinia carotovora	Soft rot	Carrot, cabbage, celery, cucumber, eggplant, iris, onion, and other plants	Affects roots, rhizomes, fruits, and fleshy stems. Gains entrance through wounds.
Erwinia phytophthora	Stem rot or blackleg	Potato, cucumber, and other vegetables	
Erwinia solanissapra	Soft rot	Potato and other vegetables	
Erwinia flavida	Soft rot	Sugar cane	
Erwinia aroideae	Soft rot	Calla lily and many vegetables	
Xanthomonas campestris	Stump rot	Cabbage	
	Soft rot	Cauliflower and rutabagas	
	Wilts		
Erwinia tracheiphila	Wilt	Cucumber, melon, squash, pumpkin, etc.	Transmitted by insect bites.
Pseudomonas solanacearum	Brown rot	Solanaceous plants, especially potato, tomato, and tobacco	
	Blights		
Erwinia amylovora	Fire blight	Pear, apple, and other pomaceous fruit plants	Attacks blossom clusters and tips of growing twigs. Then leaves, petioles, twig bark, and unripe fruit are affected.
Xanthomonas phaseoli	Bean blight	Hyacinth bean, lupine, etc. (Not soy bean)	
Pseudomonas mori	Blight	Mulberry	
Pseudomonas medicaginis	Stem blight	Alfalfa	
Pseudomonas pisi	Stem blight	Peas	
Xanthomonas juglandis	Blight	Walnut	Affects all new, tender, growing parts of tree.
Erwinia lathyri	Streak disease	Sweet pea and clover	
	Spots		
Xanthomonas cucurbitae	Leaf spot	Squash and related plants	
Xanthomonas malvacearum	Angular leaf spot	Cotton	Causes leaf spot, and stem and boll lesions.
Xanthomonas ricinicola	Leaf spot	Castor bean	
Xanthomonas vesicatoria	Spot	Tomato	Affects fruit, leaf, and stem.
Xanthomonas begoniae	Leaf spot	Begonia	
Pseudomonas angulata	Angular leaf spot	Tobacco	Leaves become injured by crawling of tobacco worm over surface. This furnishes organism portal of entry. Legs of worm may carry organism.
Pseudomonas maculicola	Leaf spot	Cauliflower and cabbage	Infection apparently through stomata.
Pseudomonas mellea	Brown rusty spot	Wisconsin tobacco	
	Galls		
Pseudomonas savastanoi	Knot	Olive	Entry is through tree wounds.
Pseudomonas tonelliana	Gall	Oleander	

3. Disinfect soil around diseased plants or allow it to lie fallow for a year or more.

4. Disinfect instruments and gloves used in handling diseased plants.

5. Take measures to control insect pests.

6. Eradicate weeds, which may harbor transmitting insects or disease organisms.

7. Spray with disinfectant as prophylaxis.

8. Propagate selectively with sturdy, resistant plants.

9. Disinfect seeds, bulbs, or cuttings used for propagation (bichloride of mercury, formaldehyde vapors, etc.).

BIBLIOGRAPHY

Burnet, F. M., and W. M. Stanley. *Plant and Bacterial Viruses.* New York: Academic Press, 1959.

Elliott, Charlotte. *Manual of Bacterial Plant Pathogens.* Waltham, Mass: Chronica Botanica Company, 1951.

Fawcett, H. S. *Citrus Diseases and Their Control.* New York: McGraw-Hill Book Company, 1936.

Heald, F. de F. *Introduction to Plant Pathology.* Second Edition. New York: McGraw-Hill Book Company, 1943.

―――. *Manual of Plant Diseases.* New York: McGraw-Hill Book Company, 1933.

Leach, J. G. *Insect Transmission of Plant Diseases.* New York: McGraw-Hill Book Company, 1940.

Medical Research Council. *A System of Bacteriology in Relation to Medicine,* Vol. III. London: His Majesty's Stationery Office, 1930.

Melhus, I. E., and G. C. Kent. *Elements of Plant Pathology.* New York: Macmillan Company, 1939.

Owens, C. E. *Principles of Plant Pathology.* New York: John Wiley and Sons, 1928.

Smith, Erwin F. *An Introduction to Bacterial Diseases of Plants.* Philadelphia: W. B. Saunders Company, 1920.

Stevens, F. L., and J. G. Hall. *Diseases of Economic Plants.* New York: Macmillan Company, 1933.

Walker, J. C. *Diseases of Vegetable Crops.* New York: McGraw-Hill Book Company, 1952.

Wormald, H. *Diseases of Fruits and Hops.* London: Crosby, Lockwood and Sons, 1939.

Yearbook, U. S. Department of Agriculture. Washington, D. C.: Government Printing Office, 1947. "Science in Farming," pp. 443–454.

PART II

MEDICAL BACTERIOLOGY

DIFFERENTIAL CHART OF PATHOGENIC BACTERIA

PRIMARY DIFFERENTIATION BY GRAM STAINING, MOTILITY RODS OR COCCI	PLATES	COLONIES	AGAR	GELATIN	LITMUS MILK	NITRATES REDUCED	BOUILLON	POTATO	DULCITOL	MALTOSE	DEXTROSE	LACTOSE	SUCROSE	XYLOSE	INOSITOL	ARABINOSE	MANNITOL	INDOL	CAPSULES	VOGES-PROSKAUER	METHYL RED TEST	CHOLERA RED TEST	SPECIAL AGGLUTINATION	EXOTOXIN
GRAM+, NON MOTILE COCCI																								
Staphylococcus aureus	Agar	Large	+	+	A.C.	+	+	+	AG	AG +	AG +	AG	AG +	AG	AG	AG	AG	AG +	–	–	+	+		
Staphylococcus albus	"	"	+	+	"	±	+	+		+	+	+							–	–	–		
Staphylococcus citreus	"	"	+	+	"	"	+	+		+	+	+					+		–				
Streptococcus pyogenes	Blood agar	Pin-point	+	–	A.	–	+	±	+	+	+	+							+	–		–	+
Streptococcus salivarius	" "	" "	+	–	A.C.	–	+	±	+	+	+	+							–	–			+
Streptococcus agalactiae	" "	" "	+	–	"	–	+	–	+	+	+	+							–	–			+
Diplococcus pneumoniae	" "	" "	+	–	"	–	+	–		+	+	+							+	–			+
GRAM+, NON MOTILE RODS																								
Corynebacterium diphtheriae	Löffler's blood serum is diagnostic	Small pin-point colonies are observed	+	–	N.C.	–	+	–	±	+	–								–	–	–		
C. pseudodiphthericum	"	"	+	–	"	+	+	+											–	–	–		
Corynebacterium xerose	"	"	+	–	"	–	+	–	+	+		+							–	–	–		
Mycobacterium tuberculosis	Glycerol agar	Cream colored	–	–	"	–	+	–			+								–				+
Bacillus anthracis	Agar	Medusa-head	+	+	A.C.P.	–	+	+		+	+								–	+	+		±
GRAM+, MOTILE RODS																								
Bacillus subtilis	Agar	Large	+	+	Alk.P.	+	+	+		+	+		+						–	+	+	+	+
GRAM–, NON MOTILE COCCI																								
Neisseria catarrhalis	Agar	Small	+	–	N.C.	+	+	+											–			–	
Neisseria intracellularis	Blood agar, ascitic agar, starch agar	Small, glistening	–	–	"	–	–	–		±	+								–			–	+
Neisseria gonorrheae			–	–	"	–	–	–			+								–			–	+
GRAM–, NON MOTILE RODS																								
Shigella dysenteriae	Eosin methylene blue agar and plain agar	Small, translucent colorless or pinkish colonies	+	–	A. to Alk.	+	+	+	–	–	+	–	–						+			–	+ +
S. paradysenteriae Flexner	"	"	+	–	A. to Alk.	+	+	+	–	+	+	–					+					–	+
S. paradysenteriae Sonne	"	"	+	–	A.C.	+	+	+	–	+	+	+	±										+
Shigella alkalescens	"	"	+	–	A. to Alk.	+	+	+	+	+	+	–	±				+	+				–	+
Pasteurella tularensis	Cystine agar	Small, grayish-white	–	–	N.C.	–	–	–	–	±	+	–					+					–	+
Pasteurella pestis	Agar	Small, colorless	+	–	A.	–	+	±	+	+	+	–					+	+	–			+	+
Brucella melitensis	Agar	Small, grayish-white	+	–	Alk.	±	+	+											–			+	+
Hemophilus influenzae	Blood agar, chocolate agar, sodium oleate agar, and	Small, colorless, transparent with entire edges	–	–	N.C.	–	+	–		±	+								+	–		–	+
Hemophilus pertussis	Bordet-Gengou agar	"	–	–	"	–	–	–											–	–			+
Hemophilus ducreyi			–	–	"	–	–	–											–	–			+
Malleomyces mallei	Agar	Tiny	+	±	A.C.	–	+	+		±									–	–		–	+
Klebsiella pneumoniae	Agar	Large, grayish-white	+	–	A.	+	+	+	+	+	+	+							±	+	–	+	+
GRAM–, MOTILE RODS																								
Escherichia coli	Endo agar and eosin methylene blue agar	On endo, red colonies. On EMB a dark with greenish sheen	+	–	A.C.G.R.	+	+	+	+	+	+	+	+	+	–	–	+	+	–	–	+	+	+
Escherichia communior	"	"	+	–	A.C.G.R.	+	+	+	+	+	+	+	+	+	–	+	+	+	–	–	+	+	+
Aerobacter aerogenes*	Agar and eosin methylene blue agar	Large, dark center, no sheen	+	–	A.C.	+	+	+	+	+	+	+	+	+	+	+	+	+	+	+	–	–	+
Aerobacter cloacae	"	"	+	+	A.C.G.R.	+	+	–	+	+	+	+	+	+	+	+	+	+	+	+	–	–	+
Eberthella typhosa	Endo agar, brilliant green agar, Wilson-Blair agar, eosin methylene blue agar, Selenite F. medium	On endo colonies are colorless. On other media growth is good.	+	–	A. to Alk.	+	+	+	–	+	+	–	–	–	–	+	+	+	–	–	+	+	+
Salmonella paratyphi	"	"	+	–	A. to Alk.	+	+	+	+	+	+	–	–	–	–	+	+	+	–	–	+	+	+
Salmonella schottmuelleri	"	"	+	–	A. to Alk.	+	+	+	+	+	+	–	–	–	–	+	+	+	–	–	+	+	+
Salmonella choleraesuis	"	"	+	–	A. to Alk.	+	+	+	+	+	+	–	–	+		+	+		–	–	+	+	+
Salmonella typhimurium	"	"	+	–	A. to Alk.	+	+	+	+	+	+	–	–	+	–	+	+	+	–	–	+	+	+
Salmonella enteritidis	"	"	+	–	A. to Alk.	+	+	+	+	+	+	–	–	+	–	+	+	+	–	–	+	+	+
Alcaligenes faecalis	Agar	Transparent	+	–	Alk.	+	+	+											–				
Proteus vulgaris	"	Gray, spreading	+	+	A. to Alk.	+	+	+		+	+	+		+	+			+	–	–		+	+
Proteus mirabilis	"	" "	+	+	A. to Alk.P.	+	+	+		–	–	+		+	+				–	–			+
Vibrio comma	"	Round, whitish-brown	+	+	A.P.R.	+	+	+		+	–	+	–		+				+	–	+	+	+
Vibrio proteus	"	Round grayish	+	+	A.C.P.	+	+	+		+		+							–	–			+
Pseudomonas aeruginosa	"	Large, spreading	+	+	Alk. C.P.R.	+	+	+		+									–	–			

KEY TO CHART

A. Acid
C. Coagulation
G. Gas
R. Reduction
P. Peptonization

Alk. Alkaline
± may or may not occur
* may be non motile
N.C. No Change

Or any combination of these.
Example:
A.C.P.- Acid, Coagulation, Peptonization.

See footnote on opposite page. (Bergey now places the Staphylococci in the genus Micrococcus and lists Neisseria intracellularis as Neisseria meningitidis.)

CHAPTER XVII

THE MICROCOCCI

The Micrococci considered in this section were placed in the genus Staphylococcus in Bergey's Fifth Edition. These organisms are most commonly found on the skin and mucous membranes. They are the organisms responsible for boils, abscesses, carbuncles, and similar suppurative processes. The genus belongs to the family Micrococcaceae and is made up of spherical organisms arranged in plates or irregular masses. Some of the species produce a yellow or, less commonly, an orange or red pigment. Outstanding members of this group are described below.

MICROCOCCUS PYOGENES VAR. AUREUS
(STAPHYLOCOCCUS AUREUS)

This is the most pathogenic of the genus. It produces a golden yellow to orange insoluble pigment.

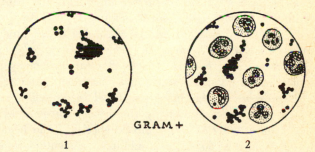

GRAM +

1   2

FIG. 41. (1) Pure culture of *Micrococcus pyogenes* var. *aureus*. (2) Smear of pus containing Micrococci and leucocytes.

Micrococcus Pyogenes Var. Aureus (Staphylococcus Aureus).
1. Group—pyogenic cocci. 2. Family—Micrococcaceae. 3. Habi-

Differential Chart: Using this chart to identify an unknown organism, practice the following: (1) Make a hanging drop and determine motility. (2) Make a smear and Gram stain. Note shape of organism and whether it is Gram-positive or -negative. (3) By elimination, the organism will be limited to a group corresponding to the known factors. *Example*, Gram +, motile rods. (4) Inoculate culture media, record reactions, and compare with chart. (5) Name the unknown organism from the data which correspond to characteristics of a particular species.

tat—skin, mucous membranes, dust, air, sputum, and pus. 4. Discoverer—Rosenbach, in 1884.

Morphological Characteristics. 1. Form—spheres. 2. Cell groupings—occurs in irregular clusters, singly, in pairs, or in short chains. 3. Size—0.8 to 1μ. 4. Staining properties—readily stained by aniline dyes. 5. Gram-positive. 6. Nonmotile. 7. Nonsporing. 8. Nonencapsulated.

Cultural Characteristics. 1. Agar cultures—smooth, circular colonies which are at first white and then turn golden. 2. Agar slant —smooth, flat, moist, plentiful, yellowish to orange growth. 3. Gelatin stab—growth, with yellowish pellicle, yellow to orange sediment, and saccate liquefaction. 4. Litmus milk—the medium is acidified and coagulated. 5. Potato—the growth is abundant and orange in color. 6. Broth—turbidity, with yellowish ring and sediment.

Physiological Characteristics. 1. Optimum temperature, 37° C. 2. Aerobic, and facultative anaerobe. 3. Nitrates are reduced to nitrites. 4. Acid is produced from dextrose, lactose, sucrose, maltose, mannitol, and glycerol, but raffinose, salicin, and inulin are not fermented. 5. Hydrogen sulfide formation is slight. 6. Pigment formation—yellow pigment is formed at 22° C. on albuminous media in which there is free access to oxygen. 7. Resistance—quite resistant to drying (for months) and antiseptics. Thermal death point (moist) varies with the strain from 60° C. for 10 minutes to 80° C. for 30 minutes.

Toxic Products. Various effects of soluble toxins produced by *Micrococcus pyogenes* var. *aureus* and some other pathogenic Micrococci have been observed. It is not yet clear how many separate toxins there are or whether a single toxin has more than one action.

1. *Hemolysin*—causes hemolysis of red blood cells. It is produced in a lesser concentration by the other pathogenic Micrococci and less frequently by the saprophytic species. The hemolysin is inactivated by heat at 60° C. for 2 minutes. Subcutaneous injection of it into rabbits results in production of an antistaphylohemolysin.

2. *Leucocidin*—destroys leucocytes. It is produced by *Micrococcus pyogenes* var. *aureus in vivo* (in animal body) and *in vitro* (in test tube), and only rarely by *Micrococcus progenes* var. *albus*.

3. *Lethal toxin* (tissue toxin)—causes death of animals when injected intravenously.

4. *Dermonecrotic toxin*—produces necrosis when injected intradermally in animals.

5. *Enterotoxin*—causes acute gastroenteritis (food poisoning) when ingested.

6. *Fibrinolysin*—digests or dissolves human blood fibrin.

7. *Coagulase*—causes coagulation of citrated blood.

Pathogenicity. *Micrococcus pyogenes* var. *aureus* enters the broken skin, and probably the unbroken skin, to produce infection, causing pus which consists of the accumulation of polymorphonuclear leucocytes in the infected area. Wounds may be infected by Micrococci, which are also found localized in boils, abscesses, carbuncles, furuncles, recurrent boils, ulcers, etc. Puerperal sepsis, periostitis, and osteomyelitis are most frequently caused by Micrococci. Septicemia of staphylococcic origin is rare; however, local peritonitis, pyemia, meningitis, and cystitis due to Micrococci have been reported as secondary infections. Micrococci are also frequently secondary invaders in tuberculosis, diphtheria, etc. Occasionally fatal Micrococci pneumonias occur. In addition, food poisoning may be caused by *Micrococcus pyogenes* var. *aureus*, often through foods containing starch thickening, such as cream puffs (see page 148).

Immunity, Therapy, and Prophylaxis. Most people have a high degree of natural immunity to the Micrococci. Stock polyvalent vaccines, autogenous vaccines, and other heat-killed preparations have been employed with some success to produce active immunity or to assist therapy. The sulfonamides, especially sulfathiazole, and penicillin appear to be very effective therapy. Some investigators have claimed good therapeutic results from the use of bacteriophage filtrates in local staphylococcic lesions. Micrococcic antiserum has also been used successfully by some workers as a local dressing, although, for the most part, reports have been negative. Micrococcic infections in surgery may be prevented by asepsis, with sterilization of instruments, skin of operator, and patient. General prophylaxis entails oral and skin cleanliness.

MICROCOCCUS PYOGENES VAR. ALBUS

Micrococcus pyogenes var. *albus*, isolated by Rosenbach in 1884, is similar to *Micrococcus pyogenes* var. *aureus* in morphological, cultural, and physiological characteristics. It produces a white growth on agar. It is also encountered in abscess and pyogenic infections but it is less pathogenic than *Micrococcus pyogenes* var. *aureus*.

MICROCOCCUS CITREUS

This organism is distinguished by its characteristic lemon-yellow pigment and is similar to the other Micrococci described, in its morphological, cultural, and physiological characteristics. It is of relatively slight pathogenicity.

MICROCOCCUS EPIDERMIDIS
(STAPHYLOCOCCUS EPIDERMIDIS)

Frequently the cause of stitch abscesses, this organism was first described by Welch in 1891. It may be a less virulent variety of *Staphylococcus albus*. It produces a white or colorless growth on culture media.

MICROCOCCUS AEROGENES
(STAPHYLOCOCCUS AEROGENES)

One of the few strict anaerobes in the genus, *Staphylococcus aerogenes* is found in natural cavities of the body and may cause puerperal fever.

GAFFKYA TETRAGENA

The genus Gaffkya of the family Micrococcaceae may be differentiated from the Staphylococci by its plane of division, for organisms of this group occur in fours or tetrads. The type species is *Gaffkya tetragena*.

Gaffkya Tetragena (**Micrococcus Tetragenus**). 1. Group—Gaffkya. 2. Family—Micrococcaceae. 3. Habitat—mucous membrane of respiratory tract, skin, air, and frequently sputum. 4. Discoverer—Gaffky, in 1883.

Morphological Characteristics. 1. Form—round or oval. 2. Cell groupings—shows typical tetrads surrounded by a pseudocapsule when present in body fluids. 3. Size—about 1μ in diameter. 4. Staining properties—easily stained by usual basic aniline dyes. 5. Gram-positive. 6. Nonmotile. 7. Nonsporing. 8. Capsule is present in the animal body.

Fig. 42. *Gaffkya tetragena* in the lymph stream. Note tetrad formation and capsules.

Cultural Characteristics. 1. Agar cultures—smooth, white, glistening colonies with an entire edge. 2. Agar slant—colonies at first transparent, later becoming grayish-white. 3. Gelatin stab—rather slow, white surface growth, with no liquefaction. 4. Litmus milk—acid formation in medium, with coagulation. 5. Potato—white, moist growth which shows a tendency to be confluent. 6. Broth—clear medium with gray sediment.

Physiological Characteristics. 1. Optimum temperature, 37° C. 2. Aerobic, and facultative anaerobe. 3. Acid is produced from dextrose, lactose, and glycerol.

Pathogenicity. In man, *Gaffkya tetragena* is usually found, without pathological significance, in sputum or saliva. It occurs in considerable numbers in some specimens of tuberculosis sputum. Rassers has suggested that in mixed infections in tuberculosis there may be a symbiotic relationship between this organism and the tubercle bacillus. *Gaffkya tetragena* has been found occasionally in the blood in septicemia, in the pus of abscesses, and in the spinal fluid in meningitis, indicating that some strains may be pathogenic for man. It is probably of low-grade virulence, and unable, as a rule, to invade the human tissues, except when resistance is lowered by some depressing influence, especially of the kind caused by the invasion of some other microorganism. *Gaffkya tetragena* is pathogenic for white mice. Only a localized reaction is produced in guinea pigs and rabbits.

Related Organisms. The other members of the Gaffkya group are without pathogenic significance to man. They include: 1. *Gaffkya tardissima,* 2. *Gaffkya verneti,* and 3. *Gaffkya anaerobia.*

BIBLIOGRAPHY

Bergey, David H. *Bergey's Manual of Determinative Bacteriology.* Seventh Edition. Baltimore: Williams & Wilkins Company, 1957.

Burnett, G. W., and H. W. Scherp. *Oral Microbiology and Infectious Disease.* Baltimore: Williams & Wilkins Company, 1957.

Jawetz, E., J. L. Melnick, and E. A. Adelberg. *Review of Medical Microbiology.* Los Altos: Lange Medical Publications, 1962. Pp. 133–136.

Marshall, M. S. *Applied Medical Bacteriology.* Philadelphia: Lea and Febiger, 1947. Pp. 1–360.

McCarty, M., ed. *Streptococcal Infections.* New York: Columbia University Press, 1954.

Rogers, D. E., ed. *Staphylcoccal Infections.* Annals of the New York Academy of Sciences, Vol. 65. New York: New York Academy of Sciences, 1956. P. 57.

Schaub, I. G., *et al. Diagnostic Bacteriology.* Fifth Edition. St. Louis: C. V. Mosby Company, 1958.

Stimson, P. M., and H. L. Hodes. *A Manual of the Common Contagious Diseases.* Fifth Edition. Philadelphia: Lea and Febiger, 1956.

Top, F. H. *Communicable Diseases.* Third Edition. St. Louis: C. V. Mosby Company, 1955.

Wilson, G. S., and A. A. Miles. *Topley and Wilson's Principles of Bacteriology and Immunity.* 2 vols. Fourth Edition. Baltimore: Williams & Wilkins Company, 1955.

Chapter XVIII

THE STREPTOCOCCI

In the genus Streptococcus (family Lactobacteriaceae, tribe Streptococceae) are included microorganisms which divide in one plane to form chains of cocci. The Streptococci are found associated with a great variety of pathological conditions, among which are erysipelas, focal infections, vegetative endocarditis, puerperal fever, septic sore throat, rheumatic fever, scarlet fever, tonsillitis, and arthritis. Members of the genus are Gram-positive, nonmotile, aerobic and facultatively anaerobic, and insoluble in bile (a property which is used to differentiate Streptococci from the pneumococcus).

CLASSIFICATION OF STREPTOCOCCI

A number of classifications of Streptococci have been propounded, but only two which have received marked attention will be presented here. They are: Smith and Brown's, and Lancefield's.

Smith and Brown's Classification. The Streptococci are classified according to their effect on blood agar plates as follows:

1. *Alpha.* In this group, known as the *Streptococcus viridans* group, are included strains which produce greenish coloration of the medium and partial hemolysis in the immediate vicinity of the colony. Examples: *Streptococcus viridans (salivarius)*, *Streptococcus bovis*, *Streptococcus equinus*.

2. *Beta.* These produce completely hemolyzed, clear, colorless zones around the colonies. They are known as the *Streptococcus hemolyticus* group. Examples: *Streptococcus pyogenes*, *Streptococcus equi*, *Streptococcus agalactiae*, *Streptococcus scarlatinae*.

3. *Gamma.* These produce no hemolysis and are known as the *Streptococcus anhemolyticus* group. Included are the lactic acid Streptococci, e.g., *Streptococcus lactis*, *Streptococcus cremoris;* and the enterococci, found in the intestinal tract of warm-blooded animals, e.g., *Streptococcus faecalis*, *Streptococcus durans*.

Lancefield's Classification. By means of specific precipitin reactions, Lancefield divided the hemolytic Streptococci into Groups A, B, C, D, and E. Since her original classification, Groups F, G, H, and the provisional K have been added.

1. *Group A.* In this, the *Streptococcus pyogenes* group, are included the Streptococci pathogenic for man. They are of etiological importance in scarlet

fever, erysipelas, tonsillitis, puerperal infection, osteomyelitis, periostitis, acute abscesses, etc.

2. *Group B.* This includes the causative agent of mastitis in cows, *Streptococcus agalactiae* (formerly *Streptococcus mastiditis*). The strains of this group are characterized by the fact that they are generally nonpathogenic for man, do not produce a toxin, do not liquefy human fibrin, are of limited and variable hemolytic power, produce yellow to brick-red chromogenesis, ferment trehalose, do not ferment sorbitol, hydrolyze sodium hippurate, and produce a pH of about 4.6 in 1% dextrose broth.

3. *Group C.* This includes animal and only occasionally human pathogens. There appear to be three biochemical groupings: *Streptococcus equi*, the cause of strangles in horses; "animal pyogenes"; and "human C," which is generally of low virulence.

4. *Group D.* These are isolated from cheese and animal and human feces. Most of the members do not hemolyze blood agar plates and are usually nonpathogenic. They include *Streptococcus faecalis*, *Streptococcus liquefaciens*, and *Streptococcus cremoris*.

5. *Group E.* Nonpathogenic Streptococci isolated from milk.

6. *Group F.* Isolated from normal throats, vaginas, and feces, these organisms are distinguished by the pin-point colonies they produce, with areas of complete hemolysis around them. They occur as harmless human parasites, but may at times be the primary cause of disease.

7. *Group G.* These are widely distributed organisms, generally not highly pathogenic.

8. *Group H.* Isolated from human feces and discharges of nose and throat, members of this group rarely cause serious human infection.

9. *Group K.* This is a provisional group whose members have not yet been proved pathogenic and may or may not be truly hemolytic.

STREPTOCOCCUS PYOGENES

Various strains of this organism have been considered specifically related to scarlet fever (*Streptococcus scarlatinae*), erysipelas (*Streptococcus erysipelatis*), and septic sore throat (*Streptococcus epidemicus*). Some investigators believe that each of these organisms is an individual

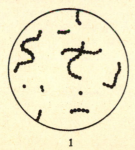 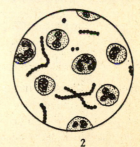

1   2

FIG. 43. (1) Pure culture of Streptococci. (2) Streptococcic pyemia, showing organisms and leucocytes.

species of Streptococcus and distinct from *Streptococcus pyogenes*. Most workers, however, consider them strains of *Streptococcus pyogenes*.

Streptococcus Pyogenes (Streptococcus Hemolyticus). 1. Group—pyogenic cocci. 2. Family—Lactobacteriaceae. 3. Habitat—widely distributed in man, animals, skin, and air. 4. Discoverers—Fehleisen, in 1882, and Rosenbach, in 1884.

Morphological Characteristics. 1. Form—small cocci. 2. Cell groupings—arranged in chains or pairs. 3. Size—0.5 to 1μ in diameter. 4. Staining properties—stains readily with basic aniline dyes. 5. Gram-positive. 6. Nonmotile. 7. Nonsporing. 8. Capsule formation variable; well developed in some strains.

Cultural Characteristics. 1. Agar cultures—small, discrete, translucent, convex colonies with entire edge. 2. Agar slant—grayish-white, opaque, more or less confluent, droplet-like, pin-point colonies. 3. Gelatin stab—slight growth resembling that on agar, with no liquefaction. 4. Blood agar—pin-point colonies surrounded by area of hemolysis (*beta* type) resulting from lysis of red blood cells and liberation of hemoglobin. 5. Litmus milk—acid is formed causing curdling. 6. Potato—delicate growth, or no visible change. 7. Broth—slight turbidity with a delicate sediment. 8. Löffler's blood serum—luxuriant growth of small isolated colonies.

Physiological Characteristics. 1. Optimum temperature, 35° to 38° C., with limits of 20° to 45° C. 2. Aerobic, and facultative anaerobe. 3. Acid is produced from dextrose, maltose, lactose, sucrose, salicin, and trehalose, but not from inulin, raffinose, arabinose, glycerol, mannitol, sorbitol, and dulcitol. 4. Insoluble in bile. 5. Final pH in dextrose broth is 4.8 to 6, with no growth at pH 9.6. 6. Nitrates are not reduced. 7. Indol is not formed.

Toxic Products. A number of toxic substances are produced as a result of the metabolism of various strains of *Streptococcus pyogenes*.

1. *Hemolysin.* This substance, which produces hemolysis of red blood cells, is of two types. One is oxygen-labile, antigenic, and resistant to heat and acid. The other is oxygen-stable, nonantigenic, and sensitive to heat and acid.

2. *Erythrogenic Toxin.* This is a rash-producing, antigenic exotoxin, which is relatively thermostable and most commonly associated with scarlet fever.

3. *Fibrinolysin.* Some members of Group A are capable of liquefying human fibrin. This property is characteristic of the *Streptococcus pyogenes* group and is helpful in differentiating human pathogens from nonpathogens.

4. *Leucocidin.* White blood cells are destroyed by filtrates prepared from Group A Streptococci.

5. *Duran-Reynals Spreading Factor.* Filtrates of pathogenic Streptococci affect the permeability of tissues, thus increasing the invasiveness (spreading power) of the organisms.

6. *Endotoxin.* It is not known whether Streptococci produce endotoxins.

Serological Reactions. Agglutination tests have been employed to differentiate the Streptococcus group serologically, but have given conflicting results. Three antigenic components have been described:

1. *C substance*—a polysaccharide haptene which is group-specific and is the common antigenic component of all Group A members.

2. *M substance*—an acid-soluble, type-specific protein which enables subdivision into immunological types. It is present in matte (rough) variants, with which virulence in the group is usually associated.

3. *T substance*—also type-specific, differing from the M substance in that it is resistant to proteolytic digestion. It seems unrelated to virulence, occurring in both matte and glossy (smooth) variants.

Pathogenicity. Man is very susceptible to invasion by *Streptococcus pyogenes*, although individual immunity varies and strains of different virulence exist. A minute skin abrasion, infected with *Streptococcus pyogenes*, may be followed by a rapidly spreading cellulitis and lymphangitis, and frequently by septicemia or blood poisoning. The following may be of streptococcic origin:

1. *Infections of the skin*—cellulitis, dermatitis, and erysipelas.
2. *Diseases of the respiratory tract*—septic sore throat, pharyngitis, tonsillitis, scarlet fever, bronchitis, empyema, and lobar pneumonia.
3. *Diseases of the ear and sinuses*—sinusitis, mastoiditis, and otitis media.
4. *Diseases of the bones and joints*—arthritis, rheumatic fever, osteomyelitis, and periostitis.
5. *Diseases of the genital tract*—cervicitis.
6. *Miscellaneous*—complication in appendicitis, peritonitis, and other infections. Secondary invaders in tuberculosis, influenza, measles, smallpox, etc.

STREPTOCOCCUS SALIVARIUS (STREPTOCOCCUS VIRIDANS)

This organism was isolated by Schottmüller in 1903 and is characterized by its growth on blood agar. The colonies are surrounded by a zone of greening and partial hemolysis. Members of the green-producing group of Streptococci are less virulent than the pyogenes group. They are constant inhabitants of the intestinal tract of man and lower animals. Some have been found to cause apical abscesses of teeth and infection of the middle ear and the sinuses, and to play an important part in focal infections. *Streptococcus salivarius* is frequently encountered in the bloodstream in cases of subacute vegetative endocarditis, which is usually fatal.

SCARLET FEVER

Scarlet fever is an acute infectious disease which is characterized by an angina of varying severity and by an exanthema. The disease occurs after an incubation period of 2 to 7 days and is usually communicable for as long a period as the Streptococci are present. It is transmissible by droplet infection from the mouth and nose, and by discharges from suppurating ears and lymph nodes. The disease is caused by a strain of *Streptococcus pyogenes*, which is sometimes termed *Streptococcus scarlatinae*. The organism produces a powerful exotoxin which causes the toxemia, rash, and fever.

Diagnosis. The disease is diagnosed by its clinical symptoms together with the isolation of hemolytic Streptococci from the throat. The *Schultz-Charlton* or *blanching reaction* is believed by some workers to be of diagnostic significance. When 1 ml. of convalescent serum or antitoxin prepared in animals is inoculated intracutaneously into the reddened skin of a scarlet fever patient, a blanching occurs at the site of injection.

The Dick Test. In 1923 the Dicks began their work on scarlet fever and succeeded in producing the disease in human beings with a Streptococcus culture isolated from a case of scarlet fever. The toxin of the organism is standardized in human beings, since no laboratory animal is available for such testing. The *skin test dose* (S.T.D.) is the unit employed. It is that amount of toxin which will produce a reaction 1.5 cm. in diameter in an average susceptible child. The unit of antitoxin is that amount which will neutralize 50 skin test doses of toxin.

The test for susceptibility to scarlet fever consists of the intradermal injection of 0.1 ml. (representing one S.T.D.) of toxin. As a control, the same amount of the same filtrate heated to 100° C. for 2 hours is similarly injected. The reactions are read at the end of 24 hours and may be classified as follows:

1. *Strongly positive*—areas of redness, swelling, and tenderness, measuring more than 3 cm. in diameter. 2. *Positive*—bright red areas from 1.5 to 3 cm. in diameter, with some swelling and tenderness. 3. *Slightly positive*—faint red areas, less than 2 cm. in diameter, without swelling or tenderness. 4. *Pseudoreactions*—these are reactions to the protein contained in the filtrate, as opposed to those in

response to the toxin, and are generally mild and less than 1 cm. in diameter. Such reactions are not an indication of susceptibility to scarlet fever. They are distinguished from a true positive reaction to the toxin by use of the control.

Active Immunization. The injection of scarlatinal Streptococcus toxin into human beings causes the Dick test to become negative. Thus, to actively immunize against scarlet fever, the individual is given a series of five injections at weekly intervals—(1) 500 S.T.D., (2) 2500 S.T.D., (3) 20,000 S.T.D., (4) 40,000 S.T.D., and (5) 80,000–100,000 S.T.D. Results have indicated that this is an effective method for prevention of scarlet fever.

Immunity, Serum Prophylaxis, and Therapy. Following an attack of scarlet fever there is usually developed an immunity against subsequent exposures, although second attacks have been known to occur. The Dicks were able to produce an antitoxin in horses by the injection of toxin. Dochez produced an antiserum which was both antitoxic and antibacterial. Horses were immunized by inoculating strains of hemolytic Streptococci into masses of agar that had been injected into the subcutaneous tissues. The microbes growing in the agar were able to stimulate the tissues of the horse to respond with antitoxin and also bactericidal substances. In emergencies following exposure to the disease a prophylactic injection of antiserum will protect for about 2 to 3 weeks. This immunity, being passive in nature, is of short duration.

Antitoxin has been employed as a therapeutic measure in scarlet fever with good results.

Complications. The incidence of complications in scarlet fever is very high, otitis media, arthritis, meningitis, and other infections being sequelae of the disease. The injection of antibacterial sera has been attended by a reduction in the percentage of complications, and treatment with sulfa drugs and antibiotics is also effective.

CHEMOTHERAPY OF STREPTOCOCCIC INFECTIONS

The introduction of prontosil by Domagk in 1935 made available a very effective drug against streptococcic infections. The sulfonamides, particularly sulfanilamide and sulfadiazine, have been found highly effective in controlling Lancefield Group A hemolytic Streptococci infections, such as puerperal sepsis, erysipelas, tonsillitis, septic sore throat, scarlet fever, etc. They are about two-thirds as effective in Group C infections, of slight value in Group B infections, without effect on Group D and enterococci infections, and of indifferent value in some of the viridans infections. (See also p. 90.)

The sulfas have also been used prophylactically. For example, an epidemic of scarlet fever was brought completely under control at a United States Naval Station by prophylactic use of 1 gm. of sulfadiazine a day.

BIBLIOGRAPHY

Bergey, David H. *Bergey's Manual of Determinative Bacteriology.* Seventh Edition. Baltimore: Williams & Wilkins Company, 1957.

Burnet, F. M. *The Natural History of Infectious Disease.* Second Edition. Cambridge, England: Cambridge University Press, 1953.

Burrows, W. *Textbook of Microbiology.* Eighteenth Edition. Philadelphia: W. B. Saunders Company, 1963. Pp. 479–506.

Dick, G. F. and G. H. *Scarlet Fever.* Chicago: Year Book Publishers, 1938.

Dubos, R. J. *Bacterial and Mycotic Infections of Man.* Third Edition. Philadelphia: J. B. Lippincott Company, 1958.

Frost, W. D., and M. Englebrecht. *The Streptococci.* Madison, Wis.: Willdof Book Company, 1940.

Holt, L. E., and R. McIntosh. *Diseases of Infancy and Childhood.* New York: Appleton-Century Company, 1940. Pp. 1215–1231.

Jawetz, E., J. L. Melnick, and E. A. Adelberg. *Review of Medical Microbiology.* Los Altos: Lange Medical Publications, 1962. Pp. 136–142.

Kolmer, J. A. *Clinical Diagnosis by Laboratory Examinations.* Second Edition. New York: Appleton-Century-Crofts, 1949.

Mackie, T. T., *et al. Manual of Tropical Medicine.* Second Edition. Philadelphia: W. B. Saunders Company, 1954.

Mellon, R. R., P. Gross, and F. B. Cooper. *Sulfanilamide Therapy of Bacterial Infections.* Springfield, Ill.: Charles C. Thomas, 1938.

Netter, E., and D. R. Edgeworth. *Medical Microbiology for Nurses.* Philadelphia: F. A. Davis Company, 1957.

Osler, W., and H. A. Christian. *Principles and Practice of Medicine.* Sixteenth Edition. New York: Appleton-Century-Crofts, 1947.

Park, William Hallock, and Anna Wessels Williams. *Pathogenic Microörganisms.* Eleventh Edition. Philadelphia: Lea and Febiger, 1939. Pp. 378–397.

Rosenau, M. J. *Preventive Medicine and Hygiene.* New York: Appleton-Century Company, 1935. Pp. 85–87, 376, 612, 712.

Sherman, James M. "The Streptococci." *Bact. Rev.* 1:1, 1937.

Smith, D. T., *et al. Zinsser's Textbook of Bacteriology.* Eleventh Edition. New York: Appleton-Century-Crofts, 1957.

Smith, Frederick C. *Sulfonamide Therapy in Medical Practice.* Philadelphia: F. A. Davis Company, 1944.

Strean, L. P. *Oral Bacterial Infections.* Brooklyn, N. Y.: Dental Publishing Company, 1949. Pp. 46–52.

Top, F. H. *Communicable Disease.* Third Edition. St. Louis: C. V. Mosby Company, 1955.

CHAPTER XIX

DIPLOCOCCUS PNEUMONIAE AND KLEBSIELLA GROUP

DIPLOCOCCUS PNEUMONIAE

Diplococcus pneumoniae, commonly referred to as pneumococcus, is the causative agent of lobar pneumonia and is most frequently encountered in this infection. The organism is a member of a group of microbes characterized by poor growth on artificial media, typical cell arrangement, and fermentation properties. The Diplococcus genus (family Lactobacteriaceae, tribe Streptococceae) is Gram-positive. Cells are arranged in pairs and are encapsulated. Most strains form acid from dextrose, lactose, sucrose, and inulin, and many are bile-soluble. The type species is *Diplococcus pneumoniae*.

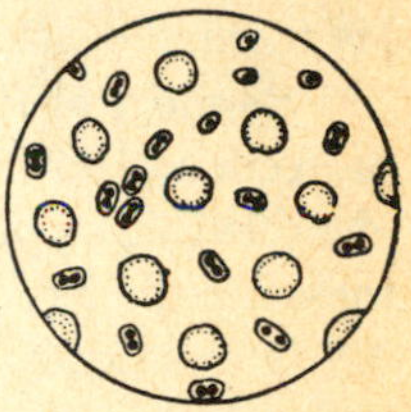

FIG. 44. *Diplococcus pneumoniae* in bloodstream, showing organisms and red blood cells.

Diplococcus Pneumoniae (Micrococcus Lanceolatus, Pneumococcus). 1. Family—Lactobacteriaceae. 2. Habitat—respiratory tract; present in sputum, blood, and exudates in pneumococcus pneumonia. 3. Discoverers—Klebs, in 1875, Pasteur, Chamberland, and Roux, in 1881, and Fränkel and Weichselbaum, in 1886.

Morphological Characteristics. 1. Form—oval-shaped. 2. Cell groupings—arranged in pairs, or occasionally in short chains. The distal ends of each pair are pointed or lancet-shaped, and each pair is enclosed in a delicate capsule. 3. Size—0.5 to 1.25μ in diameter. 4. Staining properties—stains well with basic aniline dyes and special capsule stains. 5. Gram-positive and non-acid-fast. 6. Nonmotile. 7. Nonsporing. 8. The organism has a well-marked capsule when observed in films of animal tissues; this may be retained on certain artificial media, but is generally lost on prolonged passage outside the body.

Cultural Characteristics. 1. Agar cultures—small, grayish, transparent colonies with entire edge. 2. Agar slant—transparent, raised, opaque, smooth, white colonies which are rather difficult to grow.

3. Blood agar plates—the organism grows well on this medium, producing pin-point colonies, surrounded by a zone of greening and hemolysis. 4. Gelatin stab—slow growth along line of inoculation, with no liquefaction. 5. Litmus milk—acid is produced and the medium is coagulated. 6. Potato—slight or no growth. 7. Blood broth— good growth, with turbidity.

Physiological Characteristics. 1. Optimum temperature, 37° C. 2. Aerobic, and facultative anaerobe. 3. Inulin is fermented. (Used in differentiating from green Streptococci.) 4. Bile-soluble. (Used in differentiating from green Streptococci.) 5. Nitrates are not reduced. 6. Indol is not formed. 7. Acid, but no gas, is formed in lactose, sucrose, dextrose, and inulin. 8. Resistance—very sensitive to heat, a temperature of 55° C. being destructive in 20 minutes or less. Exposure to drying, sunlight, and disinfectants kills *Diplococcus pneumoniae*, although the organism may survive in sputum for some time. Its thermal death point lies between 52° and 54° C.

Serological Reactions. By means of agglutination reactions and protection tests made with immune sera, Cooper and other workers identified 32 types of pneumococci, Types I to XXXII. Most of these are distinct types, although XXVI is closely related to VI, and XV is practically identical with XXX. Some slight cross-reactions, probably due to structural similarity of the Specific Soluble Substance, is observed between II and V, III and VIII, VII and XVIII, XV and XXX. Before Cooper separated the pneumococci into 32 types, there had been Types I, II, and III, and Group IV, to which all pneumococci which could not be identified as any of the three common types were designated. Types of pneumococci beyond the 32 reported by Cooper have been encountered, and these are becoming of increasing importance.

The serological differentiation of pneumococci is related to chemical differences in the cell constituents. The virulent pneumococcus is composed of a cell body and a capsule. The serological reactions seem to be determined by two substances, one which is type-specific and polysaccharide in nature (Specific Soluble Substance, S.S.S.) and is present in the capsular material, and the other which is group-specific and protein in nature.

1. *Specific Soluble Substance.* When isolated in purified form, these polysaccharides react with immune sera produced by injecting animals with whole pneumococci. They produce skin reactions in immunized human beings and animals. When inoculated alone, they do not form antibodies; but when combined with proteins it is possible to produce immune sera which are specific for the type of

pneumococcus from which the polysaccharide was isolated. Therefore, these pneumococcus polysaccharides belong to the group of substances termed *haptenes*, or partial antigens, by Landsteiner. The Specific Soluble Substance determines the type specificity of pneumococci.

2. *Somatic Antigen.* This is a group antigen, protein in nature, isolated from the bodies of pneumococci, and immunologically identical in all types.

Pneumococcus Typing.

Various methods are available for the determination of the type of pneumococcus present in sputum of a pneumonia patient. Most widely used are:

1. *Krumwiede-Valentine Method.* (1) About 3 to 10 ml. of sputum are coagulated by placing in a hot water bath. (2) The coagulum is broken up and saline is added. (3) The tube is replaced in the hot water bath and then centrifuged. (4) The type sera, in amounts of 0.2 ml., are added to test tubes, and the supernatant fluid from the centrifuged sputum specimens is layered on. (5) The tubes are then placed in a water bath at 50° to 55° C. for several minutes. (6) Formation of a definite contact ring indicates the type of pneumococcus present in the sputum. A positive reaction usually appears within 30 minutes.

2. *Mouse Inoculation Method.* In certain specimens of sputum, where pneumococci are present in very small amounts, about 1 ml. of the sample is injected intraperitoneally into a white mouse. After 3 to 5 hours, some peritoneal exudate is withdrawn and washed into a sterile Petri dish with 1 to 2 ml. of saline. The washings are centrifuged and the supernatant fluid used for typing.

3. *Sabin Microscopic Agglutination Method.* Small drops of peritoneal exudate of a mouse are placed on a slide. The type sera are added in a dilution of 1:10, and the mixtures spread in a thin film, allowed to dry, and stained with fuchsin. The presence of stained clumps of pneumococci indicates a positive reaction and designates the infecting type.

4. *Neufeld-Quellung Phenomenon.* This method is more frequently used as the preliminary step in typing sputum than the others. Both serum pools and monospecific antisera are used. (1) A loopful of undiluted antiserum is mixed with a fleck of sputum on a cover-slip, and then a loopful of Löffler's alkaline methylene blue is added. (2) The mixtures are examined with the oil immersion lens. (3) The homologous mixture will show pneumococci with swollen capsules and very distinct outlines.

Toxic Products.

Diplococcus pneumoniae produces hemolysin, leucocidin, necrotizing substance, and probably a purpura-producing substance. Virulence of the organism, however, is not directly dependent on the formation of such toxic products, but on the production of Specific Soluble Substance and encapsulation.

Pathogenicity.

The pneumococcus is found in at least 85% of cases of lobar pneumonia. The organism is found in the lung, in the lymph channels, and in the bloodstream in many cases. The pneumococcus may also provoke other infections, either primary or secondary, such as pericarditis, arthritis, periarthritis, meningitis, otitis media, mastoiditis, endocarditis, rhinitis, tonsillitis, conjunctivitis, osteomyelitis, and peritonitis. In children pneumococcic pleurisy may occur frequently.

Among the experimental animals, mice and rabbits are most susceptible to *Diplococcus pneumoniae.* Subcutaneous or intracutaneous injection of cultures produces death in 1 to 7 days, the disease taking the form of a septicemia and not a true pneumonia. Some investigators have produced pneumonia in dogs and monkeys by intratracheal injection of virulent pneumococci.

Variation. The organism exhibits the usual smooth-rough variation. It is virulent in the S form and avirulent in the R form. The S⟶R change involves loss of capsule and loss of ability to form S.S.S. (and, therefore, loss of type-specificity).

Immunity. Immunity following an attack of pneumonia is generally of short duration. Antibodies for pneumococci are demonstrable in the sera of normal individuals as well as in those recovering from the disease.

Prophylaxis. Use of pneumococcus vaccines for the prevention of pneumonia has given inconclusive results.

Therapy. Sulfonamides are the treatment of choice. Sulfapyridine, sulfadiazine, and sulfathiazole are most often used. Penicillin is also highly effective therapy.

Related Organisms. *Diplococcus mucosus (Streptococcus mucosus capsulatus)* is believed identical with Type III pneumococcus. It produces a heavy mucoid growth. It is found in the respiratory tract and has been isolated from the nasopharynx in cases of epidemic meningitis and cases of mastoiditis.

KLEBSIELLA GROUP

Included in the genus Klebsiella (family Enterobacteriaceae, tribe Eschericheae) are organisms characterized by the possession of a heavy gelatinous capsule which lends to their growth a mucoid consistency. They are small, plump rods with rounded ends, Gram-negative, and nonmotile. Carbohydrates are fermented with the formation of acid and gas. Nitrates are reduced to nitrites. They are generally encountered in infections of the respiratory tract, but it is probable that they are most often secondary invaders in these conditions. They cause but a small per cent of all the pneumonia cases; these are almost always fatal.

KLEBSIELLA PNEUMONIAE

Klebsiella Pneumoniae (Bacillus Pneumoniae, Bacillus Mucosus Capsulatus, Friedländer's Pneumobacillus). 1. Family—Enterobacteriaceae. 2. Habitat—respiratory tract of man, associated with inflammations. 3. Discoverer—Friedländer, in 1882.

Morphological Characteristics. 1. Form—short, thick rod possessing a capsule. 2. Cell groupings—arranged singly and in pairs end to end. 3. Size—0.5 to 0.8μ by 1 to 5μ. 4. Staining properties—stains well with ordinary aniline dyes. 5. Gram-negative. 6. Non-motile. 7. Nonsporing.

Cultural Characteristics. 1. Agar cultures—round, mucoid, translucent, grayish-white colonies with an entire edge. 2. Agar slant—luxuriant growth of whitish, sometimes viscid colonies, with a glistening surface and entire edge. 3. Gelatin stab—moderate, filiform, grayish-white growth tending toward confluence, without liquefaction. 4. Litmus milk—acid is produced, without coagulation of the medium. 5. Potato—yellowish, slimy growth, becoming café-au-lait in appearance. 6. Broth—turbidity, with the development of a slimy sediment and a pellicle.

Physiological Characteristics. 1. Optimum temperature, 37° C. 2. Aerobic, and facultative anaerobe. 3. Nitrates reduced to nitrites. 4. Acid and gas are produced in glucose, sucrose, lactose, galactose, and levulose, but not in maltose, inulin, mannitol, arabinose, and dulcitol. 5. Resistance—cultures at room temperature remain viable for months, but the organism is easily killed by exposure to direct sunlight and chemical disinfectants. A temperature of 55° C. is lethal in 30 minutes.

Serological Reactions. Like the pneumococcus, Friedländer's bacillus possesses a somatic, protein, group-specific antigen and a capsular, polysaccharide, type-specific antigen. By agglutination, absorption, and precipitin tests, *Klebsiella pneumoniae* has been differentiated into three types termed A, B, and C, and a heterogeneous group labeled X. Immunologically, the B type is similar to Type II pneumococcus.

Variation. This organism, like the pneumococcus, is virulent in the S form, and avirulent in the R form, which possesses no capsule and cannot form the capsular, type-specific antigen.

Pathogenicity. *Klebsiella pneumoniae* causes a number of cases of pneumonia which are very severe. Cases of otitis media, empyema, pericarditis, meningitis, and septicemia have been produced by it, and it is commonly associated with both acute and chronic bronchitis. It has been occasionally the cause of abscess formation in the respiratory tract, of suppuration in the sinuses, and in the antrum of Highmore. It is highly pathogenic to mice. Rabbits are less susceptible. The different types vary in pathogenicity, A and B being highly virulent for mice, whereas C is without effect.

ORGANISMS RELATED TO KLEBSIELLA PNEUMONIAE

Klebsiella Ozaenae (Bacillus Ozaenae). This organism was isolated by Abel in 1893 and is similar to *Klebsiella pneumoniae* in morphological, cultural, and physiological characteristics. It is infectious for house and field mice. In man it is considered the probable cause of ozena, a rhinitis characterized by the formation of crusts in the nose and the development of a foul odor as of putrefaction.

Klebsiella Rhinoscleromatis. Von Frisch, in 1882, described this organism, which is now believed identical with *Klebsiella pneumoniae* Type C. The organism causes in man rhinoscleroma, a disease in which there is a granulomatous inflammation of the external nares or upon the mucosa of the nose, mouth, pharynx, or larynx.

BIBLIOGRAPHY

Burnett, G. W., and H. W. Scherp. *Oral Microbiology and Infectious Disease.* Baltimore: Williams & Wilkins Company, 1957.

Burrows, W. *Textbook of Microbiology.* Eighteenth Edition. Philadelphia: W. B. Saunders Company, 1963.

Fairbrother, R. W. *A Textbook of Bacteriology.* Fifth Edition. New York: Grune & Stratton, 1948.

Holt, L. E., and R. McIntosh. *Diseases of Infancy and Childhood.* New York: Appleton-Century Company, 1940. Pp. 501–518.

Lord, Frederick T., Elliot S. Robinson, and Roderick Heffron. *Chemotherapy and Serum Therapy of Pneumonia.* New York: The Commonwealth Fund, 1940.

Park, William Hallock, and Anna Wessels Williams. *Pathogenic Microörganisms.* Eleventh Edition. Philadelphia: Lea and Febiger, 1939. Pp. 398–420.

Schaub, I. G., *et al. Diagnostic Bacteriology.* Fifth Edition. St. Louis: C. V. Mosby Company, 1958.

Society of American Bacteriologists. *Manual of Methods for Pure Culture Study of Bacteria.* Geneva, N. Y.: Biotechnical Publications, 1948. Chap. VII.

Strean, L. P. *Oral Bacterial Injections.* Brooklyn, N. Y.: Dental Publishing Company, 1949. Pp. 53–57.

Top, F. H. *Communicable Diseases.* Third Edition. St. Louis: C. V. Mosby Company, 1955.

Chapter XX

GRAM–NEGATIVE COCCI
(GENUS NEISSERIA)

The Gram-negative cocci are included in the genus Neisseria, Family Neisseriaceae, and are strict parasites. Most important among them are: 1. *Neisseria gonorrheae,* the cause of gonorrhea, 2. *Neisseria meningitidis,* the cause of epidemic meningitis, and 3. *Neisseria catarrhalis,* which is present in the nasopharynx, saliva, and respiratory tract, and is frequently associated with other organisms in inflammations of the mucous membrane. These organisms are coffee-bean-shaped, Gram-negative diplococci which are intracellular or extracellular, but generally intracellular. The points of selective action are the urethral membranes for the gonococci and the meninges of the brain and spinal cord for the meningococci.

NEISSERIA GONORRHEAE

Neisseria Gonorrheae (Gonococcus). 1. Group—Gram-negative cocci. 2. Family—Neisseriaceae. 3. Habitat—strict parasite of man found in discharges in gonorrhea and in pus from the conjunctiva in gonorrheal conjunctivitis. 4. Discoverer—Neisser, in 1879.

Morphological Characteristics. 1. Form—oval or spherical cocci. 2. Cell groupings—arranged as diplococci with adjacent sides flattened and resembling a pair of coffee beans. They are usually found intracellularly in exudates. 3. Size—0.6 to 1μ in diameter. 4. Staining properties—stains well with aniline dyes, including safranin and carbol fuchsin. 5. Gram-negative and non-acid-fast. 6. Nonmotile. 7. Nonsporing. 8. Usually nonencapsulated.

Cultural Characteristics. *Neisseria gonorrheae* grows only on special media, such as blood agar, ascitic agar, starch agar, or dextrose agar. Incubation in an atmosphere of increased CO_2 tension improves growth and is almost mandatory on primary isolation. 1. Blood agar—small, fine colonies which grow without hemolysis of the medium. 2. Serum agar—round colonies which are grayish-white and have an entire edge and a smooth glistening surface. 3. Agar plate—no growth. 4. Gelatin stab—no growth. 5. Potato—no growth. 6. Blood broth—slight growth with turbidity of the medium. 7. Chocolate (heated blood) agar—good growth, similar to that on blood agar.

Physiological Characteristics. 1. Optimum temperature, 37° C., with limits of 25°–30° to 40° C. 2. Aerobic, and facultative anaerobe. 3. Acid, but no gas, is produced from dextrose, but not from maltose, levulose, sucrose, and mannitol. 4. Methylene blue reductase is absent. 5. Resistance—only slightly resistant. Cultures kept at room temperature or in the incubator are rapidly killed. Exposure to 42° C., sunlight, desiccation, or weak disinfectants are destructive to it.

Serological Reactions. Agglutination tests and agglutinin absorption methods have indicated that strains of *Neisseria gonorrheae* fall into two main groups, but there is no clear division into separate serological types. The two immunological types appear to be correlated with two colony types, one a large, irregular, flattened, translucent colony which gives rise to surface papillae on continued incubation, and the other somewhat smaller, round, raised, opaque, slightly convex, with an irregular surface. Patients infected with gonococci produce antibodies against these organisms.

Pathogenicity. Man is the only animal naturally infected by gonococci. Infection is usually transmitted directly from individual to individual. The gonococcus attacks chiefly the urethra, both in the male and female, producing an acute catarrh. The organism quickly penetrates the surface, passing between the epithelial cells which are loosened and desquamated, and invades the tissues as far as the superficial layers of the submucous connective tissue. The disease spreads, and in the male, if untreated, the prostate may be involved, and occasionally the bladder. Orchitis and inflammation of the cord and epididymis are also fairly frequent. In the female the urethra is the most common starting point, and later the cervix is involved. Occasionally, in either sex, the mucous membrane of the rectum and anus may be affected. Arthritis also may occur as a complication. Infectious vulvo-vaginitis is a disease of children, usually caused by the gonococcus but sometimes by other Neisseria, which may spread in epidemic proportions in hospital wards and orphan homes.

Ophthalmia Neonatorum. This is a conjunctivitis in the newborn, which results from infection during parturition. If neglected, this is the most frequent cause of blindness. It is preventable by Crede's method which requires the instillation of 2% silver nitrate in the eyes of all newborn. This procedure has resulted in a great decrease in blindness attributable to gonococcal infections.

Diagnosis. 1. *Direct smears* are made of the infected urethral or conjunctival discharges, stained with Gram's stain, and examined for Gram-negative, coffee-bean-shaped, intracellular diplococci. If only extracellular diplococci are found, another specimen should be examined.

2. *Cultures* may be made from the material on the special media noted above and incubated at 37° C. in an atmosphere containing 10% CO_2.

3. *Oxidase Reaction.* This depends on the formation of indophenoloxidase, which is common to all Neisseria. The suspected material is cultured on heated blood (chocolatc) agar. After growth has occurred, the plate is flooded with a 1% solution of dimethyl or tetramethyl para-phenylenediamine, which is immediately poured off. Colonies of bacteria forming the indophenoloxidase will turn pink, then maroon, and finally black, if the dimethyl compound has been used, or magenta, with the tetramethyl. The organisms are not immediately killed, and may be subcultured within a half-hour.

4. *Carbohydrate Fermentations.* Fermentation tubes may be inoculated from oxidase-positive colonies.

Immunity, Prophylaxis, and Therapy. An attack of gonorrhea produces little, if any, immunity. Methods of prophylaxis depend upon the susceptibility of the gonococcus to silver salts, such as argyrol and protargol, in solutions which do not harm the mucous membranes but do prevent infection after exposure.

Various sulfonamides have been used very successfully in gonorrhea therapy, best results being obtained by combined oral and local administration. There are, however, some strains which are sulfaresistant or which develop drug-fastness after continued sulfa treatment. These have been effectively controlled by penicillin.

NEISSERIA INTRACELLULARIS

Neisseria Meningitidis (Meningococcus, Neisseria Intracellularis). 1. Group—Gram-negative cocci. 2. Family—Neisseriaceae. 3. Habitat—nasopharynx of man, and spinal fluid in cases of sporadic and epidemic meningitis. 4. Discoverer—Weichselbaum, in 1887.

Morphological Characteristics. 1. Form—small, ovoid cocci. 2. Cell groupings—diplococci with adjacent sides flattened, or sometimes in tetrads. 3. Size—0.6 to 0.8μ in diameter. 4. Staining properties—stains well with basic aniline dyes. 5. Gram-negative and non-acid-fast. 6. Nonmotile. 7. Nonsporing. 8. Usually non-encapsulated, but freshly isolated strains have been reported to possess a capsule.

Cultural Characteristics. *Neisseria intracellularis* does not grow well on ordinary artificial media, but requires the presence of blood, blood serum, ascitic fluid, or hydrocele fluid to which dextrose has been added. Like the gonococcus, growth of the meningococcus is favored, especially in primary isolation, by incubation in an atmosphere containing 10% CO_2. 1. Blood agar plates—these are generally employed to isolate the organism. Small, glistening, transparent, slightly convex colonies are formed. 2. Gelatin stab—growth slight, with no liquefaction of the medium. 3. Litmus milk—very little growth, with no acid formation. 4. Dextrose agar—small colonies which resemble those of Streptococcus, but are less transparent. 5. Ascitic broth—fine granular growth. 6. Potato—no growth. 7. Serum broth—slight granular growth.

Physiological Characteristics. 1. Optimum temperature, 37° C. 2. Aerobe. 3. Acid, but no gas, is produced from dextrose and maltose. 4. Catalase is present. 5. Methylene blue reductase is present. 6. Resistance—susceptible to heat and disinfectants. Drying destroys it in less than 3 hours, and moist heat at 55° C. for about 5 minutes is sufficient to kill it. It may be preserved for several weeks when dried and kept frozen. Methylene blue inhibits growth.

Serological Reactions. On the basis of agglutination reactions with immune serum, four antigenic types, I, II, IIα, and III have been proposed. Types I and III are closely related and comprise Group I. Type IV, which together with Type II comprised Group II in the original classification, seems to have disappeared. Agglutinating sera are produced in animals by repeated injections of meningococcus. The serum of persons who have recovered from meningitis contains agglutinins in dilutions up to 1:400.

The antigenic substances present in the meningococcus include:

1. *C substance*—a polysaccharide found in all types of meningococcus and also in the gonococcus.
2. *P substance*—a protein substance common to all types of meningococcus and highly toxic to rabbits.
3. *Type-specific antigen.* In Types I and III, this is a polysaccharide identical in both. In Type II, it is apparently a polysaccharide-polypeptide complex.

Pathogenicity. *Neisseria meningitidis* produces in man an infection involving the base and cortex of the brain and the surfaces of the spinal cord. The organism is initially present in the naso-pharynx from which it gains entrance to the central nervous system. The spinal fluid may be turbid or very cloudy, depending upon the number of organisms present. Blood cultures have revealed menin-gococci in the bloodstream in a considerable proportion of cases of meningococcic meningitis, and, in some cases, before symptoms of meningitis were detectable. The organisms have also been found in arthritis, pericarditis, and purpuric patches of the skin. Case-fatality was variable but high, ranging from 35% to 80% before the advent of sulfonamide and penicillin therapy, which have greatly reduced fatalities and the duration of sickness.

Experimentally, *Neisseria meningitidis* may be injected intra-peritoneally into mice, guinea pigs, and rabbits, with the production of a toxemia in one to four days when large doses are inoculated. Animals, as a rule, do not appear to be very susceptible to the men-ingococcus.

Toxic Products. Meningococcus meningitis is usually accom-panied by great toxemia. The organism, however, does not appear to form a soluble toxin, and the toxemia is believed due to an endotoxin.

Diagnosis. 1. *Microscopic.* A smear made from the sediment of centrifuged spinal fluid is stained and examined for typical Gram-negative, coffee-bean-shaped, intracellular diplococci. A differential cell count is made by preparing films with sediment from the fluid and staining with Wright's stain or methylene blue. The polymorpho-nuclear leucocytes are greatly increased in epidemic meningitis.

2. *Cultural.* Sediment of centrifuged spinal fluid is inoculated into blood agar or infusion base chocolate agar. Incubate at 37° C. for 18 to 24 hours and examine colonies for typical Gram-negative, coffee-bean-shaped diplococci.

3. *Oxidase Test.* See p. 209.

4. *Fermentation Reactions.* Pure cultures of the organism are inoculated into serum-water carbohydrate media. Incubate at 37° C. The meningococcus produces acid in dextrose and maltose, but not in levulose and saccharose.

5. *Serological Tests.* Agglutination tests are performed using anti-meningococcic sera and a suspension of a pure culture of the organism. Usually only polyvalent antiserum is employed; typing by means of type-specific sera may be performed.

Differentiation of Gonococcus from Meningococcus. These two organisms resemble each other morphologically, but may be dif-ferentiated on the basis of the following characteristics:

1. Gonococcus grows more slowly, forms smaller colonies, and grows on fewer media than does meningococcus.

2. Gonococcus produces acid in glucose only, whereas meningococcus produces acid in both glucose and maltose.

3. Gonococcus is less toxic to mice or guinea pigs by intraperitoneal injection than is meningococcus.

Therapy. Sera prepared in horses have been used effectively in the treatment of meningococcic meningitis. Reports have indicated a reduction in the mortality rate from the expected percentage of 70 to 30. When given early in the disease it may be reduced to about 18%. Serum is given intraspinally and intravenously together.

Sulfanilamide and sulfadiazine are of marked beneficial effect in the treatment of meningitis cases, reducing the fatality rate to less than 10%. When combined with serum the results have been better than with the drug alone.

Carriers. The spread of meningitis is effected by the nasal secretions of sick individuals and, to a much greater extent, by carriers—those who harbor meningococci in their nasopharynx without manifesting symptoms of the disease. Meningococci may be present in cases for a long time after convalescence and also in individuals who have never had the disease. Carriers are best identified by obtaining swabs from high up in the pharynx behind the soft palate, with the aid of the West tube.* Agglutination tests have also been used to determine carriers. Identified carriers may be subjected to treatments, mostly sulfonamides, nasal and pharyngeal sprays of chemicals, disinfectants, and antimeningococcic serum, in an effort to destroy the organisms.

NEISSERIA CATARRHALIS

Neisseria catarrhalis, isolated by Frosch and Kolle in 1896, is similar to *Neisseria meningitidis* and *Neisseria gonorrheae* in morphological characteristics, but is differentiated from these organisms by its ability to grow on ordinary artificial media without enrichment. It is further distinguished by its inability to ferment any of the carbohydrates. The organism will grow at temperatures below 20° C., a property which helps to distinguish it from *Neisseria meningitidis*, which will not grow at temperatures below 25° C. *Neisseria catarrhalis* is more resistant to drying and heat. It is pathogenic for guinea

---

* This is made by fixing a cotton swab on the end of a copper wire, about 18 cm. long, and inserting this in a glass tube bent upward at the swab end in such a way as to permit its passage upward behind the soft palate. The swab is placed in the tube, both ends are plugged with cotton, and the entire apparatus is sterilized.

pigs, which die of toxemia in 24 hours following injection of large doses of the organism. It has been isolated from the nasopharynx of healthy persons as well as of those with catarrhal inflammations of the upper respiratory tract.

RELATED ORGANISMS

The other members of the Neisseria group are, for the most part, inhabitants of the respiratory tract of man, and are usually without pathogenic significance. They include: *Neisseria sicca, Neisseria perflava, Neisseria flava,* and *Neisseria flavescens,* which are similar in morphological, cultural, and physiological characteristics.

BIBLIOGRAPHY

American Public Health Association. *Diagnostic Procedures and Reagents.* New York: 1945. Pp. 73–136.

Bergey, David H. *Bergey's Manual of Determinative Bacteriology.* Seventh Edition. Baltimore: Williams & Wilkins Company, 1957.

Dubos, R. J. *Bacterial and Mycotic Infections of Man.* Third Edition. Philadelphia: J. B. Lippincott Company, 1958.

Gay, Frederick P. *Agents of Disease and Host Resistance.* Springfield, Ill.: Charles C. Thomas, 1935. Pp. 550–588.

Jawetz, E., J. L. Melnick, and E. A. Adelberg. *Review of Medical Microbiology.* Los Altos: Lange Medical Publications, 1962. Pp. 145–148.

Marshall, M. S. *Applied Medical Bacteriology.* Philadelphia: Lea and Febiger, 1947. Pp. 1–340.

Medical Research Council. *A System of Bacteriology in Relation to Medicine,* Vol. II. London: His Majesty's Stationery Office, 1929. Pp. 239–325.

Moulton, Forest R., ed. *The Gonococcus and Gonococcal Infection.* Washington, D. C.: American Association for the Advancement of Science, 1939.

Nelson, N. K., and G. L. Crain. *Syphilis, Gonorrhea and the Public Health.* New York: Macmillan Company, 1938.

Schaub, I. G., *et al. Diagnostic Bacteriology.* Fifth Edition. St. Louis: C. V. Mosby Company, 1958.

Taber, C. W. *Cyclopedic Medical Dictionary.* Philadelphia: F. A. Davis Company, 1958. B1–B5.

Wilson, G. S., and A. A. Miles. *Topley and Wilson's Principles of Bacteriology and Immunity.* 2 vols. Fourth Edition. Baltimore: Williams & Wilkins Company, 1955.

Chapter XXI

THE CORYNEBACTERIUM GROUP

The genus Corynebacterium belongs to the family Corynebacteriaceae, order Eubacteriales. It is made up of long, slender, Gram-positive, nonmotile bacilli, with involution forms showing a decided club-shaped appearance. These organisms exhibit uneven, striated staining due to the characteristic Babes-Ernst (metachromatic) bodies present. The most important pathogenic member of this group is *Corynebacterium diphtheriae*, the diphtheria bacillus, characterized by its strong exotoxin which is responsible for the lesions peculiar to the disease caused by this organism. In this group are also certain non-pathogens which, because of similar morphological characteristics, may be mistaken for *Corynebacterium diphtheriae*.

CORYNEBACTERIUM DIPHTHERIAE

Corynebacterium Diphtheriae (Bacillus Diphtheriae). 1. Habitat—cause of diphtheria in man. May be isolated from pharynx,

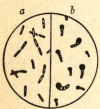

FIG. 45. *Corynebacterium diphtheriae.* (*a*) Young 12-hour culture showing Babes-Ernst bodies. (*b*) Involution forms or variants, from 5-day cultures.

larynx, trachea, and nose of cases with the disease, from healthy pharynx and nose of carriers, occasionally from the conjunctiva and infected superficial wounds. 2. Discoverers—Klebs and Löffler, in 1883.

Morphological Characteristics. 1. Form—slender rods, straight, slightly curved, or club-shaped. 2. Cell groupings—occurs singly usually, or in pairs. 3. Size—0.3 to 8μ in width and 1 to 8μ in length, but the size varies greatly. 4. Staining properties—stains readily with aniline dyes, unevenly, with forms that appear barred. The most diagnostic stains are Löffler's methylene blue and Neisser stain. 5. Gram-positive and non-acid-fast. 6. Nonmotile. 7. Nonsporing. 8. Nonencapsulated.

Cultural Characteristics. *Corynebacterium diphtheriae* grows well on artificial media, but it grows most luxuriantly on Löffler's blood serum. 1. Agar cultures—small, convex, grayish, granular colonies with irregular margin. 2. Agar slant—slight growth of grayish, granular, translucent colonies. 3. Gelatin stab—slight surface growth, with no liquefaction. 4. Litmus milk—grows readily, but the milk remains unchanged in appearance. 5. Potato—growth scanty, thin, and whitish. 6. Broth—fine granular deposit on bottom and sides of tube, with broth nearly clear and a thin pellicle formed. 7. Löffler's blood serum—small, moist, smooth, slightly raised, grayish to cream-colored colonies, with entire margin. 8. Potassium tellurite medium —this medium is used together with Löffler's blood serum for diagnostic purposes and differentiates *Corynebacterium diphtheriae* from the diphtheroid bacilli which may be confused with it. The colonies of *Corynebacterium diphtheriae* are a distinct black, due to reduction of the tellurite.

Physiological Characteristics. 1. Optimum temperature, 37° C., with a maximum of 43° C. and a minimum of 15° C. 2. Aerobic, and facultative anaerobe. 3. Indol is not produced. 4. Acid, but no gas, is produced in dextrose and levulose by all strains. Some strains also ferment galactose, maltose, dextrin, and glycerol. Fermentation reactions may be used to differentiate *Corynebacterium diphtheriae* from other members of the genus as shown below. 5. Resistance—destroyed by diffuse sunlight in 24 hours, by rapid drying over sulfuric acid in 48 hours, and by 1% phenol in 10 minutes.

FERMENTATION REACTIONS OF THE GENUS CORYNEBACTERIUM
(Acid production in Hiss's serum water with 1% sugar and indicator)

Organism	Dextrose	Maltose	Lactose	Saccharose	Dextrin	Mannitol
Corynebacterium diphtheriae	+	+	−	−	+	−
Corynebacterium xerose	+	+	−	+	−	−
Corynebacterium pseudodiphthericum	−	−	−	−	−	−
Corynebacterium pyogenes	+	+	+	+	−	−
Corynebacterium murisepticum	+	+	+	+	−	+

Toxic Products. *Corynebacterium diphtheriae* produces a powerful exotoxin, believed protein in nature, which is toxic for man, rabbits, and guinea pigs. The production of toxin *in vitro* depends upon the composition of the medium, a pH of 7.8–8.0, iron concentration (0.14 gm./ml. is optimal, 5 gm./ml. almost inhibitory), and a large surface area. Diphtheria toxin is unstable and deteriorates rapidly when exposed to light and oxidation. It is destroyed by a temperature of 75–80° C., and is rendered inactive by digestive juices and slight acidity (pH of 6 or less). It may be concentrated by salting out and dialysis and dried, in which state it keeps well even at 70° C. Toxins produced by various types of *Corynebacterium diphtheriae* seem to be identical.

Pathogenicity. Diphtheria is an acute infectious disease of children, among whom the highest incidence occurs. In man the disease is usually confined to a local infection of the mucous membranes, the bacilli affecting the upper part of the respiratory tract, especially the pharynx, where a grayish pseudomembrane or "false membrane" is formed. The pseudomembrane may extend to the entire respiratory tract, in which case the child becomes cyanotic and dyspnoeic. This pseudomembrane may also be formed in other parts of the body, as the conjunctiva, vulva, vagina, and wounds, but these are rare. To the toxin produced by *Corynebacterium diphtheriae* are ascribed the injuries which occur to the kidneys, muscles of the heart, and nerves. Heart injury and paralysis are the most dreaded sequelae of diphtheria. The disease is spread by droplet infection through discharges of the nose and throat of infected persons and carriers.

Carriers. Studies of the incidence of diphtheria bacilli in the throats of normal healthy individuals have revealed a surprisingly high number of carriers. Most of these, fortunately, do not harbor virulent strains, but the existence of carriers creates an important epidemiological problem. Convalescents from diphtheria usually harbor the bacilli in their throats for about 3 to 4 weeks, after which time they generally spontaneously get rid of them. By the end of 2 months, about 99.4% of convalescents are bacteria-free, the remainder becoming chronic carriers. Healthy persons exposed to carriers do not retain the organisms for more than a few days or, at most, a few weeks. Chronic carriers are dangerous only if the organisms are virulent. This can be determined by virulence tests, which are described under the diagnosis of diphtheria.

Various chemical sprays have been advocated to cure carriers. These include hydrogen peroxide, potassium permanganate, iodine, glycerin, formaldehyde, and acriflavine and other dyes.

Diagnosis. 1. *Collection of Material.* A sterile swab is rubbed over any visible membrane on the tonsils or pharynx, and then spread over Löffler's blood serum and/or a tellurite agar.

2. *Microscopic Examination.* Smears are made from the collected exudate and stained with Löffler's methylene blue and Neisser's stain. These are examined for characteristic barred and granular bacilli. A Gram stain shows a Gram-positive bacillus.

3. *Cultural.* The cultures are incubated for 12 hours at 37° C., after which time they are examined for characteristic colonies. These are further cultured into other tubes of Löffler's serum to obtain a pure culture. After incubation for 24 hours at 37° C., typical colonies are examined and smears made and studied. If typical diphtheria bacilli are found and the culture is from a suspected case, a presumptive diagnosis of diphtheria is made. When cultures from suspected carriers are examined, bacilli resembling *Corynebacterium diphtheriae* should be studied for fermentation reactions and by virulence tests.

4. *Fermentation Reactions.* The isolated culture is tested for fermentation reactions in dextrose, maltose, lactose, saccharose, dextrin, and mannitol. When the isolated culture fails to ferment dextrose and ferments saccharose, it is usually not *Corynebacterium diphtheriae.* (See p. 215.)

5. *Virulence Test.* This test is distinctive for *Corynebacterium diphtheriae* and is based on the characteristic picture produced in guinea pigs by the inoculation of a virulent strain of the organism which produces diphtheria toxin. Two methods are available:

(a) *Intracutaneous.* This requires experience in recognizing skin lesions. The advantage of this method is that four to six cultures can be tested with only two guinea pigs. (1) The abdomens of the animals are denuded of hair. (2) About 250 units of diphtheria antitoxin are inoculated intraperitoneally into one of the animals which is the control. (3) The growth from a 24-hour Löffler slant is rubbed up with 20 ml. of normal saline and 0.15 ml. is injected intracutaneously into each guinea pig. (4) If the culture is a true diphtheria bacillus, it will develop toxin, which in the nonimmunized guinea pig will produce a definite local inflammatory lesion in 24 hours which goes on to superficial necrosis in 48 to 72 hours. (5) The immunized guinea pig shows no lesions.

(b) *Subcutaneous.* (1) Two ml. of a pure culture of the organism grown in infusion broth for 48 hours, or twice the amount of a suspension prepared from the growth of a Löffler slant in 10 ml. of saline, is injected subcutaneously into a guinea pig about 250 grams in weight. (2) A second guinea pig is injected with the same amount of culture to which has been added 50 to 100 units of diphtheria antitoxin. (3) The procedure may be altered by giving the control animal 250 units of diphtheria antitoxin intraperitoneally 24 hours prior to the culture. (4) If the first guinea pig dies within 2 to 5 days and shows the typical engorgement of the suprarenals and local edema on post-mortem examination, and the second guinea pig lives, the test is positive and the organisms are considered virulent.

Schick Test. The reaction of the skin to the injection of diphtheria toxin gives an indication of the susceptibility of the individual to diphtheria. The toxin is diluted so that $\frac{1}{50}$ of a minimum lethal dose is contained in 0.1 ml. which is injected intradermally. A control of 0.1 ml. of heated toxin is similarly injected on the flexor surface

of the opposite arm.   Observations are made daily for several days.
The most reliable readings are those made on the fourth day.   The
reactions may be classified as follows:

1. *Positive Reaction.*   A circumscribed, slightly raised area of redness, 1 to
2 cm. in diameter, appears in 24 to 36 hours and reaches a maximum in 48 to 72
hours.   This persists for 7 to 15 days, and gradually fades to superficial scaling
and persistent brownish pigmentation.   There is no reaction in the control arm.

2. *Negative Reaction.*   There is no reaction in the test or control arm.

3. *Negative Reaction with Pseudoreaction.*   An inflammatory reaction, less
sharply circumscribed than in a true positive, develops equally on both arms within
24 hours.   Fading is rapid, and usually complete by the fourth day.

4. *Positive Reaction with Pseudoreaction.*   Control shows changes as in 3.
During the first 24 hours, the test arm also shows the same type of changes.   After
this time, the reaction in the test arm continues to develop and the one in the
control to fade.   By the fourth day, the difference is quite definite.

A positive reaction indicates lack of immunity to diphtheria.   It
is usually taken to mean that the individual has less than $\frac{1}{30}$ of a unit
of antitoxin per ml. of blood.   Recent experiments suggest, however,
that the "Schick level" of immunity may be much lower—about $\frac{1}{250}$
to $\frac{1}{500}$ unit of antitoxin per ml.

Immunity.   An attack of diphtheria usually confers lifelong immunity (which is essentially an antitoxic rather than an antibacterial
immunity).   As the age of a group increases it is found that resistance
to diphtheria increases proportionately.   This is attributed to continued exposure to the disease.   Infants up to the first year of life are
generally immune because of passive transmission of diphtheria
antitoxin through the placenta from the mother if she is immune.

Prophylaxis.   Three preparations are in use for diphtheria immunization—toxin-antitoxin, toxoid, and alum-precipitated toxoid.
The success of immunization with any of these may be followed by
studying the titer of antitoxin in the serum* of the vaccinated individual
or by the Schick test.

1. *Toxin-antitoxin.*   This mixture of toxin and antitoxin depends
for immunizing efficiency, not on the excess of toxin present, but on
the slow dissociation of the toxin-antitoxin complex to liberate free
toxin.   Immunity develops slowly; 1 to 6 months or more may be

---

* *Ramon Flocculation Test.*   Toxin and antitoxin, mixed to the point of neutralization, produce a flocculation, or precipitation.   Decreasing amounts of antitoxin are added to a
constant volume of toxin in a series of test tubes.   The tube in which flocculation first
occurs is an index of the amounts of toxin and antitoxin which are necessary to neutralize
each other.   The Ramon flocculation test is a rough method of estimating how many units
of diphtheria antitoxin are contained in a particular serum.   More exact determinations
of antitoxic potency are made on guinea pigs.

required for a negative Schick test. Three injections are reported to render 85% of individuals Schick-negative.

Toxin-antitoxin is preferred by some workers for immunization of older children and adults because it gives fewer reactions. Its great disadvantage lies in the presence of horse serum which may sensitize the individual.

2. *Toxoid*. This product, also known as anatoxin, is produced by exposing toxin to formalin, which results in loss of toxicity without alteration of antigenic power. Toxoid is said to render 95% of individuals Schick-negative. Advantages are stability, thermoresistance, and higher immunizing power than toxin-antitoxin. It also eliminates the hazard of horse serum sensitization.

3. *Alum-precipitated Toxoid*. Toxoid precipitated with alum produces an almost protein-free immunizing agent which is more effective than toxoid because of slow absorption by the tissues. Although the product does not give severe reactions in younger children, it does produce rather severe reactions in older children and adults.

Therapy. The effective therapy of diphtheria depends upon the *early* administration of diphtheria antitoxin, administered intramuscularly in less severe cases and intravenously in more severe cases. Antitoxin use can reduce mortality from 35% to zero, if given in time.

Antitoxin may also be used to protect nonimmunized individuals who have been exposed to diphtheria. This usually affords protection for 2 to 4 weeks, when the dose is repeated if danger of infection persists.

Types. Three types of *Corynebacterium diphtheriae* have been described by Anderson and his British coworkers: *gravis*, *intermedius*, and *mitis*. The *gravis* strain is said to be distinguished by (1) higher toxigenicity, (2) ability to ferment glycogen and starch, (3) short, diphtheroid morphology in contrast to the long forms with many metachromatic granules of *mitis* and barred forms of *intermedius*, (4) colonies on tellurite agar appearing irregular, striated, rough, predominantly gray in color, in contrast to small, round, smooth, flat, convex, predominantly black colonies of *mitis*, (5) pellicle formation and alkalization of broth, (6) inability to hemolyze human erythrocytes. The three types appear to be antigenically distinct, but not necessarily homogeneous. The *mitis* strains are heterogeneous, the *gravis* strains fall largely into two types, and *intermedius* strains are relatively homogeneous.

The validity of these distinctions, which show many exceptions and leave many strains untypable, and the applicability of them to the strains prevalent in this country, have as yet not been clearly defined.

RELATED ORGANISMS

Several organisms in the genus Corynebacterium are frequently found in the throat and may be confused with the diphtheria bacillus which they resemble morphologically. Granular forms, however, are very rare. They may be distinguished by their nonpathogenicity for guinea pigs and by fermentation reactions. (See p. 215.)

Corynebacterium Pseudodiphtheriae. This organism, isolated by Hoffmann in 1888, is shorter and thicker than *Corynebacterium diphtheriae* and is innocuous to laboratory animals. The organism forms no toxins and is easily differentiated from the diphtheria bacillus.

Corynebacterium Xerose. This bacillus was isolated by Kuschbert and Neisser in 1884 and is frequently found in normal as well as in inflamed eyes. It is morphologically similar to *Corynebacterium diphtheriae* but differs in pathogenicity for animals, in inability to form a toxin, and in fermentation properties.

Other Members of Corynebacterium Group. 1. *Corynebacterium pyogenes*—pathogenic for animals, causing abscesses in cattle. 2. *Corynebacterium acnes*—isolated from sebaceous glands, hair follicles, and acne pus. 3. *Corynebacterium murisepticum*—isolated in septicemia in mice. 4. *Corynebacterium gallinarum*—a nonpathogenic organism isolated from the throat of chickens.

BIBLIOGRAPHY

Andrewes, F. W., *et al*. *Diphtheria*. London: Medical Research Council, 1923.

Bergey, David H. *Bergey's Manual of Determinative Bacteriology*. Seventh Edition. Baltimore: Williams & Wilkins Company, 1957.

Gay, Frederick P. *Agents of Disease and Host Resistance*. Springfield, Ill.: Charles C. Thomas, 1935. Pp. 911–956.

Medical Research Council. *A System of Bacteriology in Relation to Medicine,* Vol. V. London: His Majesty's Stationery Office, 1930. Pp. 67–150.

Methods for Laboratory Technicians. TM 8-227. Washington, D. C.: U. S. Department of Defense, 1952.

Park, William Hallock, and Anna Wessels Williams. *Pathogenic Microörganisms*. Eleventh Edition. Philadelphia: Lea and Febiger, 1939. Pp. 442–473.

Rosebury, T. *Experimental Air-borne Infections*. Baltimore: Williams & Wilkins Company, 1947. Pp. 6–212.

Schaub, I. G., *et al*. *Diagnostic Bacteriology*. Fifth Edition. St. Louis: C. V. Mosby Company, 1958.

Chapter XXII

THE MYCOBACTERIA

The organisms classified in the genus Mycobacterium (order Actinomycetales, family Mycobacteriaceae) are characterized by the fact that they are stained with difficulty and that once stained they are not decolorized by acid. This characteristic gives them the name *acid-fast group*. They contain a waxy, granular material which accounts for the acid-fast characteristics and the necessity for special staining procedures to allow penetration of the stain. They are long, slender, Gram-positive, acid-fast rods which have a resistant cell membrane and which sometimes undergo branching. Many of the species of this group (pathogenic) grow very slowly, or not at all, on artificial media, and need specially prepared media for their cultivation. Those saprophytic or parasitic on cold-blooded animals grow rapidly on most media.

MYCOBACTERIUM TUBERCULOSIS VAR. HOMINIS

Mycobacterium Tuberculosis Var. Hominis (Bacillus Tuberculosis). 1. Group—Acid-fast. 2. Family—Mycobacteriaceae. 3. Habitat—cause of tuberculosis in man. 4. Discoverer—Koch, in 1882.

Morphological Characteristics. 1. Form—slender, beaded rods containing granules. 2. Cell groupings—occurs singly and occasion-

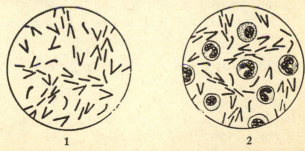

1	2

FIG. 46. (1) *Mycobacterium tuberculosis*, showing granules and tendency to V formation, with some curved forms. (2) *Mycobacterium tuberculosis* in transient tubercular septicemia, together with leucocytes.

ally in threads, with tendency toward V formation. 3. Size—0.3 to 0.6μ by 1.5 to 4μ. 4. Staining properties—irregular and difficult due to acid-fast characteristics. Stains well with Ziehl-Neelsen and Pappenheim stains.* (See p. 58.) 5. Gram-positive. 6. Nonmotile. 7. Nonsporing. 8. Fatty envelope surrounds the organism. 9. Much's granules—non-acid-fast, Gram-positive granules have been described by Much in tuberculous lesions in which acid-fast bacilli could not be found. Although Much claims that these granules are viable, infective, and virulent, and give rise to typical acid-fast bacilli, their true significance is not understood.

Cultural Characteristics. Growth of *Mycobacterium tuberculosis var. hominis* is slow in all media, requiring weeks for any visible development. 1. Nutrient agar—no growth. 2. Nutrient broth—no growth. 3. Glycerol agar—colonies are raised, cream-colored, and having a wrinkled surface. 4. Glycerol broth—after 8 weeks, a wrinkled gray pellicle is formed with no turbidity and a granular deposit. 5. Dorset's egg medium—after 3 or 4 weeks, a dull, dry, or warty film of gray color with a yellowish tinge is formed. 6. Bordet-Gengou medium—visible growth in 8 to 10 days, with maximum development in about 3 weeks. Raised, finely granular colonies, 1.1 to 3.3 mm. in diameter, with indented edges. 7. Löwenstein's medium—raised colonies which appear ball-like. 8. Litmus milk—when 2% glycerin is added, the medium is favorable for growth showing no coagulation. The reaction is at first acid, then neutral, and finally slightly basic. 9. Glycerol potato—luxuriant cream-colored growth with a nodular or warty surface after 4 weeks. 10. Petragnani special medium—luxuriant, raised, granular, cream-colored, coalesced growth.

Physiological Characteristics. 1. Optimum temperature, 37° C., with limits at 30° to 42° C. 2. Aerobe. 3. Utilizes dextrose, levulose, arabinose, and galactose, but not sucrose and lactose. 4. Golden-yellow pigment is usually produced in serum medium. 5. Resistance—cultures may remain viable for a year. Moist heat at 60° C. kills the bacilli in 15 to 20 minutes. They may survive in dried sputum even for a few months. Exposure to 5% phenol does not destroy the organism even after 24 hours. Direct sunlight kills in a few hours.

Chemical Composition. The chemical composition of the tubercle bacillus has been extensively studied by Anderson, Seibert, and others. They have shown a high lipoidal content, perhaps as high as 40%

---

* Recently use of the fluorescent auramine has been introduced. Its use is reported to be more convenient and allow easier detection of the organism in materials.

Protein, largely nucleoprotein, makes up about one-half the dry weight; polysaccharides are present in relatively small amounts. The lipoids, some of which have been shown to be physiologically active in stimulating multiplication of tissue cells, are of three types: neutral fat, phospholipoids, and waxes. The latter include a polysaccharide-containing acid-fast wax, a complex glyceride soft wax, and an acid-fast unsaponifiable wax made up of higher alcohols.

Pathogenicity. *Mycobacterium tuberculosis var. hominis* is strictly a pathogenic organism. The bacillus is found in man in tuberculous lesions, usually in the lungs. Transmission occurs through the respiratory or digestive tracts, usually by means of sputum. When the organisms have gained a foothold, the tissues where they localize immediately begin to react. Numbers of tissue cells grow around the bacilli to wall them off, forming the characteristic tubercle. With further growth of the organisms, the surrounding cells are killed. Adjacent tubercles may coalesce. Dead tissue at the center of such masses becomes yellowish and cheesy—"caseated." Localized skin infections may be contracted by handling the organism. Tuberculous lesions may involve the lungs, skin, central nervous system, digestive tract, liver, bones, joints, glands, etc.

Although predisposition to the disease may run in families, it is not hereditary. Occupational tuberculosis occurs among workers in the glass, asbestos, and granite industries as a sequela of silicosis or anthracosis.

Toxic Products. *Mycobacterium tuberculosis* does not produce a true exotoxin. Filtrates of glycerin broth cultures of the organism contain certain proteinaceous products, which when inoculated into normal healthy animals result in a reaction differing from that in a tuberculous animal. These substances, termed *tuberculins*, were found by Koch to produce in tuberculous animals a local inflammation followed by necrosis, without an increase in the original infection. In a few instances, however, fever may be produced, and the animal may become severely ill. This is the *Koch phenomenon* and is interpreted as a state of hypersensitivity induced by previous exposure to the tubercle bacillus and its products. When controlled doses of tuberculin are inoculated into normal animals, there is no reaction. However, when tubercle bacilli are injected into healthy animals, tubercles develop and the infection spreads to various organs, until the animal finally succumbs.

Bacteriological Diagnosis. The material collected, sputum, urine, feces, cerebrospinal fluid, stomach contents, or serous exudates, is examined for tubercle bacilli.   Sputum is concentrated by adding to one part of sputum two parts of antiformin (sodium hypochlorite) and shaking for 10 minutes.   The sediment obtained after centrifuging is spread on a slide, dried, and stained with Ziehl-Neelsen stain.   This sediment is also inoculated into specific culture media, and into proper laboratory animals.   Other material, like urine, feces, cerebrospinal fluid, etc. may be concentrated in a similar manner.

1. *Staining Prepared Films.*   When acid-fast bacilli stained red against a blue background are found in sputum or other exudates and excretions, this is a presumptive positive test for tubercle bacilli.

2. *Animal Inoculation.*   Simultaneous with the preparation of a film for microscopic examination, guinea pigs should be inoculated subcutaneously or intramuscularly in the inguinal region.   The animals die at the end of 3 weeks to 4 months, the inguinal nodes and spleen teeming with bacilli.

3. *Inoculation of Culture Media.*   Some investigators advocate culturing on media such as the Bordet-Gengou or Löwenstein, instead of inoculation into guinea pigs.   Others suggest inoculation of guinea pigs and then cultivation of minced infected tissue from these animals. However, because the methods take such a long time for results, initial bacteriological examination revealing characteristic acid-fast bacilli, differentiated from *Mycobacterium lacticola* (*Bacillus smegmatis*) with which they may be confused, may be considered as an almost positive diagnosis in the majority of cases.

Tuberculin Diagnostic Methods.   The specific response obtained upon the injection of tuberculin has been utilized in diagnosis.   Because of the wide prevalence of some degree of tubercular infection in human beings, however, tuberculin testing is of diagnostic value only in young children.

1. *Mantoux Intracutaneous Test.*   The test consists of graded intracutaneous injections of 0.1 ml. of tuberculin diluted 1:10,000 and 1:100.   A positive reaction appears in 6 to 8 hours, reaches its maximum in 24 to 48 hours, and generally subsides in 6 to 10 days.   It is characterized by infiltration, hyperemia, and, in rare instances, vesiculation.   The width of the area of infiltration and the degree of inflammation are noted.   This is the most accurate of the human tuberculin tests.

2. *Von Pirquet Cutaneous Test.*   A drop of tuberculin, diluted to 25%, is injected into a small scarification made on the skin.   A similar scarification is made about three inches away from the first one.   This is kept as the control.   Both areas should be examined at the end of 12, 24, and 36 hours.   A positive reaction appears after 3 to 24 hours and is usually at its height at 36 to 48 hours.   It consists of itching and reddening of the skin about 10 mm. in diameter.

3. *Moro Percutaneous Test.* A small amount of ointment made of equal parts of lanolin and tuberculin is rubbed into the skin on the chest. The development of reddening and papules indicates a positive reaction.

4. *Vollmer Patch Test.* Small squares of filter paper, impregnated with tuberculin and dried, are taped on the cleansed skin over the sternum. Reddening indicates a positive reaction.

Immunity. Tuberculosis infection confers a definite protection against reinfection. Active immunization by means of suspensions of living attenuated bacilli or of killed bacilli have not proved of any definite use. Calmette and Guérin, by carrying a strain of bovine tubercle bacillus through several hundred transfers, obtained the B.C.G. (*Bacille Calmette-Guérin*) *vaccine.* This is believed to be harmless, and although it has been used fairly widely, especially in Europe, since its introduction in 1924, evidence is lacking that it is of any value, at least in this country.

Prophylaxis. The prevention of tuberculosis is in large measure dependent upon personal and community hygiene. Fresh air, sunshine, and nourishing food are important factors in building up a resistance to the disease. Prevention of infection from animal sources involves tuberculin-testing all animals used for food and destroying the reactors, thorough cooking of meats, and pasteurization of milk.

Therapy. Treatment of tuberculosis is for the most part systemic, fresh air, sunlight, rest, and nourishing food being advised. The use of tuberculin in therapy has been proposed. Although the results have been favorable in a few cases, it is not advocated for widespread use because of the severe reactions resulting from its haphazard injection. The best results have been obtained with small doses and its use has been limited to early cases of the disease. Some encouraging results have been obtained by the use of the sulfonamides promin, promizole, and diazone, and by combined use of promin and streptomycin, the latter being indicated in tubercular meningitis, miliary genito-urinary tuberculosis, chronic tuberculous empyema, etc.

MYCOBACTERIUM TUBERCULOSIS VAR. BOVIS

The bovine type of tuberculosis is highly pathogenic for cattle and all mammals except man. It is pathogenic for man, but not to the degree of the human tubercle bacillus. Mostly children are infected, and the portal of entry is usually the alimentary tract by the ingestion of contaminated raw milk. Bovine tuberculosis in man is usually nonpulmonary, tuberculosis of the bones and joints being the most common.

It is among cattle that tuberculin testing for diagnosis finds its most extensive application.   Three types of test may be used:

1. *Thermal Test.*   Constitutive reaction involving rise in temperature is observed after subcutaneous injection.   Formerly widely used, this is being replaced by shorter methods.

2. *Intradermal Test.*   Extensively used for cattle, swine, and chickens, this is similar to the Mantoux test for human beings.   Local swelling indicates positive reaction.

3. *Calmette Ophthalmic Test.*   Inflammation of the conjunctiva about 6 hours after tuberculin has been dropped into it constitutes a positive test.

Mycobacterium tuberculosis var. bovis may be differentiated from *Mycobacterium tuberculosis var. hominis* as follows:

CHARACTERISTIC	BOVINE TYPE	HUMAN TYPE
Morphological	Tends to be short, thick, and solidly stained.	Tends to be long, slim, slightly bent, and may show beading and irregular staining.
Cultural	Grows poorly or not at all when freshly isolated on media such as egg, glycerin egg, glycerin potato, and glycerin broth. Glycerol had little effect on growth.	Grows well on all of these media. Growth greatly enhanced by glycerol.
Acid production in glycerin broth	Renders bouillon less and less acid.	Causes preliminary fall in acidity. As growth progresses acidity is increased and may exceed original acidity.
pH of tuberculin	Alkaline or slightly acid.	Markedly acid.
Virulence for rabbits	Very virulent for rabbits by any method of inoculation. Rabbits usually die before 40 to 50 days after injection.	Slightly virulent for rabbits. Animals survive 40 to 50 days, and on autopsy show only lesions in lungs or kidneys, or both.
Virulence for calves	Causes generalized tuberculosis.	Causes local lesion or, at most, a spreading to the nearest lymph node.

MYCOBACTERIUM LEPRAE

Mycobacterium leprae has characteristics very similar to those of *Mycobacterium tuberculosis var. hominis.*   The organism is numerous in lesions of leprosy, often occurring as coarsely-beaded rods which stain more readily than does the tubercle bacillus.   Histopathological sections from lesions show millions of typical acid-fast, beaded rods, usually intracellular and arranged in packets.

Cultivation.   It has not been definitely determined whether *Mycobacterium leprae* has been cultivated.   A bacillus was described by Hansen as acid-fast, Gram-positive, nonmotile, nonsporing, and

1 to 7μ in length by 0.3 to 0.5μ in breadth, with tendency to occur in clumps or packets. None of the organisms described has been accepted as the cause of leprosy, for it has been impossible to produce typical symptoms in laboratory animals.

Diagnosis. The bacteriological diagnosis of leprosy at the present time consists in making smears from material obtained from the lesions (discharges, ulcerated nodules, or scrapings from excised tissue), and in finding typical acid-fast bacteria.

Pathogenicity. Leprosy is a chronic disease which usually occurs in two forms, the anesthetic or nerve type and the nodular. Infection results in the disfigurement of the body and destruction of tissue with scar formation. Death is usually due to tuberculosis or other coincident infection. The infectivity of the disease is slight. Both intimate and prolonged contact are believed necessary for the spread of infection, and workers in leper colonies do not necessarily contract the disease. The incubation period of the disease is unknown, and it has been estimated to be 3 to 6 years.

The organism enters via the mucous membranes of the nose and pharynx, the respiratory and gastrointestinal tracts, and the skin. Early localization is the nose along the septum.

Immunological Reactions. Persons with leprosy respond to injections of extracts of leprous tissue and tuberculin with a febrile reaction. The serum of such patients fixes complement with antigens prepared from cultures of organisms in the genus Mycobacterium and also gives a positive Wassermann reaction.

Treatment and Prevention. Chaulmoogra oil has been beneficial in the treatment of leprosy. Isolation and segregation of lepers are employed in the prevention and control of the disease.

RELATED ORGANISMS

A number of organisms in the acid-fast group resemble the tubercle bacillus in morphological characteristics. A large majority of these are saprophytes or parasites on cold-blooded animals and grow rapidly on most media. These include: 1. *Mycobacterium tuberculosis avium*—cause of tuberculosis in domestic fowls and other birds. 2. *Mycobacterium tuberculosis piscium*—cause of nodule and tumor-like formations in carp and other fish. 3. *Mycobacterium lacticola* (*smegmatis*)—found on skin and genitalia of man. 4. *Mycobacterium phlei*—found in hay, grass, soil, and dust. 5. *Mycobacterium thamnopheos*—parasite in the garter snake. 6. *Mycobacterium friedmannii*

—parasite of turtles.   7. *Mycobacterium paratuberculosis*—cause of Johne's disease in cattle.   (See p. 159.)

BIBLIOGRAPHY

American Association for the Advancement of Science. *Tuberculosis and Leprosy, The Mycobacterial Diseases.* Washington, D. C.: 1938.

American Public Health Association. *Diagnostic Procedures and Reagents.* New York: 1945. Pp. 305–318.

Anderson, W. A. *Pathology.* Second Edition. St. Louis: C. V. Mosby Company, 1953.

Darzins, E. *The Bacteriology of Tuberculosis.* Minneapolis: University of Minnesota Press, 1958.

Diagnostic Standards and Classification of Tuberculosis. New York: National Tuberculosis Association, 1940.

Gardner, A. D. *Bacteriology for Medical Students and Practitioners.* New York: Oxford University Press, 1944. Pp. 133–148.

Gershenfeld, L. *Bacteriology and Allied Subjects.* Easton, Pa.: Mack Publishing Company, 1945. Pp. 104–169.

Koch, R. *The Aetiology of Tuberculosis.* (Translation of Koch's original paper.) New York: National Tuberculosis Association, 1932.

Marshall, M. S. *Applied Medical Bacteriology.* Philadelphia: Lea and Febiger, 1947. Pp. 1–340.

Park, William Hallock, and Anna Wessels Williams. *Pathogenic Microörganisms.* Eleventh Edition. Philadelphia: Lea and Febiger, 1939. Pp. 568–617.

Pollitzer, R. *Plague.* WHO Monograph Series, No. 22. New York: Columbia University Press, 1954.

Rosebury, T. *Experimental Air-borne Infections.* Baltimore: Williams & Wilkins Company, 1947. Pp. 6–212.

———. *Microörganisms Indigenous to Man.* New York: McGraw-Hill Book Company, 1962. Pp. 82–87.

Rosenthal, S. R. *BCG Vaccination against Tuberculosis.* Boston: Little, Brown and Company, 1957.

Strean, L. P. *Oral Bacterial Infections.* Brooklyn, N. Y.: Dental Publishing Company, 1949. Pp. 68–72.

Strong, R. P. *Stitt's Diagnosis, Prevention, and Treatment of Tropical Diseases.* Philadelphia: The Blakiston Company, 1945. Pp. 541–813.

Sulzberger, M. B., and R. L. Baer. *Office Immunology.* Chicago: Yearbook Publishers, 1947. Pp. 95–129.

Chapter XXIII

THE PASTEURELLA GROUP

The genus Pasteurella (family Parvobacteriaceae, tribe Pasteurelleae) comprises the Gram-negative, bipolar-stained, nonmotile rods of the plague and hemorrhagic septicemia group of organisms that are of human and veterinary pathogenic significance. The members of this genus of importance in veterinary medicine have been discussed. (See pp. 164 to 169.) In this chapter the causative agent of plague, *Pasteurella pestis*, will be described in its relation to the disease.

PASTEURELLA PESTIS

Pasteurella Pestis (Bacillus Pestis). 1. Group—hemorrhagic septicemia. 2. Family—Parvobacteriaceae. 3. Habitat—causes plague in man and in rats, ground squirrels, and other rodents. 4. Discoverers—Kitazato and Yersin, in 1894.

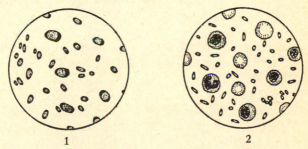

Fig. 47. (1) Involution forms of *Pasteurella pestis*. (2) *Pasteurella pestis* in the blood in plague septicemia.

Morphological Characteristics. 1. Form—short, plump rods, rounded at both ends, slightly thicker in the middle. 2. Cell groupings—occurs singly. 3. Size—0.5 to 0.7μ in breadth and 1.5 to 2μ in length. 4. Staining properties—stains well with ordinary aniline dyes, methylene blue, or strong carbol fuchsin; bipolar staining. 5. Gram-negative. 6. Nonmotile. 7. Nonsporing. 8. Gelatinous capsule. 9. Tendency to form involution forms, especially in media containing 3–4% sodium chloride.

Cultural Characteristics. *Pasteurella pestis* grows well in ordinary artificial media. 1. Agar slant—viscid, slimy, grayish-white, expansive growth. 2. Gelatin stab—slow arborescent growth in stab, with no liquefaction. 3. Litmus milk—slow growth with slight acidity of the medium and no coagulation. 4. Potato—slight grayish growth. 5. Agar cultures—grayish-yellow colonies with raised center, coarsely granular, and jagged edge. 6. Broth—delicate, whitish, flocculent sediment with stalactite growth.

Physiological Characteristics. 1. Optimum temperature, 25° to 30° C., with minimum of 0° C. and maximum of 45° C. 2. Aerobic, and facultative anaerobe. 3. Indol is not produced. 4. Nitrates are reduced to nitrites. 5. Acid, but no gas, is formed from glucose, levulose, maltose, galactose, and mannite. No fermentation of lactose, saccharose, dulcite, raffinose, and inulin. 6. Resistance— quite resistant to cold; killed in 2 to 3 hours by drying, in 3 to 5 hours by sunlight, in 5 minutes by 1% Lysol and chloride of lime.

Pathogenicity. Plague is one of the oldest diseases of man, dating back to Biblical times. Known as the "Black Death," it swept through Europe in 1348 killing 25,000,000 people—about one-fourth the population.

It is primarily a disease of rodents, and only secondarily of man. In rats, generalized blood infection occurs. Plague in man may be of three types:

1. *Bubonic or Glandular*—This type is spread by the flea. The organisms enter through the skin and may produce a local lesion at the point of infection. Then they spread and infect the lymph glands which swell and often suppurate, forming the characteristic buboes. Hemorrhages from mucous surfaces and beneath the skin may occur. Bacilli may pass into the blood from the buboes, and multiply there. Mortality for this type is 50-80%.

2. *Pneumonic*—This type may occur secondarily to the bubonic form, or primarily. It involves the lungs and is spread by contact. There is the characteristic picture of lung inflammation or bronchopneumonia. The sputum is laden with blood and organisms. Mortality is nearly 100%.

3. *Septicemic*—This rapidly fatal type is also transmitted by fleas. Numerous hemorrhages occur under the skin, which turns black.

Mode of Transmission. Plague is transmitted from rat to rat and from rat to man by infected fleas. Healthy rats become infected most usually by the bite of a flea which has previously bitten a rat with plague. An infected flea, biting a man or other animal, deposits on the skin of the new host regurgitated material or feces or both, which contain the organism. Bacilli are rubbed farther into the skin by scratching. Pneumonic plague may be spread from man to man by droplet infection.

Diagnosis. 1. *Smears* from buboes or other plague lesions, although these may not contain many organisms. Internal organs, especially the spleen, are often more fruitful. In the pneumonic form, smears of sputum yield good results.

2. *Cultures* from material from buboes, blood, or internal organs.

3. *Post-mortem* diagnosis from investigation of heart's blood and spleen.

4. *Guinea pig inoculation* subcutaneously or, with material that has undergone gross contamination, by rubbing on freshly shaved abdomen. Plague bacilli will penetrate the minute abrasions, but the contaminants will not. The animals die within 2 to 5 days, when the organism can be isolated from the heart's blood.

5. *Agglutination tests*, not proved practicable.

Immunity and Prophylaxis. One attack of plague usually protects human beings against subsequent attack, and second attacks are extremely rare. Experimental animals have been immunized with suspensions of attenuated or killed plague bacilli. Use of the latter has not been satisfactory in man. Possible danger has severely limited use in human beings of living attenuated organisms, which in animals produce a more solid immunity. Some recent experiments, however, have given encouraging results. Intramuscular administration of antiplague serum has also been suggested for prophylaxis.

The United States Public Health Service has instituted energetic methods of rat extermination: rat-proofing of houses and other rat-control measures. Quarantine regulations and supervision of ships and sleeping cars have been enacted.

Therapy. Antiplague serum produced in animals is claimed by some to have great value as therapy, but most workers question its efficiency.

Related Organisms. (See pp. 164–169.) Among the organisms also classified in the genus Pasteurella are: 1. *Pasteurella avicida*—the cause of fowl cholera. 2. *Pasteurella cuniculicida*—the cause of hemorrhagic septicemia of rabbits. 3. *Pasteurella tularensis*—the cause of tularemia, a rodent disease transmissible to man. 4. *Pasteurella suilla*—the cause of swine plague. 5. *Pasteurella bollingeri*—the cause of hemorrhagic septicemia in domestic cattle, hogs, and horses.

BIBLIOGRAPHY

Anderson, W. A. *Pathology.* Second Edition. St. Louis: C. V. Mosby Company, 1953.

Bergey, David H. *Bergey's Manual of Determinative Bacteriology.* Seventh Edition. Baltimore: Williams & Wilkins Company, 1957.

Dorland, W. A. N. *American Illustrated Medical Dictionary.* Philadelphia: W. B. Saunders Company, 1951.

Gay, Frederick P. *Agents of Disease and Host Resistance.* Springfield, Ill.: Charles C. Thomas, 1935. Pp. 725–752.

Hirst, L. F. *The Conquest of Plague.* Oxford: Clarendon Press, 1953.

Marshall, M. S. *Applied Medical Bacteriology.* Philadelphia: Lea and Febiger, 1947. Pp. 1–340.

Medical Research Council. *A System of Bacteriology in Relation to Medicine,* Vol. III, pp. 137–224, Vol. IV, pp. 446–473. London: His Majesty's Stationery Office, 1929.

Methods for Laboratory Technicians. TM 8-227. Washington, D. C.: U. S. Department of Defense, 1952.

Parish, H. J. *Antisera, Toxoids, Vaccines and Tuberculins in Prophylaxis and Treatment.* Third Edition. Baltimore: Williams & Wilkins Company, 1955.

Smith, Geddes. *Plague on Us.* New York: The Commonwealth Fund, 1941.

Society of American Bacteriologists. *Manual of Methods for Pure Culture Study of Bacteria.* Geneva, N. Y.: Biotechnical Publications, 1949. Chap. VII.

Taylor, B. T. *Stedman's Medical Dictionary.* Baltimore: Williams & Wilkins Company, 1957. Pp. 923–924.

Top, F. H. *Communicable Diseases.* Third Edition. St. Louis: C. V. Mosby Company, 1955.

Wu Lien Teh. *A Treatise on Pneumonic Plague.* Geneva: League of Nations, Health Organization, 1926.

Chapter XXIV

THE HEMOPHILUS GROUP

Members of the genus Hemophilus (family Parvobacteriaceae, tribe Hemophileae) are strictly parasitic, nonmotile, pleomorphic, Gram-negative, minute rods, which require for growth certain accessory substances found in blood and some plant tissues. Included at the present time are certain organisms, such as *Hemophilus pertussis* and *Hemophilus ducreyi*, which satisfy most of the generic criteria, but not all the generic growth requirements.

HEMOPHILUS INFLUENZAE

Hemophilus Influenzae (Bacillus Influenzae, Pfeiffer Bacillus). 1. Group—hemoglobinophilic. 2. Family—Parvobacteriaceae. 3. Habitat—found in the respiratory tract of man and was at one time regarded by Pfeiffer and other investigators as the cause of human influenza. 4. Discoverer—Pfeiffer, in 1892.

Morphological Characteristics. 1. Form—minute rods. 2. Cell groupings—occurs singly and in pairs, occasionally in short chains, and at times in long thread forms. 3. Size—0.2 to 0.3μ broad and 0.5 to 2μ long. 4. Staining properties—shows tendency to bipolar staining; stains well with carbol fuchsin and alkaline methylene blue. 5. Gram-negative. 6. Nonmotile. 7. Nonsporing. 8. Some strains are encapsulated.

Cultural Characteristics. *Hemophilus influenzae* is a strict parasite. It requires two factors for its growth, which have been termed X and V. Both are present in fresh whole blood. The heat-stable X factor is replaceable by hemoglobin or hematin, with which it may be identical. The heat-labile V factor is extractable from yeast, certain other vegetable cells, and many bacteria, notably certain strains of Staphylococcus. The V factor may be replaced by coenzyme I or II. *Hemophilus influenzae* grown together with cultures of Staphylococci and other bacteria exhibits larger colonies in the region of the other bacteria, owing to their synthesis of coenzyme, than those in the area farther removed from these bacteria. This has been termed the *satellite phenomenon.*

1. No growth on ordinary culture media.   2. Blood agar—clear, droplike colonies after 24 hours.   3. Chocolate agar—growth similar to that on blood agar.   4. Glycerin-potato-blood agar—arched colonies resembling drops of quicksilver; area of hemolysis appears on second or third day.

Physiological Characteristics.   1. Optimum temperature, 37° C. 2. Strict aerobe.   3. Forty to 50% of the strains produce indol. 4. Nitrates are reduced to nitrites.   5. Some strains attack carbohydrates, but mannitol and lactose are not fermented.   6. Resistance— killed by exposure to heat and antiseptics, and by drying.

Serological Reactions.   By precipitation tests, six types of *Hemophilus influenzae*, a, b, c, d, e, and f, have been identified.

Pathogenicity.   Whereas the etiology of human influenza was formerly attributed to *Hemophilus influenzae*, present evidence points toward virus causation.   (See p. 295.)   The bacillus is frequently but not always present in influenza, probably as a secondary invader as it is in cases of scarlet fever, measles, chickenpox, whooping cough, etc. A severe throat infection of children, meningitis, endocarditis, and sinusitis have also been attributed to the organism.

HEMOPHILUS PERTUSSIS

Hemophilus Pertussis (Bacillus Pertussis, Bordet-Gengou Bacillus).   1. Group—hemoglobinophilic.   2. Family—Parvobacteriaceae. 3. Habitat—respiratory tract in cases of whooping cough.   4. Discoverers—Bordet and Gengou, in 1906.

Morphological Characteristics.   1. Form—small, ovoid rods, with a tendency to pleomorphism in fluid media.   2. Cell groupings—occurs singly and sometimes in pairs; appears in masses and clumps in exudates.   3. Size—0.2 to 0.3μ broad and 0.5μ long.   4. Staining properties—frequently shows bipolar staining; stains well with alkaline methylene blue, dilute carbol fuchsin, or aqueous carbol toluidine. 5. Gram-negative.   6. Nonmotile.   7. Nonsporing.

Cultural Characteristics.   1. Glycerin-potato-blood agar (Bordet-Gengou medium)—colonies barely visible after 24 hours; these become visible in 48 to 72 hours and are small, glistening, grayish, and rather thick.   2. Blood agar—the blood is hemolyzed and the colonies are small, transparent, and with entire edges.   3. Gelatin stab—no growth.   4. Potato—light yellow streak of growth, becoming tan. 5. Litmus milk—the medium becomes brownish in color, with no other visible changes.   6. Fluid media—viscid sediment formed.

Physiological Characteristics. 1. Optimum temperature, 37° C., but may grow at 5° to 10° C. 2. Strict aerobe. 3. Nitrates are not reduced to nitrites. 4. No indol is formed. 5. No carbohydrates are fermented. 6. Resistance—easily killed by drying, disinfectants, sunlight, and heat.

Variant Forms. Virulent and avirulent forms of *Hemophilus pertussis* have been identified, as well as smooth, rough, and intermediate types. By means of serological reactions four phases have been recognized, Phases I, II, III, and IV. The organisms isolated from patients with whooping cough are usually in Phase I of the groups described by Leslie and Gardner. They are smooth, encapsulated, virulent for guinea pigs, mice, and rabbits, hemolytic, and very fastidious about X and V factors. After artificial cultivation, the organisms lose these properties and enter Phases II, III, and IV, which probably represent stages in the smooth——>rough transition. Only virulent organisms in Phase I are capable of stimulating antigens for immunization of man.

Clinical Aspects. Whooping cough is an acute, specific, infectious disease of the respiratory passages, characterized by recurring series of convulsive coughs which terminate in a long-drawn inspiration. The disease manifests itself as an ordinary cold after an incubation period of usually 7 days, and almost uniformly within 10 days. The points of predilection are the laryngeal and bronchial mucous membranes. The organism, *Hemophilus pertussis*, is found in large numbers in the sputum in the early stages of the disease, but disappears as the disease progresses. The disease is infectious and communicable in the early catarrhal stages before characteristic whoops make clinical diagnosis possible. After the characteristic whoops appear, the infecting organism can be isolated in discharges by the use of cough plates. The cough may be due to a toxin which has recently been demonstrated present in the organism and culture filtrates.

Bacteriological Diagnosis. 1. *Cough plate method* of Chievitz and Meyer—open plate containing Bordet-Gengou medium is exposed to one or more explosive coughs, covered, incubated, and examined for characteristic colonies after 2–5 days.

2. *Nasopharyngeal swabs* on culturing may give higher percentage of positive findings than the cough plate.

Pathogenicity. The organisms produce a mild toxin which may be carried by leucocytes to the trachea and bronchi. Slight changes in the lymph nodules of the spleen and gastrointestinal tract are noted, with lymphocytosis.

The greatest incidence of whooping cough occurs under 5 years of age, and the death rates for whooping cough are highest under 1 year of age. Children having whooping cough become predisposed to infection with Streptococci, Micrococci, pneumococci, and *Mycobacterium tuberculosis*. Death from pneumonia in the lower age group is the greatest danger, and otitis media and other streptococcic infections also claim a high percentage of children.

Serological Reactions. Agglutinins are present in the serum of persons with whooping cough, but are not of diagnostic significance. Agglutinins have also been found in the sera produced by immunizing animals. The complement fixation test has been used diagnostically. Convalescent serum has some anti-endotoxic properties.

Immunity, Prophylaxis, and Therapy. An attack of whooping cough usually confers lifelong immunity. Vaccines made of suspensions of heat-killed and chemically treated cultures of *Hemophilus pertussis* have been used for the prevention and treatment of whooping cough. Sauer's vaccine has been used prophylactically with good results and as a therapeutic agent has been claimed to lessen the number and severity of the paroxysms and to shorten the duration of the disease slightly. The disease is appreciably milder in immunized children. Results with serum in the prophylaxis of whooping cough have been conflicting.

HEMOPHILUS DUCREYI

Hemophilus Ducreyi (Bacillus of Soft Chancre). 1. Group—hemoglobinophilic. 2. Family—Parvobacteriaceae. 3. Habitat—cause of soft chancre (chancroid). 4. Discoverer—Ducrey, in 1889.

Morphological Characteristics. 1. Form—small rods with rounded ends. 2. Cell groupings—occurs singly or in chains of from 3 to 20 bacilli; solitary bacilli may be seen free or within pus cells. Filaments in pathological material are characteristically seen in "schools." 3. Size—0.2μ broad and 1.5 to 2μ long. 4. Staining properties—exhibits polar staining; stains well with aniline dyes. 5. Gram-negative. 6. Nonmotile. 7. Nonsporing.

Cultural Characteristics. *Hemophilus ducreyi* does not grow on ordinary media. The most favorable medium is either a rich blood agar, or heated, clotted blood.

Physiological Characteristics. 1. Optimum temperature, 37° C. 2. Strict aerobe. 3. Resistance—the organism is delicate and is easily killed by exposure to light.

Pathogenicity. *Hemophilus ducreyi* produces soft chancre or chancroid, an acute inflammatory lesion that occurs usually upon the genitals or skin surrounding them. It is transmitted by direct contact, or indirectly in some cases by infected dressings, towels, or instruments. The lesion starts as a small pustule, which ruptures, leaving a ragged ulcer with necrosis that spreads rapidly. The disease differs from the syphilitic chancre in its lack of induration and in the severe inflammatory response. Although the disease has been experimentally produced in man by cultures of *Hemophilus ducreyi*, attempts to infect animals have been unsuccessful.

Immunity and Therapy. Infection results in little or no immunity. Chancroid is frequently multiple and auto-inoculable. Antisera and autogenous vaccines seem to be of value.

Differential Diagnosis. Soft chancre caused by *Hemophilus ducreyi* may be differentiated from syphilitic chancre by (a) examining chancre fluid by dark field illumination, (b) Wassermann or Kahn serological tests, and (c) use of specific *Hemophilus ducreyi* vaccine.

Related Organisms. Among the organisms also classified in the genus Hemophilus are: 1. *Hemophilus hemoglobinophilus*, associated with preputial secretions of dogs; 2. *Hemophilus duplex*, the cause of subacute infectious or angular conjunctivitis; 3. *Hemophilus suis*, which together with a filterable virus causes swine influenza (see p. 307); 4. *Hemophilus hemolyticus*, a nonpathogen found in the upper respiratory tract of man; and 5. *Hemophilus parainfluenzae*, also nonpathogenic and found in the upper respiratory tract of man and cat.

BIBLIOGRAPHY

Dubos, R. J. *Bacterial and Mycotic Infections of Man.* Third Edition. Philadelphia: J. B. Lippincott Company, 1958.

Fairbrother, R. W. *A Textbook of Bacteriology.* Fifth Edition. New York: Grune & Stratton, 1948.

Netter, E., and D. R. Edgeworth. *Medical Microbiology for Nurses.* Philadelphia: F. A. Davis Company, 1957.

Parish, H. J. *Antisera, Toxoids, Vaccines and Tuberculins in Prophylaxis and Treatment.* Third Edition. Baltimore: Williams & Wilkins Company, 1955.

Schaub, I. G., *et al. Diagnostic Bacteriology.* Fifth Edition. St. Louis: C. V. Mosby Company, 1958.

Strean, L. P. *Oral Bacterial Infections.* Brooklyn, N. Y.: Dental Publishing Company, 1949. Pp. 95–99.

Tanner, F. W., and L. P. Tanner. *Food-borne Infections and Intoxications.* Second Edition. Champaign, Ill.: Garrard Press, 1953.

CHAPTER XXV

THE COLIFORM OR COLON–TYPHOID–DYSENTERY GROUP AND INTERMEDIATE FORMS

The members of the coliform or colon-typhoid-dysentery group, classified in the genera Escherichia, Aerobacter, Shigella, and Salmonella of the family Enterobacteriaceae, have stimulated more research in determinative methods than any other group of bacteria because of the biological and morphological similarity which make their differentiation and identification often very difficult. Special culture media, fermentation reactions, and specific agglutination tests are the most valuable means of identification. (See Chapter V on Culture Media.) Some species are widely distributed in nature and are solely saprophytic, others are specifically pathogenic, and some are occasionally pathogenic. A primary differentiation of practical but not absolute value is made on the basis of lactose fermentation. Formation of acid and gas within 24 hours usually indicates a nonpathogen colon-aerogenes organism. Those which do not ferment lactose are essentially pathogens.

In general, the coliform bacilli are Gram-negative, nonsporing, usually motile rods which ferment carbohydrates to a varying degree. These organisms may be normally present in the gastrointestinal tract or may invade it upon infection. The symptoms of infection with the pathogenic forms are gastrointestinal disturbances, and the organisms may be isolated from the feces. Specific agglutinins are produced against the infecting species.

COLON–AEROGENES GROUP
(GENERA ESCHERICHIA AND AEROBACTER)

Classified in the tribe Eschericheae, the genera Escherichia and Aerobacter are similar in morphological characteristics, but may be distinguished by physiological tests. *Escherichia coli* gives a positive methyl red test and a negative Voges-Proskauer test, whereas *Aerobacter aerogenes* gives a negative methyl red test and a positive Voges-Proskauer test. *Escherichia coli* produces indol and cannot utilize citrates as sole carbon source, whereas the reverse holds true for

Aerobacter aerogenes. Both organisms ferment dextrose and lactose with acid and gas formation and grow aerobically.

A. **Escherichia Coli (Bacillus Coli Communis)**. 1. Group—colon or Escherichia. 2. Family—Enterobacteriaceae. 3. Habitat—intestinal tract of man and animals; also water, milk, and soil. 4. Discoverers—Buchner, in 1885, and Escherich, in 1886.

Morphological Characteristics. 1. Form—short, plump rods, sometimes coccus-like. 2. Cell groupings—occurs singly, in pairs, or in short chains. 3. Size—0.5μ broad and 1 to 3μ long. 4. Staining properties—stains well with aniline dyes. 5. Gram-negative. 6. Nonmotile or motile with peritrichous flagella. 7. Nonsporing. 8. Usually nonencapsulated.

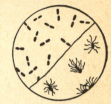

Fig. 48. *Escherichia coli*, showing peritrichous flagella.

Cultural Characteristics. 1. Agar cultures—circular colonies, which are smooth and white to yellowish-white; cultures have a peculiar fetid odor. 2. Agar streak—grayish colonies become visible in 12 to 18 hours; these gradually become more and more opaque. 3. Eosin methylene blue agar—moist, circular colonies about 2 to 3 mm. in diameter after 24 hours incubation at 37° C. These colonies have dark centers when examined by transmitted light. 4. Endo medium —circular, smooth, red colonies which have a metallic sheen when viewed by reflected light. The growth has a fetid odor. 5. Krumwiede's triple sugar medium—acid and gas are formed in the butt, and acid in the slant. 6. Wilson and Blair medium—no growth because of presence of brilliant green. 7. Selenite F and desoxycholate media —no growth. 8. Litmus milk—acid and gas formed, with coagulation of the medium. 9. Potato—profuse grayish to yellowish-brown growth. 10. Broth—rapid growth, with formation of a pellicle and a slight, slimy sediment. 11. Gelatin stab—good growth along line of inoculation, without liquefaction of the medium.

Physiological Characteristics. 1. Optimum temperature, 37° C. 2. Aerobic, and facultative anaerobe. 3. Nitrates are reduced to nitrites. 4. Indol is formed in peptone broth. 5. Methyl red test, positive. 6. Voges-Proskauer reaction, negative. 7. No hydrogen sulfide is produced. 8. Acid and gas ($CO_2:H_2::1:1$) are produced from dextrose, levulose, galactose, lactose, maltose, arabinose, xylose, rhamnose, and mannitol; variable in sucrose, raffinose, salicin, and dulcitol. Sucrose-fermenting strains are known as *Escherichia coli communior*. 9. Cannot use citrates as sole carbon source. 10. Eijkman test (fermentation of glucose at 46°C.), positive. 11. Cultures are characterized by fetid odor. 12. Resistance—60° C. kills in about 15 to 30 minutes.

Toxic Products.   *Escherichia coli* owes its toxicity, it is believed, to endotoxins, one of which may be enterotoxic.

Pathogenicity.   Ordinarily, *Escherichia coli* is present in the alimentary tract of man and other animals without pathogenic significance.   The organism has been known, however, to invade other parts of the body, most frequently the urinary tract where it causes a cystitis. The colon bacillus has been found to be of only slight pathogenicity for animals.

Serological Reactions.   Organisms of this genus are immunologically heterogeneous.

B. Aerobacter Aerogenes (Bacillus Aerogenes).   This organism was described by Escherich in 1885.   It is widely distributed in nature, found in milk, on grains, in sewage and water, and in the intestinal tract.   It is considered nonpathogenic.   It is frequently nonmotile and usually encapsulated.   In contrast to *Escherichia coli*, *Aerobacter aerogenes* is methyl red test negative, Voges-Proskauer test positive, can utilize citrates as the sole source of carbon, generally does not produce indol from peptone broth, and gives a negative Eijkman test. The organism ferments dextrose, galactose, lactose, levulose, arabinose, maltose, raffinose, cellobiose, salicin, starch, dextrin, glycerol, mannitol, sorbitol, and inositol with the production of acid and gas (CO_2:H_2::2:1).   Nitrates are reduced to nitrites.   Gelatin is not liquefied.   *Aerobacter aerogenes* grows aerobically at an optimum temperature of about 30° C.

C. Aerobacter Cloacae (Bacillus Cloacae, Aerobacter Liquefaciens).   In cultural, morphological, and physiological characteristics *Aerobacter cloacae* resembles *Aerobacter aerogenes*.   It is differentiated from the latter by its ability to liquefy gelatin.   It is found in human and animal feces, sewage, soil, and water.   The organism is of minor pathogenic significance.

D. Escherichia Freundii.   This organism, common in soil and water and frequently found in the normal human intestine, holds a position between *Escherichia coli* and Aerobacter.   It resembles the former in some respects, and the latter in being able to utilize citrates as the sole source of energy or carbon in an otherwise inorganic medium. It is sometimes given the generic name Citrobacter.   Like other organisms intermediate between the *coli* and *aerogenes* groups, it has been isolated in pathological processes.

DYSENTERY GROUP (GENUS SHIGELLA)

The genus Shigella, tribe Salmonelleae, is differentiated by its nonmotility and characteristic fermentation reactions. A number of the members cause intestinal disturbances, ranging from a mild, subclinical diarrhea to severe inflammation and ulceration of the large bowel, often with scar formation and stricture of the bowel after recovery (bacillary dysentery).

Shigella Dysenteriae (Bacillus Dysenteriae). 1. Group—dysentery or Shigella. 2. Family—Enterobacteriaceae. 3. Habitat—intestinal tract of bacillary dysentery cases and carriers. 4. Discoverer—Shiga, in 1898.

Morphological Characteristics. *Shigella dysenteriae* is a Gram-negative, nonmotile rod that is 0.4 to 0.6μ broad and 1 to 3μ long. It is nonsporing and is easily stained by aniline dyes.

Cultural Characteristics. The organism grows well on ordinary culture media. Its growth is inhibited by brilliant green.

Physiological Characteristics. 1. Optimum temperature, 37° C. 2. Aerobic, and facultative anaerobe. 3. Nitrates are reduced to nitrites. 4. Acid, but no gas, is produced from dextrose, levulose, raffinose, glycerol, and adonitol. The organism does not ferment arabinose, xylose, rhamnose, maltose, lactose, sucrose, salicin, mannitol, dulcitol, and isodulcitol.

Toxic Products. *Shigella dysenteriae* produces a powerful exotoxin. According to Olitsky and Kligler, this poison in small amounts produces typical paralysis and severe nerve lesions in rabbits after an incubation period of a few hours to 4 days. These investigators also isolated an endotoxin from the dysentery bacillus. This endotoxin when inoculated into animals produces loss of weight and diarrhea but no paralysis. The exotoxin of *Shigella dysenteriae*, when inoculated intravenously into rabbits in large doses, produces a rapid fall in temperature, marked respiratory failure, and severe diarrhea. It is believed that most of the lesions observed in the gastrointestinal tract of patients with dysentery are due to the toxin rather then to the direct local action of the bacilli.

Pathogenicity. *Shigella dysenteriae* produces in man a gastrointestinal disease accompanied by abdominal pain and diarrhea. As the disease progresses, the pains and diarrhea increase, and the stools become small in quantity and filled with mucus and flakes of blood. Marked nervous symptoms, attributable to the absorption of toxic products, are noted. The pathological findings are extensive ulcerations in the mucous membranes of the intestine. The bacteria may be found in the depths of the mucosa and the submucosa. Organisms are not found in the bloodstream as often as in typhoid fever.

Carriers. Persons may continue to excrete *Shigella dysenteriae* after convalescence. There are also normal, healthy carriers. Most individuals recovering from bacillary dysentery continue to excrete bacilli for a number of weeks; in a few instances they become chronic carriers. Sulfonamides have proved useful in curing carriers.

Prophylactic Vaccination. A number of vaccines have been employed as a prophylactic measure against bacillary dysentery. These consist of heat-killed preparations and serum sensitized cultures. Oral vaccination has also been advocated. Although favorable results have been reported, further experimentation is required.

Serum Treatment. Some investigators have advocated the use of monovalent and polyvalent sera in the treatment of dysentery, and have claimed good results. Bacteriophage, also, has been tried in a number of instances with a good outcome.

Related Organisms. In addition to *Shigella dysenteriae* there are a number of other organisms which have been isolated in cases of human dysentery. These microbes are morphologically related to, but much less toxic than, *Shigella dysenteriae* and can be differentiated by fermentation reactions. They include *Shigella paradysenteriae*, *Shigella ambigua*, *Shigella alkalescens*, *Shigella ceylonensis*.

TYPHOID AND PARATYPHOID GROUP (GENUS SALMONELLA)

Included in the genus Salmonella (tribe Salmonelleae) are Gram-negative, aerobic, motile rods, which ferment dextrose and many other carbohydrates with the production of acid and gas. They are lactose negative and salicin negative. The use of agglutination tests for analysis of antigenic structure and classification has been highly developed in this genus. The organisms have O and H antigen; some species have Vi antigen.

They are of particular importance to man because of their etiological role in food infections, producing a sudden, usually transient, stormy gastrointestinal disturbance after a short incubation period. Contaminated food, water, or milk may serve as the mode of transmission. Infection of the food may occur by means of human carriers, excreta of mice and rats, or flesh of infected animals which is eaten raw or improperly cooked. (See Chapter XIII.) Salmonella species may also produce paratyphoid fever, a slow, continued fever of the typhoid type. The typhoid organisms are also included in the Genus

Salmonella. These organisms were previously put in a separate genus, Eberthella. *Salmonella typhosa* (*Eberthella typhosa*) produces two types of disease: (1) a gastroenteritis similar to that produced by the pathogenic Shigella and Salmonella, and (2) a particular and generalized infection of the small intestine, typhoid fever, involving invasion of the blood stream.

Salmonella Typhosa (Eberthella Typhosa). 1. Group—typhoid or Eberthella. 2. Family—Enterobacteriaceae. 3. Habitat—stools of patients suffering from typhoid fever or carriers of the organism. 4. Discoverers—Eberth and Gaffky, in 1884.

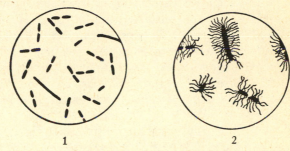

FIG. 49. (1) *Salmonella typhosa*, showing some long variant forms. (2) *Salmonella typhosa*; note peritrichous flagella.

Morphological Characteristics. 1. Form—short, plump rods. 2. Cell groupings—occurs singly, in pairs, occasionally in short chains. 3. Size—0.6 to 0.7μ in breadth, and 2 to 3μ in length. 4. Staining properties—stains well with ordinary aniline dyes. 5. Gram-negative. 6. Motile with peritrichous flagella, usually about 12 arranged peripherally. 7. Nonsporing. 8. Nonencapsulated.

Cultural Characteristics. 1. Agar cultures—colonies are grayish in color and transparent to opaque. 2. Agar slant—thin, glistening, bluish-gray growth. 3. Gelatin stab—thin, white growth, with no liquefaction. 4. Litmus milk—the medium is at first acid and then alkaline. 5. Potato—slight, moist growth, barely visible. 6. Broth —uniform turbidity, with a moderate sediment and a delicate pellicle in old cultures. 7. Endo medium—colonies are colorless. 8. Brilliant green agar—colonies have a snowflake appearance. 9. Desoxycholate citrate agar—colonies are gray, translucent, round, and glistening. 10. Selenite F medium—inhibits growth of all coliform organisms except the typhoid group. 11. Wilson-Blair medium— colonies of *Salmonella typhosa* appear black. 12. Krumwiede's triple sugar medium—acid is produced in the butt only.

Physiological Characteristics. 1. Optimum temperature, 37° C., with poor growth at 4° C. and no growth beyond 46° C. 2. Aerobic, and facultative anaerobe. 3. Nitrates are reduced to nitrites. 4. Hydrogen sulfide is produced. 5. Indol is not formed. 6. Acid is produced from dextrose, levulose, galactose, xylose, maltose, raffinose, dextrin, glycerol, mannitol, and sorbitol, but no fermentation of lactose, sucrose, inulin, rhamnose, inositol, and salicin is observed. 7. Resistance—the organism may be cultivated for a long period of time in artificial media. It is destroyed by a temperature of 56° C. in 20 minutes, by 5% phenol or 1:500 bichloride of mercury in 5 minutes.

Serological Reactions. Three antigens have been identified in *Salmonella typhosa*: (1) O or somatic group antigen, (2) H or flagellar specific antigen, present in motile varieties only, and (3) Vi or virulence antigen, which appears to have some relation to virulence. Isolated from active cases and carriers, Vi antigen is gradually lost on artificial media. Demonstration of agglutinins, as in the Widal test described below, in the serum of cases or convalescents is an important diagnostic procedure.

Widal Test. Originally a slide agglutination test, this is now performed macroscopically with both H and O antigens. The reaction usually becomes positive during the third week of typhoid fever; it is therefore of no value during the early days of the disease. Agglutination of a suspension of typhoid bacilli by serum in a dilution of 1:40 or 1:80 has usually been considered diagnostic. Recently, it has been suggested that a significant titer should be at least 1:100 for O antigen and 1:200 for H antigen. Positive reaction has been observed from three months to a year after convalescence or after vaccination against typhoid fever, resulting in confusing interpretations in vaccinated persons. The test may be applied to detection of carriers; at any one time about 75% of chronic carriers will give a positive reaction.

Pathogenicity. *Salmonella typhosa* is the cause of typhoid fever, an acute, generalized infection transmitted by water, milk, food, and flies. The organisms enter through the alimentary canal, and the body tries to localize them in Peyer's patches, lymph tissue nodules in the small intestine. The invaders, however, usually escape into the bloodstream and the small intestines. An irritation of the walls of the gastrointestinal tract is caused, with the formation of ulcers and sometimes the production of diarrhea. The organism may be isolated from the feces during and after the second week. It is also demonstrable in large numbers in the spleen and liver of those dying of typhoid fever. During the first week of the disease a blood culture is usually positive, and cultures made from the rose spots characteristic of the disease reveal numerous typhoid bacilli.

When inoculated intraperitoneally or intravenously into laboratory animals *Salmonella typhosa* may produce a fatal bacteremia. The disease itself has never been transmitted to a laboratory animal.

Carriers. Typhoid fever is an outstanding example of disease which may be transmitted by individuals who to all appearances are normal and healthy, but are dangerous because they harbor infectious agents. Many epidemics of typhoid fever have been traced to persons who excrete typhoid bacilli in their feces, although they show no outward manifestations of typhoid fever. The most notable was Typhoid Mary, whose death on November 11, 1938, brought to an end a long series of outbreaks of the disease which followed wherever she was employed. About one-third of the cases discharge bacilli for 3 weeks after onset; 10%, for 8–10 weeks (convalescent carriers). Some may continue to excrete the organisms for several months, years, or their lifetime.

Repeated fecal examination, Widal testing, and determination of Vi agglutinins in the blood are used to detect carriers. Of the methods that have been proposed for their cure, chemotherapy and phage therapy have been unsuccessful; vaccine therapy and removal of the gall bladder (believed to be a typhoid bacillus reservoir) have been moderately successful. Chloromycin is the first antibiotic to prove effective against typhoid fever.

Immunity. Following an attack of typhoid fever, there is usually an immunity which lasts for a period of years. About 2% are subject to a second attack.

Prophylactic Vaccination. Since its introduction by Wright in 1898, vaccination against typhoid fever by parenteral inoculation of killed organisms has been widely used with favorable results. Compulsory vaccination in the United States Army since 1911 has resulted in the virtual disappearance of the disease among our troops.

Combined vaccines against paratyphoid fever and typhoid fever have been used, especially during wartime when the danger of exposure to enteric fevers is very great. These vaccines, of which T.A.B. or triple vaccine has been used most extensively, consist of various proportions of *Salmonella typhosa*, *Salmonella paratyphi* (paratyphoid fever type A), and *Salmonella schottmuelleri* (paratyphoid fever type B).

Mode of Transmission. Outbreaks of typhoid fever have been traced most often to water and milk supplies contaminated with

Salmonella typhosa. Other means of transmission are flies, infected shellfish, infected food, fomites, and carriers.

Public Health. The most important measure in preventing spread of the disease is immediate isolation of the patient and proper disposal of articles handled by him and any of his dejecta. Persons exposed to typhoid fever should be held in quarantine. For those having the disease the period of isolation extends until three successive negative stools have been obtained. The stools of convalescent patients and carriers must be properly disinfected before disposal, and care must be taken to prevent them from contaminating food, milk, or water supply.

Salmonella paratyphi (Bacillus paratyphosus A) produces paratyphoid fever. The organism is differentiated by its fermentation reactions, especially inability to ferment xylose, and serological specificity.

Salmonella schottmuelleri (Bacillus paratyphosus B) also causes paratyphoid fever, indistinguishable from the disease caused by *Salmonella paratyphi* except by identification of the invading organism by specific agglutination tests.

Salmonella hirschfeldii (Bacillus paratyphosus C) is found in enteric fevers in Africa and Asia. The disease is essentially a bacteremia without intestinal involvement. The organism is closely related biologically to *Salmonella choleraesuis.*

Salmonella typhimurium (Salmonella aertrycke) is the species most commonly isolated from food infection outbreaks. It is also pathogenic for laboratory and domestic animals and birds.

Salmonella enteritidis (Gärtner's Bacillus) was the first Salmonella species to be linked to food infection, in which it is now known to play an important role. It resembles *Salmonella schottmuelleri,* from which it may be differentiated by serological reactions.

Salmonella choleraesuis (Bacillus suipestifer) was once believed to be the cause of hog cholera, now known to be due to a virus. The bacillus is normally present in the intestinal tract of pigs, and becomes a secondary invader during the disease. It is sometimes isolated from cases of human gastroenteritis.

Salmonella pullorum causes an enteric disease of young chicks called white diarrhea. It may also infect the ovaries and eggs of adult birds, effecting transmission of the disease to the offspring.

Salmonella abortivoequina causes abortion of pregnant mares, guinea pigs, rabbits, goats, and cows.

Salmonella abortusovis produces abortion in sheep.

A Method for Differentiation of Coliform and Related Organisms

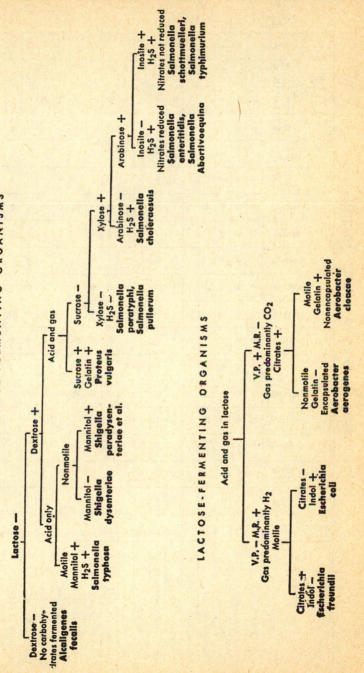

NON-LACTOSE-FERMENTING ORGANISMS

LACTOSE-FERMENTING ORGANISMS

Lactose —

Dextrose —
No carbohy-
drates fermented
**Alcaligenes
fecalis**

Dextrose +

Acid only

Motile
Mannitol +
H₂S +
**Salmonella
typhosa**

Nonmotile

Mannitol —
**Shigella
dysenteriae**

Mannitol +
**Shigella
paradysen-
teriae et al.**

Acid and gas

Sucrose +
Gelatin +
**Proteus
vulgaris**

Sucrose —

Xylose —
H₂S —
**Salmonella
paratyphi,
Salmonella
pullerum**

Xylose +

Arabinose +
H₂S +
**Salmonella
choleraesuis**

Arabinose +

Inosite —
H₂S —
Nitrates reduced
**Salmonella
enteritidis,
Salmonella
Abortivoequina**

Inosite +
H₂S +
Nitrates not reduced
**Salmonella
schottmuelleri,
Salmonella
typhimurium**

Acid and gas in lactose

V.P. — M.R. +
Gas predominantly H₂
Motile

Citrates —
Indol —
**Escherichia
freundii**

Citrates —
Indol +
**Escherichia
coli**

V.P. + M.R. —
Gas predominantly CO₂
Citrates +

Nonmotile
Gelatin —
Encapsulated
**Aerobacter
aerogenes**

Motile
Gelatin +
Nonencapsulated
**Aerobacter
cloacae**

DIFFERENTIAL REACTIONS OF ENTERIC ORGANISMS

Organism	M.R.	V.P.	Indol Production	Citrates	Lactose	Dextrose	Sucrose	Mannitol	Xylose	Arabinose	Inositol	Salicin	Maltose	Nitrate Reduction	Gelatin Liquefaction	Milk	Double Sugar Medium Butt	Double Sugar Medium Slant	Motility	H₂S Production
Escherichia coli	+	-	+	-	●	●	⊕	●	●	●	-	⊕	●	+	-	⊕c	⊕	+	+	-
Aerobacter aerogenes	-	+	-	+	●	●	●	●	●	●	⊕	⊕	●	+	-	⊕c	⊕	+	-	-
Aerobacter cloacae	-	+	±	+	⊕	⊕	⊕	⊕	⊕	⊕	+	⊕	⊕	+	+		⊕	+k	+	-
Escherichia freundii	+	-	-	+	⊕	+		+	±				⊕		-		+		+	
Eberthella typhosa	+		-		-	+	-	+	±	-	-	-	+	+	-	+k	+	k	+	+
Shigella dysenteriae	+		-	-	-	+	-	-	-	-	-	-	-	+	-	+	+	k	-	-
Shigella paradysenteriae	+		±	-	-	+	±	±	±	±	±	-	±	+	-	+	+	k	-	-
Shigella ambigua	+		+	-	-	+	-	-	-	-	-	-	-	+	-	+	+	k	-	-
Shigella alkalescens			+		-	+	-	+	-	-	-	-	+	+	-		+	k	-	-
Salmonella paratyphi			-		-	⊕	-	⊕	⊕	⊕	-	-	⊕	+	-		⊕	k	+	-
Salmonella schottmuelleri			-		-	⊕	-	⊕	⊕	⊕	⊕	-	⊕	-	-	+k	⊕	k	+	+
Salmonella hirschfeldii			-		-	⊕	-	⊕	⊕	⊕	-	-	⊕		-	+k	⊕	+k	+	+
Salmonella enteriidis			-		-	⊕	-	⊕	⊕	⊕	⊕	-	⊕	+	-	+k	⊕	+k	+	+
Salmonella choleraesuis			-		-	⊕	-	⊕	⊕	⊕	-	-	⊕	-	-	+k	⊕	+k	+	±
Salmonella typhimurium			-		-	⊕	-	⊕	⊕	-	-	-	⊕	-	-	+k	⊕	+k	+	+
Salmonella abortus equi			-		-	⊕	-	⊕	⊕	⊕	-	-	⊕	+	-	+k		+k	+	+
Salmonella pullorum			-		-	⊕	-	⊕	-	⊕	-	-	-		-	+k			-	-

Legend: +, acid formation or positive reaction. ●, acid and gas formation. ⊕, variable (acid and gas or negative). c, coagulation. k, alkaline. + k, acid to alkaline. -, no change. ±, variable (acid or negative).

BACTERIOLOGICAL DIAGNOSIS

Laboratory diagnosis of typhoid, dysentery, and related diseases usually involves some or all of the following procedures.

Blood Culture. An invasion of the bloodstream occurs regularly in the early stages of typhoid fever, less often in Shigella and Salmonella infections, and probably only rarely with paracolon organisms. Blood cultures are routinely made in the diagnosis of typhoid fever during the first week after onset. The specimen is taken in bile broth, from which selenite agar plates are seeded.

Feces and Urine Culture. 1. *Selective Media*—Fluid selenite media may be used for partial inhibition of large numbers of organisms such as *Escherichia coli*. *Salmonella typhosa* grows well, but the medium is somewhat toxic to Salmonella and Shigella. From the growth obtained, selective agar plates are seeded.

2. *Direct Culture*—Fresh stool is streaked on selective agar medium, e.g., desoxycholate citrate agar, S–S agar, MacConkey's agar.

Metabolic tests, e.g., fermentation reactions and H_2S and indol production, and motility tests are made on subcultures of pure cultures obtained from above.

Agglutination studies are also made on the patient's serum. A marked increase of O agglutinins is especially characteristic of the group. If there is also a marked increase in H agglutinins, it is usually possible to determine the species.

PARACOLON GROUP

This is a large, heterogeneous group of organisms, culturally and serologically complex, occupying a position intermediate between the Escherichia-Aerogenes and the Salmonella-Shigella groups. Sometimes referred to as "aberrant coliforms," their position is not now clearly understood, and they have not attained generic status. A frequent property is the inability to ferment lactose, or a much delayed fermentation. The paracolon organisms are frequently found in food infections and are often mistaken for Salmonella. They have been placed in a separate genus, Paracolobactrum, by Borman, Wheeler, and Stuart, who reported the type species *Paracolobactrum aeronogenoides*, and other species *Paracolobactrum intermedium* and *Paracolobactrum coliforme*. They have been isolated in human gastroenteritis and in the urine of patients with genito-urinary infections. Habitats include intestinal tracts of animals, soils, surface water, and grains.

BIBLIOGRAPHY

Anderson, G. W., and M. G. Arnstein. *Communicable Disease Control.* Third Edition. New York: Macmillan Company, 1953.

Anderson, W. A. *Pathology.* Second Edition. St. Louis: C. V. Mosby Company, 1953.

Bigger, Joseph W. *Handbook of Bacteriology.* Sixth Edition. Baltimore: Williams & Wilkins Company, 1950.

Burrows, W. *Textbook of Microbiology.* Eighteenth Edition. Philadelphia: W. B. Saunders Company, 1963. Pp. 538–593.

Dack, G. M. *Food Poisoning.* Third Edition. Chicago: University of Chicago Press, 1956.

Dubos, R. J. *Bacterial and Mycotic Infections of Man.* Third Edition. Philadelphia: J. B. Lippincott Company, 1958.

Frazier, W. C. *Food Microbiology.* New York: McGraw-Hill Book Company, 1958.

Gardner, A. D. *Bacteriology for Medical Students and Practitioners.* New York: Oxford University Press, 1944. Pp. 74–93.

Jawetz, E., J. L. Melnick, and E. A. Adelberg. *Review of Medical Microbiology.* Los Altos: Lange Medical Publications, 1962. Pp. 168–177.

Kolle, W., and H. Hetsch. *Experimental Bacteriology in Its Application to the Diagnosis, Epidemiology, and Immunology of Infectious Diseases,* Vol. I. New York: Macmillan Company, 1935. Pp. 235–309.

Ordinance and Code Relative to Eating and Drinking Establishments. U. S. Public Health Service Publication No. 37. Washington, D. C.: Government Printing Office, 1950.

Parish, H. J. *Antisera, Toxoids, Vaccines and Tuberculins in Prophylaxis and Treatment.* Third Edition. Baltimore: Williams & Wilkins Company, 1955.

Siler, J. F. *et. al. Immunization to Typhoid Fever.* Baltimore: Johns Hopkins Press, 1941.

Standard Methods for the Microbiological Examination of Foods. New York: American Public Health Association, 1958.

Strean, L. P. *Oral Bacterial Infections.* Brooklyn, N. Y.: Dental Publishing Company, 1949. Pp. 74–83.

Suckling, E. V. *Examination of Waters and Water Supplies.* Philadelphia: The Blakiston Company, 1943. Pp. 413–595.

Tanner, F. W., and L. P. Tanner. *Food-borne Infections and Intoxications.* Second Edition. Champaign, Ill.: Garrard Press, 1953.

Umbreit, W. W. *Modern Microbiology.* San Francisco: W. H. Freeman and Company, 1962. Pp. 448–457.

Walter, W. G., and R. H. McBee. *General Microbiology.* Second Edition. Princeton, N. J.: D. Van Nostrand Company, 1962. Pp. 161–162.

CHAPTER XXVI

THE ANAEROBIC SPORE–FORMERS
(GENUS CLOSTRIDIUM)

The members of the genus Clostridium, family Bacillaceae, are the obligate anaerobic spore-formers, which readily decompose protein media and ferment carbohydrates. They are usually Gram-positive and non-acid-fast. This group includes many soil anaerobes, some of which are pathogenic, such as *Clostridium botulinum*, *Clostridium tetani*, *Clostridium perfringens*, *Clostridium novyi*, and *Clostridium feseri*. These organisms are of primary importance in food poisoning, wound infections, and gas gangrene. The soil anaerobes which cause putrefaction include *Clostridium sporogenes* and *Clostridium lentoputrescens*.

METHODS OF CULTIVATING ANAEROBES

Exclusion of Air.

1. *Heat.*

(a) Drive off oxygen from medium by boiling or autoclaving for 10 minutes.

(b) Cool rapidly in ice water to keep oxygen absorption to a minimum.

(c) Inoculate; cover with layer of petroleum jelly or melted agar.

This method does not provide complete anaerobiosis, and is messy if gas-formers are present and blow the jelly or agar plug out of the tube.

2. *Roux's Method.*

(a) Suck inoculated gelatin or agar into narrow tubes and then close both ends by fusing in the flame.

(b) Incubate until growth occurs, break, and recover organism by fishing.

3. *Wright's Method.*

(a) Insert a short piece of glass tubing, constricted at both ends and fitted at each end with a small piece of soft rubber tubing, into a test tube of broth.

(b) Connect upper end of inserted glass tubing by the rubber with a pipette passed through the cotton plug in the tube.

(c) After inoculation of the medium, suck the fluid up until the glass tubing is completely filled.

(d) Prevent downflow of fluid by placing finger over pipette or by constricting the small piece of rubber tubing attached to upper end of pipette.

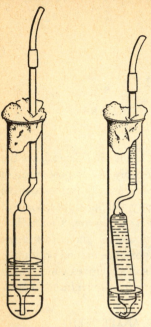

FIG. 50. Wright's method.

(e) Push entire system of tubes downward so that both pieces of rubber tubing attached to the ends of the glass chamber are bent.

This is a simple and effective method for cultivating anaerobic bacteria in fluid media.

4. *Deep Agar.*

(a) Melt dextrose infusion agar in tubes 8–10 cm. deep; cool to 50° C.

(b) Inoculate and mix thoroughly.

(c) Solidify rapidly in cold water and incubate.

(d) To isolate colonies: heat tube carefully in hot flame to melt layer of agar in direct contact with glass, place flamed mouth of tube in sterile covered dish, flame bottom of tube to force agar out of tube by steam, and slice agar to reach desired colonies.

Strict anaerobes grow only in the depths. Less strict anaerobes grow in the depths and also somewhat nearer to the surface. Facultative anaerobes grow on the surface and in the depths. Microaerophils may grow in a narrow zone some distance below the surface. This method is quite useful, but the method of removal of the column is awkward.

5. *Deep Agar—T'ung's Modification.*

(a) Close vertical glass cylinder (about 15 cm. × 1.8 cm.) at bottom end with rubber stopper into which hook-shaped glass rod has been inserted.

(b) Place agar in tube, plug open end with cotton, and autoclave upright.

(c) Inoculate as shake tube.

(d) To remove column: withdraw rubber stopper gently.   Glass hook withdraws the agar with it.

Displacement of Air.

1. *Novy Jar Method*

(a) Set inoculated plates on wire frame about an inch above bottom of jar; set cover tightly in place.

(b) Exhaust air by suction pump.

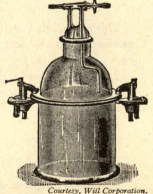

Courtesy, Will Corporation.

FIG. 51.   Anaerobic culture jar (Novy, Improved). For plate cultures of anaerobic bacteria by vacuum, gas, or pyrogallate methods.

(c) By means of double stopcock at top, admit hydrogen from Kipp generator.

(d) Repeat process of alternate exhaustion and admission of hydrogen several times.

Reducing Agents or Oxygen Exhaustion.

1. *McIntosh and Fildes' Method.*

(a) Place cultures in a sealed jar in which a small amount of asbestos impregnated with palladium black (catalyst) is suspended.

(b) Heat palladinized asbestos immediately before closing the jar by passing an electric current through a surrounding coil of nichrome wire. (*Brewer's modification* eliminates danger of explosion from sparks by enclosing heating element in gas-tight tube inside the catalytic mass.)

(c) Slowly introduce hydrogen, which combines with the oxygen present to form water. This is absorbed by calcium chloride or other drying agent.

This method provides complete anaerobiosis.

2. *Buchner's Method*, based on absorption of oxygen by alkaline solutions of pyrogallol.

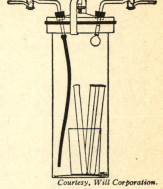

Courtesy, Will Corporation.

Fig. 52. Anaerobic culture apparatus (Smillie). The oxygen is completely removed by means of the catalytic action of palladinized asbestos upon oxygen and hydrogen.

(a) Place small wire or glass holder within a large test tube.
(b) Drop dry pyrogallol into bottom of tube.
(c) Add 30% sodium hydroxide solution.
(d) Insert smaller test tube containing inoculated culture medium into the larger tube and stopper latter tightly.

Courtesy,
Arthur H. Thomas Co.

Fig. 53. Spray's anaerobic apparatus.

3. *Spray's Method.*

(a) Place pyrogallol and sodium hydroxide solution in separate shallow compartments in bottom of Spray apparatus.
(b) Invert inoculated agar plate over top.
(c) Seal joint with paraffin or plastic.
(d) Tilt vessel slightly to mix solutions.

4. *Tissue Method.*

(a) Addition of chopped beef heart, brain, fish, or other tissues which act as reducing agents to a liquid medium permits cultivation of anaerobes without precautions for excluding air.

5. *Brewer's Method.*

(a) Infusion or blood agar base with 0.2% sodium thioglycollate (or similar compound having affinity for oxygen) is inoculated in bottom half of ordinary Petri dish.

FIG. 54. Cross section of Brewer cover for anaerobic Petri dish culture.

(b) Cover with Brewer cover, which has most of the central portion depressed so that there is only a very thin layer of air above the agar. This air is sealed in by a deeper, ringed depression in the cover which actually contacts the agar.

6. *Use of Aerobic Bacteria.*

(a) Streak one-half of an agar plate with a culture of an anaerobe; the other half, with a culture of a facultative aerobe.

(b) Replace cover, and seal edges with modeling clay. Incubate in inverted position.

Utilization of oxygen and elimination of carbon dioxide by the facultative aerobe reduces oxygen tension to a level that permits growth of the anaerobe, usually within 24–48 hours.

CLOSTRIDIUM TETANI

Clostridium Tetani (Bacillus Tetanus). 1. Group—anaerobic spore-former. 2. Family—Bacillaceae. 3. Habitat—widely distributed in upper layers of soil and rich cultivated lands, also in feces of herbivora, particularly horses. 4. Discoverer—Nicolaier, in 1884.

Morphological Characteristics. 1. Form—straight, round-ended bacilli. 2. Cell groupings—occurs singly, in pairs, and in long chains and filaments. 3. Size—0.4 to 0.6μ in breadth and 2 to 5μ in length. 4. Spores—spherical, terminal, and broader than the cell, giving appearance of drumsticks. 5. Staining properties—stains well with ordinary aniline dyes and spore stains. 6. Gram-positive and non-acid-fast. 7. Motile, with peritrichous flagella. 8. Nonencapsulated.

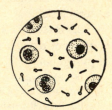

FIG. 55. *Clostridium tetani.* Tack-head bacillus with and without enlarged terminal spores, taken from an infected wound.

Cultural Characteristics. The organism grows well in ordinary culture media under anaerobic conditions. 1. Agar streak—thin transparent film. 2. Gelatin—the medium is liquefied and blackened. 3. Serum agar—small, villous, transparent colonies. 4. Anaerobic agar plate cultures—colonies give appearance of light fleecy clouds. 5. Blood agar—hemolysis occurs. 6. Litmus milk—casein is slowly precipitated. 7. Cooked meat medium—slightly turbid growth, with some gas formation. 8. Coagulated albumin—the medium is slowly liquefied. 9. Brain medium—slow digestion and blackening of the medium.

Physiological Characteristics. 1. Optimum temperature, 37° C. 2. Anaerobe. 3. Indol is formed. 4. Nitrates are not reduced to nitrites. 5. Proteolytic enzymes are produced. 6. No acid or gas is formed in carbohydrate media. 7. Resistance—in dry state the spores can withstand a temperature of 80° C. for 1 hour, and boiling for 15 to 70 minutes. All spores are killed by exposure to dry heat at 160° C. for 1 hour or to steam under pressure at 120° C. for 20 minutes. A solution of 1% silver nitrate is lethal in 1 minute; 5% carbolic acid, in 12 hours; and a 1% solution of bichloride of mercury, in 2 or 3 hours.

Toxic Substances. *Clostridium tetani* produces two poisonous substances: (1) tetanospasmin, a true toxin, one of the most potent poisons known, affecting the nervous tissue, and (2) tetanolysin, a hemolysin.

Tetanospasmin is best produced in a neutral or slightly alkaline beef infusion bouillon to which 0.5% sodium chloride and 1% peptone have been added. The toxin is destroyed by heat, sunlight, and oxidation. A powerful antitoxin may be produced in horses by the injection of the toxin.

Pathogenicity. Likelihood of infection with *Clostridium tetani* depends on the nature of the wound and the presence of other microorganisms. Thus, deep, lacerated wounds in which there has been considerable tissue destruction predispose toward tetanus. Compound fracture and gunshot wounds are especially suitable for the development of tetanus bacilli, since in these instances much tissue has been destroyed and the spores can gain a foothold. Tetanus is essentially an intoxication. The bacilli do not circulate in the blood, but elaborate toxins at the point of entry, and these find their way into the blood.

Following the injection of tetanus toxin subcutaneously into guinea pigs there is usually an incubation period of 8 to 24 hours before toxic spasms occur. Spasms are first noted in the muscles nearest the point of inoculation. When the toxin is given intravenously, there is general tetanus of all the muscles.

In man, following an incubation period which may vary from 5 to 50 days, the first symptoms appear. There is usually headache and general depression followed by difficulty in swallowing and in opening the mouth, due to spasms of the masseters. There is slight stiffness of the neck and gradually a spasm of the muscles of the cheeks develops. The spasms spread to the trunk and back.

Prophylaxis, Therapy, and Immunity. When a wound is much lacerated or contused and does not bleed freely, an effective protection against the development of tetanus is afforded by the use of antitoxin.

Although antitoxin is more effective as prophylaxis than as therapy, it probably is of some value when administered early in clinical cases.

The addition of 0.4% formalin to tetanus toxin produces, after incubation for 4 to 6 weeks at 37° C., a toxoid similar to diphtheria toxoid, which has been found to produce an effective immunity against tetanus. Three doses each of 1 ml. given at intervals of 3 weeks are considered capable of producing a high antitoxin titer in the blood of human beings. The injections are harmless and the immunity is considered to last from 1 to 5 years. The injection of a supplementary dose of 2 ml. of anatoxin a year after vaccination is advocated to reinforce the titer of antitoxin.

It has been reported that alum-precipitated toxoid given in two injections of 1 ml. each, 2 to 4 weeks apart, produces a greater and more rapid immunity than does toxoid, and with fewer reactions.

Bacteriological Diagnosis. 1. *Film preparations* are made from suspected materials and examined under the microscope for typical spores of *Clostridium tetani*.

2. *Media are inoculated* with the material and incubated at 37° C. for 72 hours under anaerobic conditions. The cooked meat medium, and blood and plain agar plates, are examined for the isolation of pure colonies.

3. *A guinea pig is inoculated* subcutaneously with a portion of the original material, or of the heated culture. Sterile emery dust is mixed with the inoculum in order to produce suitable conditions for the development of the tetanus bacilli if present. The development of tetanus in the inoculated guinea pig and its death in 1 to 4 days indicates the presence of *Clostridium tetani*, if a control animal protected by an intraperitoneal injection of tetanus antitoxin remains well.

CLOSTRIDIUM BOTULINUM

Clostridium Botulinum (Bacillus Botulinus). 1. Group—anaerobic spore-former. 2. Family—Bacillaceae. 3. Habitat—soil, decayed vegetables. 4. Discoverer—Van Ermengem, in 1896.

Morphological Characteristics. 1. Form—large bacillus with rounded ends. 2. Cell groupings—occurs singly, in pairs, and in short chains; occasionally long chains. 3. Size—0.5 to 0.8μ in breadth and 3 to 8μ in length. 4. Spores—oval and subterminal. 5. Staining properties—stains well with ordinary aniline dyes and spore stains. 6. Gram-positive and non-acid-fast. 7. Motile, with peritrichous flagella.

Cultural Characteristics. The organism grows well anaerobically. 1. Gelatin—liquefaction of the medium. 2. Agar cultures—grayish-yellow, flat, filamentous colonies. 3. Agar slant—translucent growth, grayish-yellow in color. 4. Glucose agar shake—opaque colonies with brownish centers, and much gas formation. 5. Coagulated albumin—poor growth without liquefaction. 6. Brain medium—not blackened or digested. 7. Cooked meat medium—good growth, with production of gas and blackening and digestion of the medium. 8. Milk—slight increase in acidity with no coagulation or gas formation. 9. Blood agar—alpha type of hemolysis.

Physiological Characteristics. 1. Optimum temperature, 35° to 37° C. 2. Anaerobe. 3. Acid and gas are produced from dextrose, levulose, maltose, dextrin, glycerol, and inositol, but sucrose, lactose, inulin, and galactose are not fermented. 4. Indol is not produced. 5. Nitrates are not reduced to nitrites. 6. Ammonia is produced. 7. Hydrogen sulfide is produced. 8. Resistance—the spores are highly resistant; they withstand dry heat at 180° C. for 15 to 30 minutes and moist heat at 100° C. for 3 to 5 hours.

Toxic Substances. *Clostridium botulinum* produces hemolysin active against human and horse red blood cells. At a temperature of 37° C. under anaerobic conditions a powerful toxin, the most potent bacterial toxin known, is produced. This toxin is not affected by gastric juices and requires a temperature of 80° C. for 30 minutes or 100° C. for 10 minutes for destruction. The toxin is found in infected meats, sausage, and canned goods, and the ingestion of such contaminated foodstuffs results in a food poisoning disease of human beings and animals called botulism. The toxin is so potent that less than 0.00001 ml. is a fatal dose for a 250-gram guinea pig.

Pathogenicity. Believed due to ingestion of preformed toxin, the disease is highly fatal. Toxin is absorbed directly from the stomach and intestine, and affects the nerve-muscle complex. Usually within 24 hours after eating the contaminated food, the individual complains of general weakness, lassitude, and headache. Ocular paresis and pharyngeal paralysis usually develop. (See also Chapter XIII.)

Various Types. There are five types of *Clostridium botulinum*, each producing an immunologically different toxin. Types A, B, and E affect man; Types C and D, lower animals.

Prevention and Therapy. Botulism may be prevented through sterilization, processing, or cooking of all canned, smoked, or dried foods. The most important precaution is boiling for 20 minutes all home-canned foods after opening. Antitoxin appears to be of little therapeutic value.

CLOSTRIDIUM PERFRINGENS

Clostridium Perfringens (Clostridium Welchii, Bacillus Welchii, Bacillus Aerogenes Capsulatus). 1. Group—anaerobic spore-former. 2. Family—Bacillaceae. 3. Habitat—dust, soil, sewage, water, fish, mollusks, milk, cheese, and feces of animals and man. 4. Discoverers—Welch and Nuttall, in 1892.

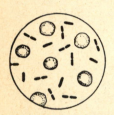

Fig. 56. *Clostridium perfringens* in blood, together with red blood corpuscles.

Morphological Characteristics. 1. Form—plump, thick-set rods. 2. Cell groupings—occurs singly or in pairs. 3. Size—1 to 1.5μ in breadth and 4 to 8μ in length. 4. Spores—oval and centrally located, rarely subterminal, and do not swell the rod in which they are formed; they are formed sparingly and only in the absence of fermentable carbohydrates. 5. Staining properties —stains readily with ordinary aniline dyes and spore stains. 6. Gram-positive and non-acid-fast. 7. Nonmotile. 8. Encapsulated.

Cultural Characteristics. 1. Agar cultures—moist, circular, slightly raised colonies with opaque center and entire edge. 2. Agar streak—compact, opaque, white or grayish-white, biconvex discs. 3. Gelatin stab—liquefaction and blackening of the medium. 4. Blood agar—zone of hemolysis in 24 hours. 5. Potato—thin, grayish-white growth, with gas formation. 6. Broth—marked turbidity, with abundant gas formation. 7. Cooked meat medium— no digestion of the medium but rapid growth with profuse gas formation. 8. Brain medium—not blackened or digested. 9. Litmus milk—produces characteristic stormy fermentation with evolution of large amount of gas and coagulation.

Physiological Characteristics. 1. Optimum temperature, 35° to 37° C. 2. Anaerobe. 3. Indol is not produced. 4. Nitrates are reduced to nitrites. 5. Acid and gas are produced from dextrose, levulose, galactose, mannose, maltose, lactose, sucrose, starch, and xylose. 6. Methyl red test, positive.

Toxic Substances. An exotoxin, which stimulates the production of a specific antitoxin, is formed by *Clostridium perfringens*. The toxin is thermolabile and is destroyed by exposure to 70° C. for 30 minutes.

Various Types. There are four types of *Clostridium perfringens*. This division is made on the basis of fermentation reactions in inulin and glycerol. Inulin and glycerol are fermented by Type I, only inulin by Type II, only glycerol by Type III, and neither inulin nor glycerol by Type IV. Antitoxin against one type may be used effectively against the toxins of the other types.

Pathogenicity. *Clostridium perfringens* is the most frequent cause of gas gangrene in man and plays a significant role in war wound complications. Tissue injury is a usual and perhaps essential preliminary to infection. Once the organisms are established, however, they rapidly invade the surrounding tissue. Injected intravenously into a rabbit which is killed after 5 minutes and kept in the incubator for 5 to 8 hours, *Clostridium perfringens* produces marked gas formation which is characteristic for the organism. The gas bubbles will be found distributed throughout the organs, especially in the liver. The organism can be isolated from the liver and heart's blood. The intramuscular injection of 1 ml. of a broth culture of *Clostridium perfringens* into guinea pigs causes death of the animals in 18 to 24 hours. At autopsy there is marked necrosis of the tissue with extensive gas formation and foul acid odor. The organism may be isolated from the heart's blood.

Prophylaxis and Therapy. Various antitoxic sera have been prepared, and claims are made that these are of therapeutic and prophylactic value.

CLOSTRIDIUM HISTOLYTICUM

Clostridium Histolyticum (Bacillus Histolyticus). 1. Group—anaerobic spore-former. 2. Family—Bacillaceae. 3. Habitat—feces and soil. 4. Discoverers—Weinberg and Seguin, in 1916.

Morphological Characteristics. 1. Form—rods with rounded ends. 2. Cell groupings—occurs singly and in pairs. 3. Size—0.5 to 0.7μ in breadth and 3 to 5μ in length. 4. Spores—oval and subterminal, causing swelling. 5. Gram-positive and non-acid-fast. 6. Motile, with peritrichous flagella.

Cultural Characteristics. 1. Agar slant—slight growth of tiny, smooth, discrete colonies. 2. Gelatin—liquefied. 3. Blood agar—hemolysis of the medium. 4. Broth—turbid growth with slight precipitate. 5. Coagulated albumin—liquefied. 6. Brain medium—digestion and blackening of the medium. 7. Cooked meat medium—digestion of medium with production of offensive odor. 8. Litmus milk—slow digestion, with coagulation.

Physiological Characteristics. 1. Grows well at 37° C. 2. Indol is not produced. 3. Nitrates are not reduced to nitrites. 4. Carbohydrates are not fermented. 5. Hydrogen sulfide is produced.

Pathogenicity. The organism has been found in mixed infections in war wounds. *Clostridium histolyticum* is intensely proteolytic, producing active necrosis of muscle tissue. A toxin is formed which is responsible for the lesions. The intramuscular injection of 2 to 3 ml. of culture into guinea pigs causes rapid digestion of the muscle tissue. Although there is marked destruction of tissue the animal appears well.

CLOSTRIDIUM LENTOPUTRESCENS

Clostridium Lentoputrescens (Clostridium Putrificum, Bacillus Putrificus). 1. Group—anaerobic spore-former. 2. Family—Bacillaceae. 3. Habitat—intestinal canal of man; widely distributed in soil. 4. Discoverer—Bienstock, in 1884.

Morphological Characteristics. 1. Form—rods with rounded ends. 2. Cell groupings—occurs singly, in pairs, and in chains. 3. Size—0.4 to 0.6μ in breadth and 7 to 9μ in length. 4. Spores—spherical, terminal, and large, causing swelling. 5. Gram-positive and non-acid-fast. 6. Motile, with peritrichous flagella.

Cultural Characteristics. 1. Agar cultures—small, circular colonies which develop a ground-glass appearance. 2. Gelatin stab—liquefied in 5 to 6 days. 3. Blood agar—hemolysis of medium. 4. Litmus milk—acid coagulation, followed by slow peptonization. 5. Broth—turbid growth, with putrid odor. 6. Brain medium—blackened and digested. 7. Cooked meat medium—digestion of medium with production of bad odor. 8. Blood serum—liquefied, and gas produced.

Physiological Characteristics. 1. Optimum temperature, 35° to 37° C. 2. Anaerobe. 3. No action on carbohydrates. 4. Hydrogen sulfide is produced. 5. Indol may or may not be formed. 6. Nitrates are not reduced to nitrites.

Pathogenicity. The organism is usually nonpathogenic. It was first isolated from the intestine of a cadaver, and it has also been associated with cases of gas gangrene.

RELATED ORGANISMS

The anaerobic spore-formers may be divided into three groups:

1. **Pathogenic for Man.** These include: (a) *Clostridium perfringens*, cause of gas gangrene; (b) *Clostridium tetani*, cause of tetanus;

(c) *Clostridium botulinum,* cause of botulism; (d) *Clostridium histoly-ticum,* cause of necrosis of war wounds; (e) *Clostridium septicum,* cause of malignant edema of man and also invader of war wounds.

2. **Pathogenic for Animals.** These include: (a) *Clostridium septicum,* cause of malignant edema in horses; (b) *Clostridium novyi,* cause of gas gangrene; (c) *Clostridium chauvoei,* cause of blackleg in cattle; (d) *Clostridium fallax,* pathogenic to guinea pigs, and possibly cause of blackleg of sheep. (See also Chapter XV.)

3. **Nonpathogenic Putrefactive Organisms.** Included in this group are: (a) *Clostridium mucosum,* (b) *Clostridium butyricum,* (c) *Clostridium sporogenes,* (d) *Clostridium lentoputrescens,* and (e) *Clostridium alcaligenes.*

BIBLIOGRAPHY

Dack, G. M. *Food Poisoning.* Revised Edition. Chicago: University of Chicago Press, 1949.

Dubos, R. J. *The Bacterial Cell.* Cambridge, Mass.: Harvard University Press, 1945. Pp. 3–352.

Gardner, A. D. *Bacteriology for Medical Students and Practitioners.* New York: Oxford University Press, 1944. Pp. 149–163.

Society of American Bacteriologists. *Manual of Methods for Pure Culture Study of Bacteria.* Geneva, N. Y.: Biotechnical Publications, 1949. Chap. VII.

Sulzberger, M. B., and R. L. Baer. *Office Immunology.* Chicago: Yearbook Publishers, 1947. Pp. 95–129.

Taber, C. W. *Cyclopedic Medical Dictionary.* Philadelphia: F. A. Davis Company, 1958. B1–B5.

Taylor, B. T. *Stedman's Medical Dictionary.* Baltimore: Williams & Wilkins Company, 1957. Pp. 308–309.

Chapter XXVII

VIBRIO COMMA AND ALLIED ORGANISMS

The Vibrio genus (family Pseudomonadaceae, tribe Spirilleae) includes short, curved, rigid rods with one to three polar flagella, making the organisms actively motile. Many species liquefy gelatin. They are usually Gram-negative. Generally non-pathogenic water forms, the group also includes *Vibrio comma*, the cause of Asiatic cholera.

VIBRIO COMMA

Vibrio Comma (Vibrio Cholerae, Spirillum Cholerae Asiaticae, Comma Bacillus). 1. Group—Vibrio. 2. Family—Pseudomonadaceae. 3. Habitat—feces of patients infected with Asiatic cholera; water. 4. Discoverer—Koch, in 1883.

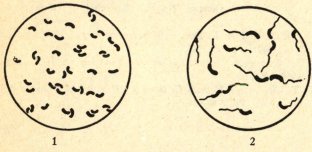

Fig. 57. (1) *Vibrio comma* in pure culture; note comma shape. (2) *Vibrio comma*, showing flagellum at one pole.

Morphological Characteristics. 1. Form—short, slightly bent, rigid rods. 2. Cell groupings—occurs singly and in spiral chains. 3. Size—0.4 to 0.6μ in breadth and 1.5 to 3μ in length. 4. Staining properties—stains well with ordinary aniline dyes. 5. Gram-negative and non-acid-fast. 6. Actively motile with one polar flagellum. 7. Nonsporing. 8. Cultures grown for some time on artificial media lose original uniformity of size and shape and tend to become less curved. Old cultures exhibit involution forms which are irregularly swollen or even coccoid in outline. Bacillary or even true spirillum forms are not uncommon.

Cultural Characteristics. *Vibrio comma* grows rapidly on all ordinary artificial media, especially if these are alkaline. 1. Agar cultures—round, thin, opalescent colonies, with entire edge; the surface is sometimes studded with small, knoblike secondary colonies. 2. Agar streak—in 24 hours there is a good, raised, translucent, grayish-yellow layer of growth, with smooth, glistening surface and edge formed of single colonies. 3. Gelatin stab—small, turnip-shaped area of liquefaction at the surface, which upon evaporation of the liquid leaves a bubble-like depression. 4. Blood agar plates—good growth; some strains show hemolysis.* 5. Potato—whitish, moist, spreading growth. 6. Litmus milk—rapid growth with slow peptonization, but generally no coagulation. 7. Broth—slight turbidity, with formation of a fragile, wrinkled pellicle and a flocculent precipitate.

Physiological Characteristics. 1. Optimum temperature, 37° C., with limits of 14° to 42° C. 2. Aerobic, and poor facultative anaerobe. 3. Indol is formed. 4. Nitrates are reduced to nitrites. 5. Cholera red reaction. (See p. 40.) 6. Hydrogen sulfide is formed. 7. Acid, but no gas, is produced from dextrose, levulose, galactose, maltose, sucrose, and mannitol, but there is no fermentation of lactose, arabinose, inulin, or dulcitol. 8. Resistance—the organism is rapidly destroyed by exposure to sunlight and drying. A solution of 5% phenol kills it in 10 minutes; it is also very sensitive to mercuric chloride, mineral acids, and very weak solutions of chlorine gas. It is killed in 10 minutes at 55° C.

Toxic Substances. The toxic symptoms of infection with the cholera vibrio have been attributed by Pfeiffer and others to endotoxins liberated upon the disintegration of the bodies of the organisms. This endotoxin is heat-stable, seems to have an affinity for epithelium, and causes shredding of the epithelium of the intestine and gall bladder in human cholera. Latest investigations indicate that no exotoxin is formed.

Transmission. Like all intestinal diseases, cholera is transmitted through infected feces. These find their way to the mouths of susceptible persons most often through water, and also by means of contaminated food eaten raw, of flies, and of personal contact. Organisms disappear rapidly from the feces, and it is believed that convales-

---

* By the *Grieg test* (mixture of 3% suspension of goat erythrocytes with an equal quantity of a 24 hour broth culture of *Vibrio comma* and reading after 2 and 4 hours incubation at 37° C.), the organism is nonhemolytic. A proposed explanation for the apparent contradiction is that blood plate hemolysis in this case is a hemodigestive process and basically different from hemoglobin liberation from a suspension of erythrocytes.

cent carriers play a small role in transmission.   Cases, especially in the incubation stage, are the most important sources.

Pathogenicity.   Cholera is primarily or entirely a toxemia. After an incubation period of from 1 to 5 days, the disease manifests itself by profuse vomiting, diarrhea, and "rice-water" stools which are watery and colorless, containing large numbers of epithelial cells, mucus flakes, and organisms.   Marked dehydration is observed, with collapse, subnormal temperature, anuria, and emaciation.   The organisms penetrate the mucosa of the intestine and lie in the layers nearest to the submucosa.

The disease runs a short course, resulting in death in some cases within 12 hours after the onset of the disease.   Fatality runs above 50% in untreated cases and about 30% in treated cases.   Animals are not spontaneously infected with the cholera vibrio.

Serological Reactions.   A specific agglutination reaction with homologous immune serum is used for identifying *Vibrio comma*. The cholera organism belongs to Group I of the six immunological groups into which the genus has been divided on the basis of specific O antigen.   The antigenicity of the cholera toxin has been open to question, but some workers have reported preparation of an effective antitoxin.

Diagnosis.

1.   Smears are prepared from a flake of mucus in the stools.   The preparations are examined by Gram's stain for typical Gram-negative, comma-shaped organisms.   A hanging drop is made and studied for motility.

2.   The stools are inoculated into several tubes of alkaline peptone water and agar plates and incubated at 37° C.   Under these conditions, the cholera organism grows more rapidly than other intestinal bacteria.   After 6 to 8 hours, the growth is smeared, stained, and examined for typical organisms, and also streaked on agar.

3.   Typical colonies are picked and identified by slide agglutination with monospecific antiserum.

The above tests, together with the clinical picture, are usually sufficient for identification.   If further substantiation is needed, the following may be employed:

4.   Negative Grieg test.   (See footnote p. 263.)

5.   Fermentation of sucrose and mannitol and failure to ferment arabinose.

6.   Positive cholera red reaction.   (See p. 40.)

7.   Pfeiffer's bacteriolysis reaction, which is relatively specific but does not distinguish between the hemolytic El Tor strains and the nonhemolytic *Vibrio comma*.   To 1 ml. of broth containing 0.001 ml. of high titer cholera serum, a loopful of an 18 to 24 hour agar culture of the organism is added.   The mixture is

injected intraperitoneally into a guinea pig. Another guinea pig is similarly injected with the same amount of bacterial suspension and 0.01 ml. of normal serum. Immediately after injection and at intervals of 20, 40, and 60 minutes, a drop of the peritoneal exudate is removed from each of the animals and examined under the microscope. A positive reaction is indicated by granular degeneration, swelling, and loss of motility of the vibrios. The peritoneal fluid of the control animal will contain many actively motile, comma-shaped vibrios.

Immunity, Prophylaxis, and Therapy. Recovery from cholera is usually said to confer immunity to subsequent infection. Many, however, believe that this immunity may not be of a high order nor of more than a year or two duration.

Various vaccines, including heat-killed, sensitized, and mixed vaccines containing *Salmonella typhosa*, *Salmonella paratyphi*, *Salmonella schottmuelleri*, and *Vibrio comma*, have been used in the prevention of cholera with somewhat favorable results. The injection is attended by certain local reactions of pain and some swelling of more or less severity, and general reactions of temperature, malaise, headache, some diarrhea in rare instances. The vaccines are generally given in two doses. The immunity conferred is not absolute and is of short duration. It develops after the fifth day and persists for a period varying from 6 months to 1 year. It is advised that vaccination be repeated annually.

Some investigators have used bacteriophage prophylactically and therapeutically in cholera, but there is not sufficient evidence upon which to base any conclusions. There is some indication, however, that the early administration of bacteriophage may be of value in reducing the severity and mortality of the disease. Therapeutic use of antisera is ineffective. Sulfaguanidine has been reported of value. Saline infusions are, of course, mandatory.

ALLIED ORGANISMS

Related to *Vibrio comma* and classified in the same genus are the following organisms: 1. *Vibrio metchnikovii*, recovered from the feces of fowl suffering from a cholera-like disease; 2. *Vibrio proteus*, found in human feces; 3. *Vibrio piscium*, cause of an epidemic infection in fish; 4. *Vibrio fetus*, isolated from cows and sheep infected with contagious abortion; 5. *Vibrio berolinensis*, *Vibrio liquefaciens*, and *Vibrio aquatilis*, found in water; 6. *Vibrio neocistes*, *Vibrio cuneatus*, *Vibrio cyclosites*, and *Vibrio agarliquefaciens*, all of which are found in soil; and 7. *Vibrio wolfii*, which have been isolated from cervical secretions in chronic endometritis.

BIBLIOGRAPHY

Anderson, W. A. *Pathology*. Second Edition. St. Louis: C. V. Mosby Company, 1953.

Burnet, F. M. *The Natural History of Infectious Disease*. Second Edition. Cambridge, England: Cambridge University Press, 1953.

Chambers, J. S. *The Conquest of Cholera*. New York: Macmillan Company, 1938.

Dubos, R. J. *Bacterial and Mycotic Infections of Man*. Third Edition. Philadelphia: J. B. Lippincott Company, 1958.

Gay, Frederick P. *Agents of Disease and Host Resistance*. Springfield, Ill.: Charles C. Thomas, 1935. Pp. 589–611.

Leifson, E. *Bacteriology for Students of Medicine and Public Health*. New York: Harper and Brothers, 1942. Pp. 301–378.

Marshall, M. S. *Applied Medical Bacteriology*. Philadelphia: Lea and Febiger, 1947. Pp. 1–340.

Park, William Hallock, and Anna Wessels Williams. *Pathogenic Microörganisms*. Philadelphia: Lea and Febiger, 1939. Pp. 369–644.

Tanner, F. W., and L. P. Tanner. *Food-borne Infections and Intoxications*. Second Edition. Champaign, Ill.: Garrard Press, 1953.

Wilson, G. S., and A. A. Miles. *Principles of Bacteriology and Immunity*. Baltimore: Williams & Wilkins Company, 1955.

Zinsser, H., and J. T. Enders. *Immunity*. New York: Macmillan Company, 1939. Pp. 440–721.

Chapter XXVIII

THE ACTINOMYCETES

The organisms to be considered in this chapter, belonging to the family Actinomycetaceae, order Actinomycetales, have characteristics intermediate between the true bacteria (order Eubacteriales) and higher organisms, particularly molds. They form long filaments which are often branched. They are mostly nonmotile, aerobic, soil saprophytes, attacking a wide variety of complex organic substances. Their classification is still quite confused. For the most recent Bergey classification and characteristics of the family and genera, see Chapter II. Explanation of some of the terms used will be found in Chapter VII on Yeasts and Molds.

GENUS ACTINOMYCES

Found in soil and dust, these are important, often highly-pigmented scavengers that decompose dead plant and animal matter. They include the most moldlike members of the family, often forming conidia-bearing, aerial hyphae.

Cultural Characteristics. The Actinomyces may be cultivated quite easily from soil, but grow slowly and are often overrun by bacteria. Best results with soil forms are obtained with Czapek-Dox and similar largely inorganic media. Agar colonies are usually flat, tough, and dry-looking, adhering to the medium and frequently exhibiting radial folds. Older colonies may be powdery or woolly, due to aerial hyphae and conidia. Growths give off a musty odor.

Physiological Characteristics. 1. Optimum temperature for saprophytes usually 25° C., parasites 37° C. 2. Soil species strictly aerobic; some pathogens are microaerophilic. *Actinomyces bovis*, once believed anaerobic, grows best in atmosphere containing 10% CO_2. 3. Grow best in alkaline material.

Pathogenicity. The actinomycosis caused by *Actinomyces bovis* ("lumpy jaw" in cattle) and *Actinomyces hominis* (in man) is an infectious, granulomatous disease, characterized by destruction of tissue, suppuration, and overgrowth of fibrous tissue. It is usually chronic and is characterized by systemic involvement. The organisms are probably introduced by thorns, splinters, etc., contaminated with soil containing virulent Actinomyces. or by insect bites. In the pus are found the so-called "sulfur granules"—large, yellow masses of matted growth with projecting tips which become enlarged and club-shaped.

GENUS NOCARDIA

In the first period of their growth the Nocardias resemble the Actinomyces in forming a typical mycelium with true branching of hyphae. Shortly thereafter, however, the filaments form transverse walls, whereupon the entire mycelium fragments into bacillary and then into coccoid cells. When cultivated on fresh media the latter are found to germinate into mycelia.

Pathogenicity. *Nocardia madurae* (*Actinomyces madurae*) causes madura foot or mycetoma, which is encountered mostly in tropical regions. *Nocardia asteroides* (*Actinomyces asteroides*) and *Nocardia farcinicus* (*Actinomyces farcinicus*) produce tuberculosis-like diseases of man and cattle, respectively. The former organism produces aerial mycelium only occasionally and forms conidia rarely or not at all. Its appearance is similar to the Mycobacterium species. Some Nocardia are plant parasites, causing such diseases as common potato scab.

Related Organisms. 1. *Nocardia leishmannii* causes disease of the lungs and pericarditis in man. 2. *Nocardia pulmonalis* causes certain lung infections in cows. 3. *Nocardia ragonensis* causes streptothricosis in man.

BIBLIOGRAPHY

Anderson, W. A. *Pathology*. Second Edition. St. Louis: C. V. Mosby Company, 1953.

Bergey, David H. *Bergey's Manual of Determinative Bacteriology*. Seventh Edition. Baltimore: Williams & Wilkins Company, 1957.

Burrows, William. *Jordan-Burrows Textbook of Bacteriology* Eighteenth Edition. Philadelphia: W. B. Saunders Company, 1963. Pp. 775–832.

Cumberland, M. C. *A Classification of the Actinomycetes and Related Branches of Bacteria*. Thesis. Baltimore: Johns Hopkins University, 1944.

Dubos, R. J. *Bacterial and Mycotic Infections of Man*. Third Edition. Philadelphia: J. B. Lippincott Company, 1958. Pp. 582–614.

Erikson. *Pathogenic Anaerobic Organisms of the Actinomyces Group*. London: Medical Research Council, Special Report Ser. No. 240, 1940.

Henrici, Arthur T. *Molds, Yeasts and Actinomycetes.* Second Edition. Revised by C. E. Skinner and H. M. Tsuchiya. New York: John Wiley and Sons, 1947.

Jawetz, E., J. L. Melnick, and E. A. Adelberg. *Review of Medical Microbiology.* Los Altos: Lange Medical Publications, 1962. Pp. 170–171.

Porter, J. R. *Bacterial Chemistry and Physiology.* New York: John Wiley and Sons, 1946. Pp. 686–716.

Smith, David T., and Norman F. Conant. *Zinsser's Textbook of Bacteriology.* Twelfth Edition. New York: Appleton-Century-Crofts, 1960.

Society of American Bacteriologists. *Manual of Methods for Pure Culture Study of Bacteria.* Geneva, N. Y.: Biotechnical Puplications, 1949. Chap. V.

Strean, L. P. *Oral Bacterial Infections.* Brooklyn, N. Y.: Dental Publishing Company, 1949. Pp. 141–147.

Walter, W. G., and R. H. McBee. *General Microbiology.* Princeton, N. J.: D. Van Nostrand Company, 1958. Pp. 140–143.

Chapter XXIX

THE SPIROCHETES

Members of the order Spirochaetales, protozoan-like in some respects, are long, spiral, cylindrical, flexible organisms without a nucleus. They do not form spores, conidia, branches, or filaments. They reproduce by transverse fission and are actively motile, but without polarity. They have been considered flagellaless, but the electron microscope has recently revealed apparent flagella on some Treponema. The spirochetes are difficult to cultivate and stain. The order contains one family, Spirochaetaceae, with six, or perhaps seven, genera.

GENERAL CHARACTERISTICS OF GENERA

Genera Spirochaeta, Saprospira, and Cristispira. The members of these genera are without pathogenicity for man. The cell bodies are 45 to 500μ long and 0.5μ broad. They are resistant to the action of a 10% solution of bile salts. (Family Spirochaetaceae)

1. *Spirochaeta.* The members of this genus have a well-defined axis filament, but no crista or ridge, and no periplast membrane. They are usually present in sewage and foul waters.

1a. *Cytophaga.* This group is tentatively included here, but also bears some relation to the Myxobacteriales. The organisms are long, flexuous rods with pointed ends, inhabiting soil and sea water, and actively hydrolyzing cellulose.

2. *Saprospira.* These organisms have a distinct periplast membrane, but lack a crista and axis filament. They have a chambered structure. Some species are found in molluscs, some free-living in water.

3. *Cristispira.* These organisms have a crista or ridge and a periplast membrane, but no axis filament. The ridge runs the length of the body and has been falsely called an undulating membrane. These spirochetes are parasitic in molluscs. The cell is divided into chambers by transverse septa.

Genera Borrelia, Treponema, and Leptospira. Members of these genera are parasitic and sometimes pathogenic. They are disintegrated by solutions of 10% bile salts. (Family Treponemataceae)

4. *Borrelia.* These organisms are small, flexible filaments, with three to five large, wavy spirals. They have an axis filament and a periplast membrane. Cause of relapsing fever is included here.

5. *Treponema.* This group is characterized by close coils and by bending and rotating movements. They have an axis filament but no crista. Causes of syphilis and yaws are included here.

6. *Leptospira.* These have tightly wound coils with flagelliform, tapering ends. One or both ends are sharply curved to form a hook. They possess neither axis filament nor periplast membrane. Cause of infectious jaundice is included here.

BORRELIA RECURRENTIS

Borrelia Recurrentis (Spirochaeta Recurrentis, Spirochaeta Obermeieri). This organism was first described by Obermeier in 1873. An anaerobic blood parasite, it causes European relapsing fever, characterized by a pyrexia of 3 to 5 days, and an equal period of apyrexia, often relapsing through as many as ten such cycles. There are marked enlargement of the spleen and slight enlargement of the liver. The spirochetes are demonstrable in the blood during the period of fever.

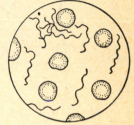

FIG. 58 *Borrelia recurrentis.* Dark field illumination of blood, showing organisms and red blood cells.

Morphological Characteristics. *Borrelia recurrentis* is very active, spiral, having from three waves up to twenty. It is 8 to 15μ in length and 0.5μ in breadth, with tapering body ends. It is best observed by dark field illumination in fresh blood. In stained, dried, blood smears there is much distortion; the spirals are often obliterated.

Transmission. The spirochete may be tick-borne or louse-borne. In the former case, it passes through an animal reservoir, such as the rat or other small animal; in the latter case, the insect carries the parasite from man to man. The spirochetes survive in ticks and lice for a considerable time, although often in an apparently unrecognizable, filterable form. This, together with the fact that the filtered blood of relapsing fever patients is capable of transmitting the disease, supports the view that the organisms pass into a filterable, granular stage.

Immunity. As an immunity is developed, the relapses occur less and less frequently. Chronic infection, however, persists for a long time, maintaining immunity by the presence of the organisms in the body.

Therapy. As in syphilis and yaws, arsenicals are used as therapy. Only one or two injections are usually required.

Related Organisms. *Borrelia duttonii* causes South African relapsing fever; *Borrelia kochii*, African relapsing fever; *Borrelia novyi*, American relapsing fever; *Borrelia carteri*, Indian relapsing fever. *Borrelia anserina* causes septicemia in chickens. The nonpathogenic *Borrelia refringens* is found on the human genitalia.

FIG. 59. *Borrelia refringens.* Occasional nonpathogenic habitant of genitalia.

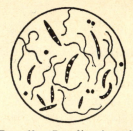

FIG. 60. *Borrelia vincentii* and fusiform bacilli. Note the irregular undulations.

BORRELIA VINCENTII

Borrelia vincentii is usually associated with the fusiform bacilli described by Vincent, Plaut, and others as present in Vincent's angina. The bacilli are 3 to 10μ in length, and 0.5 to 0.8μ in breadth. They are readily stained by Löffler's methylene blue, carbol fuchsin, and Giemsa's stain. The fusiform bacilli are classified as *Fusobacterium plauti-vincenti.* Their relationship to *Fusiformis dentium* and *Bacillus fusiformis* has as yet not been ascertained.

Vincent's angina, or trench mouth, which is highly contagious, is characterized by a false membrane in the tonsillar region, similar to the croupous diphtheritic membrane. The infection differs from diphtheria by an absence of severe systemic disturbances. The membrane sloughs off, leaving a sore mouth, bleeding gums, and a deep ulcer. Smears from the deeper parts show the spirochete and the bacilli in large numbers. The bacillus may be cultivated on blood or serum agar under strict anaerobiosis.

BORRELIA DUTTONII

Borrelia duttonii causes Central and South African relapsing fever. The organism is closely related to *Borrelia recurrentis*, but it is antigenically distinct from other causes of relapsing fever and so plastic that many serological types have been identified.

Borrelia duttonii is transmitted to man through the bite of a tick

(*Ornithodorus moubata*) by fecal contamination of the bite. The organism produces a severe disease in rodents and guinea-pigs, being found in the brains of experimentally infected animals.

Early administration of bismarsen, neoarsphenamine, or novarsenobillon constitutes accepted therapy.

TREPONEMA PALLIDUM

Treponema Pallidum (Spirochaeta Pallida). 1. Group—Treponema. 2. Family—Treponemataceae. 3. Habitat—cause of syphilis in man. 4. Discoverers—Schaudinn and Hoffmann, in 1905.

Morphological Characteristics. 1. Form —delicate, slender, undulating filaments with pointed ends. 2. Cell groupings—consist of 3 to 12 convolutions. 3. Size—4 to 14μ in length and up to 0.5μ in thickness. 4. Staining properties—stained only with great difficulty, and then with special stains like Giemsa, Burri India ink method (negative staining), Fontana-Tribondeau method (silver impregnation), and Noguchi's method. 5. Motility— active backward and forward movements, rotating about long axis and occasionally bending.

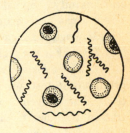

FIG. 61. *Treponema pallidum*, best seen under dark field illumination in syphilitic chancre fluid. Note regular spirals.

Cultural Characteristics. *Treponema pallidum* has never been cultivated in artificial media.

Physiological Characteristics. *Treponema pallidum* is a strict anaerobe. It is killed by 50–55° C. in 30 minutes, or by moist heat at 39° C. in 5 hours, and is quickly destroyed by drying, sunlight, and mild disinfectants.

Pathogenicity. Syphilis normally affects only man. In 1903, however, Metchnikoff and Roux succeeded in transmitting the disease to monkeys; others have since infected rabbits.

Syphilis or lues is a disease most commonly acquired by sexual contact. Congenital syphilis in which the disease is acquired *in utero* from an infected mother is frequently encountered. The primary lesion of syphilis is the hard chancre which develops after an incubation period of 2 to 4 weeks. The lesion starts as a papule at the point of infection (usually the genitalia), enlarges, becomes indurated, and ulcerates. The serous exudate from the chancre, when examined under dark field illumination, reveals many spirochetes. The lymphatics, which drain the area of the chancre, soon carry the *Treponema pallidum* to other parts of the body, and the systemic invasive or secondary stage follows, during which any tissue may become involved. Eruptions on skin and mucous membranes, headaches, and many vague symptoms appear. When this stage develops, serologic diagnostic tests become strongly positive. The tests are usually negative during the chancre stage. A lengthy latent period may precede the third stage. The characteristic lesion of the tertiary stage is the gumma, or tubercles with centers of necrosis, which break down to form ulcers and contain the *Treponema pallidum*. There may be central nervous system involvement with paresis, general paralysis, and insanity. Spinal cord infection results in locomotor ataxia. Lesions may also occur in other organs.

Serological Diagnostic Tests. 1. *Wassermann Complement Fixation Test.* For the technic see page 345.

2. *Kahn Precipitation Test.* This is a macroscopic flocculation or precipitation test, using a nonspecific antigen as in the Wassermann test. The antigen is an alcoholic cholesterinized extract of ether extracted beef heart. The three agents needed are patient's serum inactivated by heat at 56° C. for one-half hour, a 0.85% solution of sodium chloride, and the antigen.

(a) Ascertain the titer of the antigen by determining the smallest amount of salt solution in which the precipitate formed when salt is added to the antigen is dissolved on adding more salt solution.

(b) Mix the antigen with salt solution according to the titer, i.e., 1 ml. of antigen with 1.2 ml. saline if this has been found to be the titer.

(c) Allow to stand for 10 minutes.

(d) Arrange three tubes one behind the other for each serum to be tested. To the front row of tubes add 0.05 ml. of the diluted antigen, to the middle row add 0.25 ml., and to the back row add 0.0125 ml.

(e) Immediately add 0.15 ml. of each patient's serum which has been inactivated and shake the whole rack vigorously for 10 seconds.

(f) If the temperatures are low, place the racks in a hot water bath for 10 minutes.

(g) Shake racks for 3 minutes.

(h) Add 1 ml. of saline to all the front tubes, and 0.5 ml. to each of the tubes in the middle and back rows.

(i) The reactions are read as follows:

(1) A strongly positive reaction $++++$ is indicated by flocculation or precipitation in which the particles are visible to the eye.

(2) Various other degrees of positivity depend upon the degree of precipitation, $+++$ and $++$.

(3) A reaction of $+$ is visible only by magnification.

(4) Negative reactions are uniformly opalescent. (5) The final result is obtained by reading the three tubes and dividing by 3.

3. *Kline Test*. This is a microscopic precipitation reaction which can be made in a few minutes and is a useful check. The procedure is shown in Fig. 62.

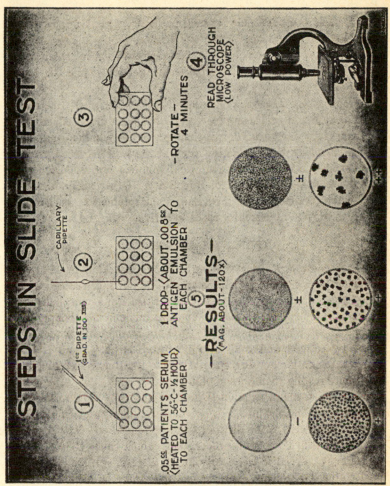

Courtesy, Dr. B. S. Kline, Mount Sinai Hospital, Cleveland, Ohio.

FIG. 62.   Kline test.

The tests may be negative during the first two weeks of the disease, and in latent syphilis following treatment. Although these tests do detect a high percentage of syphilis cases, without a history of possible exposure a positive test does not always indicate syphilis and a negative test does not always exclude the disease. Positive

Wassermann reactions have been known to occur in certain other infections, including leprosy, yaws, sleeping sickness, malaria, and tuberculosis. The ingestion of alcohol or the taking of an anesthetic interferes with the test, the former giving a negative reaction and the latter a false positive. The Wassermann reaction is nearly always positive in the secondary stage of the disease.

Immunity. One attack of syphilis does not usually confer lasting immunity. However, immunity to superinfection, while the organisms are in the body, is afforded.

Prophylaxis. Local application of 33% calomel (mercurous chloride) ointment or 2% or 10% argyrol has been found effective.

Therapy. Early administration of arsphenamine and other organic arsenicals in conjunction with bismuth or mercury preparations is widely accepted therapy. Penicillin has revolutionized treatment by reducing the period of infectivity. Other antibiotics, such as chlortetracycline, oxytetracycline, chloramphenical carbomycin, and erythromycin appear also to be effective.

TREPONEMA PERTENUE

Treponema pertenue was described by Castellani in 1905, who isolated it from cases of frambesia tropica or yaws, a tropical, ulcerative disease similar to syphilis. The organism is morphologically similar to *Treponema pallidum*. Yaws is less chronic than syphilis, and one attack confers immunity. It is not a venereal disease. It is spread by contact and may also be insect-borne. Yaws is found most often in children. It responds well to treatment with arsphenamine, more readily than does syphilis.

The Wassermann reaction is positive in yaws. The relationship between yaws and syphilis has not been definitely determined, and they are considered to be distinct from one another. The organism may be found in the papular ulcerations which cause the ugly body mutilations observed so frequently in the disease.

Related Organisms. *Treponema mucosum* is associated with pyorrhea, which is a mixed infection. *Treponema microdentium* and *Treponema macrodentium* are nonpathogenic inhabitants of the mouth of man. *Treponema cuniculi* causes rabbit spirochetosis.

LEPTOSPIRA ICTEROHAEMORRHAGIAE

This organism causes Weil's disease, or infectious jaundice. The disease is characterized by an incubation period of 5 to 7 days, followed by fever, nausea, vomiting, headache, and muscular pain, lasting 2 to 3 days. Then jaundice is usually exhibited, with edema of the liver and enlargement of the spleen. The blood is highly infectious during the febrile period and contains many tightly-wound, spiralled, microaerophilic organisms about 6 to 9μ in length, with ends curved in the shape of hooks. The organisms are present in the urine towards the end of the disease.

Fig. 63. *Leptospira icterohaemorrhagiae*. Note irregular undulations.

Pollution by the urine of infected rats, and probably dogs, of water, mines, trenches, etc., serves to transmit the disease to man. The mode of entry is still uncertain and may involve minute cuts in the skin or the alimentary tract.

As in cholera, cytolysis—the Pfeiffer phenomenon (see p. 264)—can be well illustrated in infectious jaundice. The lytic antibodies appear in the blood shortly after infection. The agglutinin titer is markedly increased and is useful diagnostically.

Active artificial immunization is said to be effective. Immune sera have given favorable therapeutic results.

Related Organisms. *Leptospira hebdomadis* is the cause of seven-day fever in Japan. Many similar infections are caused by Leptospira in other parts of the Far East. *Leptospira pomona* causes swineherd's disease throughout the world. It is transmitted to man by infected swine and cattle.

BIBLIOGRAPHY

American Association for the Advancement of Science, Symposium No. 18. *Relapsing Fever in the Americas.* Washington, D. C., 1942.

Boyd, W. C. *Fundamentals of Immunology.* New York: Interscience Publishers, 1947. Pp. 369–392.

Cowdry, E. V. *Microscopic Technique in Biology and Medicine.* Baltimore: Williams & Wilkins Company, 1943. Pp. 193–194.

Jawetz, E., J. L. Melnick, and E. A. Adelberg. *Review of Medical Microbiology.* Los Altos: Lange Medical Publications, 1962. Pp. 196–203.

Manson-Bahr, P. H. *Manson's Tropical Diseases.* Baltimore: Williams & Wilkins Company, 1940.

Medical Research Council. *A System of Bacteriology in Relation to Medicine.* Vol. VIII. London: His Majesty's Stationery Office, 1931. Pp. 101–333.

Moore, Joseph Earle. *Penicillin in Syphilis.* Springfield, Ill.: Charles C. Thomas, 1948.

Netter, E., and D. R. Edgeworth. *Medical Microbiology for Nurses.* Philadelphia: F. A. Davis Company, 1957.

Raper, K., and C. Thom. *Manual of the Penicillia.* Baltimore: Williams & Wilkins Company, 1949. Pp. 84–104.

Stokes, John H., H. Beerman, and N. R. Ingraham, Jr. *Modern Clinical Syphilology.* Third Edition. Philadelphia: W. B. Saunders Company, 1944.

Swingle, D. B. *General Bacteriology,* rev. by G. W. Walter. New York: D. Van Nostrand Company, 1947. Pp. 41–69.

Taber, C. W. *Cyclopedic Medical Dictionary.* Philadelphia: F. A. Davis Company, 1958. B1–B5.

Taylor, N. B., and A. E. Taylor. *Stedman's Medical Dictionary.* Baltimore: Williams & Wilkins Company, 1957. Pp. 1324–1325.

Walter, W. G., and R. H. McBee. *General Microbiology.* Princeton, N. J.: D. Van Nostrand Company, 1958. Pp. 150–152.

Chapter XXX

THE RICKETTSIAE

The rickettsia bodies were first described by Ricketts in 1909, who found them in the blood of persons with Rocky Mountain spotted fever. They are intracellular parasites whose exact nature is unknown, but they are generally considered intermediate between the bacteria and filterable viruses. They resemble bacteria in shape and may be cocci, diplococci, or short bacilli. They are Gram-negative, nonmotile, nonsporing, nonencapsulated, probably all nonfilterable, and are difficult to stain with ordinary aniline dyes. The Giemsa stain is good for demonstrating these bodies. It is difficult to cultivate the rickettsia bodies, and therefore diagnosis is hampered. The diseases caused by rickettsia bodies are all vectored by intermediate insect hosts of the phylum Arthropoda, such as ticks and lice. The rickettsia bodies are $0.3–0.5\mu$ or more in length and about 0.3μ in diameter. They have electrophoretic characteristics similar to the bacteria. They are thermolabile and killed by dehydration or chemical antiseptics. Electron microscopic examination of the rickettsia bodies reveals a homogeneous or slightly granular internal structure closely resembling that of the bacteria. Like the viruses, the pathogenic Rickettsiae are true parasites and can be cultivated in the intestines of rats, ticks, and lice, in tissue cultures, and in the membranes of chick embryos. The natural habitat of the Rickettsiae appears to be the cells lining the intestine and other tissues of insects, both blood-sucking and non-blood-sucking. Their classification is given on pages 33 and 34.

Diseases. The majority of known rickettsial diseases fall into three classes: typhus fevers, spotted fevers, and tsutsugamushi. Clinically they are similar in many respects—fever, skin rashes or dark blotches due to multiplication of organisms within the endothelial cells of small blood vessels, and brain involvement.

TYPHUS FEVERS

Typhus fever, a disease of filth and famine, is known by various names in different parts of the world. It has been generally believed to manifest itself in two main types: (1) *Murine, rat,* or *endemic typhus,* caused by *Rickettsia prowazeki var. mooseri* and transmitted from rat

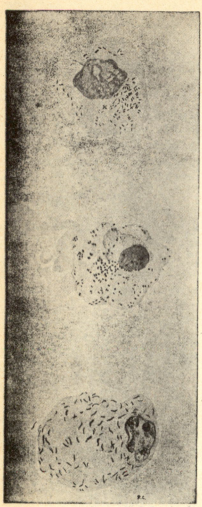

to rat by the rat flea and rat louse and from rat to man usually by the rat flea. This is the form that prevails in the southern United States and in Mexico, where it is known as *tabardillo.* It has also been termed *shop, Toulon, Moscow,* and *Manchurian typhus.* It appears sporadically and has a mortality below 5%. (2) *Human, European,* or *epidemic typhus,* caused by *Rickettsia prowazeki var. prowazeki* and spread from man to man by the body louse and possibly the head louse. It is highly communicable and has a mortality of 20–70%. *Brill's disease* is a mild type of louse-borne European typhus endemic in Atlantic coast cities.

The disease is characterized by initial violent headache, nausea, dizziness, alternate chills and fever, and a typical rash. Complications include typhus gangrene, severe bronchopneumonia, otitis media, and typhus encephalitis. Clinically the two forms are not very different; both may exist in endemic and epidemic form with corresponding mildness and fatality. Typhus is also a disease of lice, causing them to sicken and die.

The organisms occur in the cytoplasm but not the nucleus of invaded cells. Blood of the patient is infectious, but organisms have not been seen in the blood.

Courtesy, W. B. Saunders Co.

Fig. 64.   Rickettsiae in the cells of a guinea pig. (Redrawn from Monteiro, in Martin Frobisher, Jr., *Fundamentals of Bacteriology,* 1949.)

Weil-Felix Reaction. The serum of persons with typhus fever contains agglutinins against certain Proteus strains, the most famous of which is Proteus OX19, isolated by Weil and Felix from the urine of typhus fever patients. The agglutination reaction of the serum in dilutions of 1:50 to 1:50,000 is diagnostic for typhus fever. The Proteus X strains are not related to typhus fever in an etiologic sense, for they are rarely found in cases of the disease and experimental inoculation of these organisms does not produce the disease. Proteus strains do not immunize against Rickettsiae, nor does recovery from rickettsial infection immunize against experimental Proteus infection. Zinsser and associates have suggested that the rickettsia bodies and certain Proteus strains contain a common antigen component.

Prophylaxis. The prevention of typhus fever lies in the eradication of lice, especially when people are confined in close quarters such as jails and trenches, and particularly in keeping lice away from known typhus cases. Cleanliness, repellents, and delousing procedures are useful. DDT (dichloro-diphenyl-trichloroethane) appears promising. (See following section on vaccines.)

Immunity and Therapy. Recovery from either form of typhus fever results in lasting immunity to both. Massive doses of vaccines of killed organisms have been fairly successful in producing active immunity. The vaccine, formerly prepared by growing the organisms in lice and using the ground intestine by tissue culture in guinea pig testes, or made from the lungs of infected rats, has been recently produced from growth of the organism in developing chick embryos. By the latter means, Bengston and Dyer of the United States Public Health Service have obtained a preparation free from bacteria and insect material.

The serum of immunized goats or rabbits is being tried therapeutically and may prove useful. Chloromycetin and aureomycin are proving clinically effective according to widespread reports.

Diagnosis. The Weil-Felix reaction is the main diagnostic tool. Complement fixation is also used in differentiating typhus, especially the murine type, from spotted fever. The U.S.S.R. Institute of Experimental Medicine has recently reported a complement fixation test claimed to be very sensitive, strain-specific, and applicable as early as the first and second days of illness. Animal inoculation is also employed.

SPOTTED FEVERS

These are diseases similar to typhus fever and characterized by petechial hemorrhages into the skin together with a rash and an incubation period of 3 to 9 days. The disease is transmitted from tick to tick hereditarily and to man by the tick. It is not transmitted from man to man, as is typhus fever (see p. 280). The disease is innocuous for ticks. It is most prevalent in the northwestern United States. The causative agent, *Rickettsia rickettsi* (also known as *Rickettsia dermacentroxenus*), can be transmitted experimentally to monkeys, guinea pigs, rabbits, and dogs. It is found within the nucleus of the invaded cell.

The disease has, with questionable correctness, usually been classified into three very closely related types, all immunologically identical and caused by the same organism. One occurs in the eastern states and is transmitted chiefly by the dog tick, *Dermacentor variabilis;* one in the western states, particularly Bitter Root Valley, Montana, and transmitted by the sheep or wood tick, *Dermacentor andersoni;* and one in Brazil (*typho-exanthematico*) transmitted by *Amblyomma cajennense.* In the United States, the disease is known as *Rocky Mountain spotted fever.* The mortality is very variable but runs nationally about 19%. The mortality in Brazil is about 70%. Other spotted fevers include *Fièvre Boutonneuse, Kenya fever, South African tick-bite fever, São Paolo typhus.*

After the first week of the disease, the Weil-Felix agglutination test is positive. The serum usually has a titer ranging from 1:20 to 1:200, not as high as in typhus fever.

Immunity and Prophylaxis. One attack of spotted fever usually confers lifelong immunity. There is a slight cross protection between typhus and spotted fevers. Immunization with massive doses of vaccines of killed organisms has been fairly successful. As with typhus, these were formerly prepared from infected insects and now from tissue cultures or preferably from chick embryos. Measures to minimize contact with ticks are, of course, mandatory. These include wearing of tick-proof clothing, eradication of ticks by brush clearing, and removal of ticks from the body as quickly as possible.

TSUTSUGAMUSHI FEVERS

This is a group of similar febrile diseases resembling typhus fever and occurring in Japan and adjacent lands. The causative agent is *Rickettsia tsutsugamushi* (also known as *Rickettsia orientalis* or *Rickettsia nipponica*), which is transmitted to man by the bites of mite larvae occurring in swampy areas. The larvae become infected from infected adults through the eggs. The adult mites do not bite mammals. In addition to the typical rickettsial symptoms of headache,

INSECT-BORNE RICKETTSIAL DISEASES

Organism	Disease	Vector	Prevention	Treatment and Control	Symptoms	Weil-Felix Reaction		
						OX19	OX2	OXK
Rickettsia prowazekii (var. *prowazekii*)	Human or epidemic typhus, goat fever (South America)	Body lice, possibly head lice	Eradication of lice, cleanliness, repellents, delousing procedures	Vaccines consisting of killed organisms, chloramphenicol, chlortetracycline hydrochloride	Violent headache; dizziness; alternate chills, fever, rash	3+	2+	–
Rickettsia prowazekii (var. *mooseri*)	Endemic typhus, urban typhus, murine typhus (southern U. S.)	Rat fleas	Eradication of rats, cleanliness, proper sanitary laws	Vaccines consisting of killed organisms (same as for var. *prowazekii*); serum of immunized goats or rabbits may prove useful	Same as for var. *prowazekii*	3+	2+	–
Rickettsia rickettsii	Rocky Mountain spotted fever (eastern and northwestern U. S.)	Dog and wood ticks	Tick-proof clothing, eradication of ticks by brushing from body as soon as possible	Vaccines consisting of killed organisms	Petechial hemorrhages into skin, rash	3+	2+	–
Rickettsia tsutsugamushi (R. orientalis)	Tsutsugamushi fever, scrub typhus (Asiatic-Pacific area)	Mite larvae, chiggers	Destruction of infected mites and rodents, immunization, chemoprophylaxis	Chloramphenicol or one of the tetracyclines	Headache, nausea, chills, fever, sore site of bite	–	–	3+
Rickettsia wolhynica	Trench fever (central Europe)	Body lice	Eradication of lice, cleanliness, repellents, delousing procedures, vaccines	The tetracyclines	Sudden onset, with fever, headache, and pains in bones and muscles, especially in the legs	–	–	–
Rickettsia burnetii (Coxiella burnetii)	Q fever (western U.S., Italy, Greece)	Wood and cattle ticks	Inactivated vaccine, milk pasteurization, prevention of consumption of milk or meat of infected wild animals	The tetracyclines	Malaise, chills (sudden onset)	–	–	–

nausea, and chills and fever, there is a primary sore at the site of the bite. The animal reservoirs are mice and rats. Mortality may run to 50%.

One attack confers prolonged but not absolute immunity. Serum of tsutsugamushi patients gives a high-titer agglutination Weil-Felix reaction with the Proteus OXK strain, which is derived from OX19 by variation.

Besides the tsutsugamushi disease, there are also included in this group the probably identical *mite fever* of Sumatra and the milder *rural* or *scrub typhus* of Malaya.

MISCELLANEOUS DISEASES

Trench Fever. This disease, also known as Wolhynian fever, is characterized by sudden onset with fever, headache, and pains in the muscles and bones, especially in the legs. There are frequent remissions and relapses. The disease is caused by *Rickettsia quintana* (*Rickettsia wolhynica*, *Rickettsia pediculi*) and is transmitted from man to man by lice. It is rarely fatal.

Q Fever. This is caused by a filterable *Coxiella burneti* (*Rickettsia diaporica*), with transmission possibly by tick or by air. There is high fever, no rash, no Weil-Felix reaction. There have been no fatalities. The disease has been found especially in slaughterhouse workers and dairy farm workers in Australia. *Nine-mile fever*, found in Montana, is believed identical.

"Heartwater" Disease. Also known as *Veldt disease*, this is a highly fatal, economically important, tick-borne disease of cattle, sheep, and goats in South Africa. The causative organism is *Cowdria ruminantium*.

Bullis Fever (Lone Star or Tick Fever). This is a relatively mild, Weil-Felix negative disease which has been observed in soldiers stationed in Texas. It is believed to be caused by Rickettsiae and transmitted by ticks.

BIBLIOGRAPHY

Bergey, David H. *Bergey's Manual of Determinative Bacteriology*. Seventh Edition. Baltimore: Williams & Wilkins Company, 1957.

Parish, H. J. *Antisera, Toxoids, Vaccines and Tuberculins in Prophylaxis and Treatment*. Fourth Edition. Baltimore: Williams & Wilkins Company, 1958.

Rivers, T. M., and F. L. Horsfall, eds. *Viral and Rickettsial Infections of Man*. Third Edition. Philadelphia: J. B. Lippincott Company, 1959.

CHAPTER XXXI

THE VIRUSES

The viruses are minute disease-producing agents, usually termed filterable because they pass through fine filters capable of holding back bacteria, and ultramicroscopic because they are not visible with the light microscope. No acceptable definition of these bodies, whose existence was first indicated in 1892 by Iwanowski's work on tobacco mosaic disease, can yet be advanced. Their nature, structure, and mode of activity are still incompletely known, although the sizable body of knowledge concerning them is constantly being enlarged.

GENERAL CHARACTERISTICS

Visibility. Viruses cannot be seen with the light microscope. Some can be photographed under the electron microscope (Figs. 65 and 66). These pictures have revealed bodies which may be the

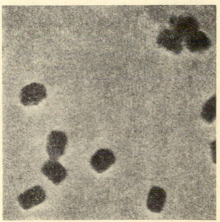

Courtesy, R.C.A.

Fig. 65. Electron micrograph of elementary bodies of vaccinia virus. (× 23,000.)

vaccinia virus as uniform rectangular particles, containing five areas of condensation. Needle-like crystals, inseparable from the ability to produce the disease, probably represent the virus of tobacco mosaic.

Filterability. Viruses generally pass through filters capable of holding back bacteria. This ability, however, is relative and dependent on the size of the virus and the filter pore, the nature of the filter, and undoubtedly other factors, such as the medium, temperature, electrical charge, etc.

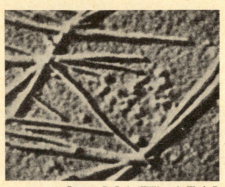

Courtesy, R. C. A. (Williams & Wyckoff).

FIG. 66. Electron micrograph of shadowed tobacco mosaic suspension. (× 60,000.)

Bacterial filters are usually made of unglazed porcelain, kieselguhr (diatomaceous earth), or asbestos, and are available in different sizes and degrees of porosity. **1.** *Berkefeld filters*, made of kieselguhr, asbestos, and organic matter, are available in three grades: V (viel), relatively coarse, approximate porosity 8–12μ; N (normal), intermediate, approximate porosity 5–7μ; W (wenig), fine, approximate porosity 3–4μ. **2.** *Pasteur-Chamberland filters*, made of unglazed porcelain from kaolin and sand, in the form of candles, are marketed in grades L₁, L₂, L₃, L₅, L₇, L₉, L₁₁, and L₁₃, running from coarsest to finest. **3.** *Mandler filters*, made of diatomaceous earth, plaster of Paris, and asbestos, are available in preliminary, regular, and fine grades. **4.** *Seitz filters* are asbestos discs.

Besides being used to separate filterable viruses, bacterial filters are used to separate exotoxins or other filterable material from bacterial cultures and to sterilize certain preparations easily destroyed by heat, such as sugar solutions, antitoxins, etc. Extreme care must be used to guard against contamination.

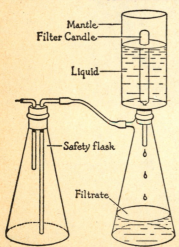

FIG. 67. Berkefeld filter set up for use.

Parasitism. All known viruses are obligate parasites requiring living cells for their cultivation. Some have been cultivated outside the body in the developing chick embryo or in tissue culture.

Viruses usually exhibit a marked but not absolute selectivity for certain tissues. On the basis of their *primary* tissue predilection, the viruses may be classified as follows:

1. *Neurotropic* — predilection for central nervous system, such as viruses of rabies, poliomyelitis, encephalomyelitis, lymphocytic choriomeningitis.

2. *Dermotropic* — predilection for the skin, such as viruses of vaccinia, variola, alastrim, varicella, molluscum contagiosum, verruca, herpes zoster, herpes simplex, trachoma.

3. *Pneumotropic* — predilection for respiratory tract, such as viruses of influenza, rubella, rubeola, psittacosis, common cold.

4. *Viscerotropic* — predilection for abdominal or thoracic viscera or produce signs of generalized infection, such as virus of yellow fever.

5. *Miscellaneous* — such as viruses of lymphogranuloma inguinale, parotitis, etc.

Inclusion Bodies. In the lesions produced by some viruses there are certain intracellular masses, termed *inclusion bodies*, which are often of diagnostic value. They may be cytoplasmic (Fig. 68) or intranuclear (Fig. 69). Typical are the *elementary bodies* of vaccinia and the *Negri bodies* of rabies. Their exact nature is disputed. Some believe them to be aggregations of the virus itself, and others consider them to be the reaction of the cell towards the virus.

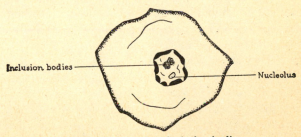

FIG. 68. Cytoplasmic inclusion bodies.

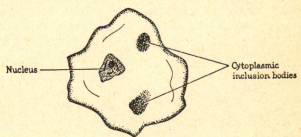

FIG. 69. Intranuclear inclusion bodies.

Immunity. Recovery from a virus infection produces usually a solid and lasting, but sometimes a transient, immunity. Second attacks of yellow fever, poliomyelitis, hog cholera, smallpox, or measles are very rare. Immunity of about a year results from foot-and-mouth disease, dengue fever, and pappataci fever. In fever blisters and common colds, immunity is very short. There is some evidence to indicate that lasting immunity when present is an infection-immunity or immunity to superinfection. Viruses have been shown to persist in the body a long time after recovery.

Various familiar immunizing substances—agglutinins, precipitins, neutralizing antibodies—have been identified in circulating blood, and the general nature of the immunity is believed the same as in other infections. However, the effectiveness of circulating antibodies against intracellular viruses has not been fully determined. Active artificial immunity is highly successful with many virus diseases. The *serum neutralization test* is often used for study and diagnosis, e.g., poliomyelitis, influenza, vaccinia. This determines the protection afforded susceptible animals by mixtures of immune serum and virus in contrast to the development of disease in animals inoculated with normal serum and virus or virus alone.

Adaptability. This ability, characteristic of living things, is also an attribute of viruses. For example, rabies virus grown in dogs is very virulent for both dog and man. If cultivated in rabbits, however, the virus becomes attenuated for dog and man and virulent for rabbits. Similarly, the selective tissue action of viruses can be modified by inoculation into an abnormal portal of entry.

Resistance. A striking characteristic of most viruses is resistance to 50% glycerin, a property shared by only a very few bacteria. Viruses may be preserved in 50% glycerin at icebox temperatures or by the lyophil process (drying and freezing). They are heat labile and also readily inactivated by direct sunlight.

Size. By means of collodion membranes (gradacol membranes or ultrafilters) of graduated and standardized porosities, electron microscope studies, and ultracentrifugation, viruses have been shown to vary greatly in size but to be relatively uniform within each "species." They range from about 275 mμ* for psittacosis, 175 mμ for vaccinia to 22 mμ for yellow fever and 12 mμ for poliomyelitis. (The diameter of Micrococcus is about 800 mμ.)

Composition. In 1935, Stanley succeeded in preparing crystalline, disease-producing nucleoproteins, considered to be the viruses of

* mμ—millimicron or the thousandth part of a micron.

tobacco and cucumber mosaic diseases. From this and other work, the belief is now held that the smaller viruses are nucleoproteins and the larger viruses like vaccinia contain nucleoprotein, fat, and carbohydrate.

NATURE OF VIRUSES

The challenge of the nature of the material known as viruses has been met with a number of suggested explanations. The first-discovered virus of tobacco mosaic disease was believed by Beijerinck (1899) to be a living contagious fluid, by Wood (1899) an oxidizing enzyme, by Goldstein (1927) a protozoan parasite, by Vinson (1931) an inanimate chemical substance. Others have postulated that viruses are infinitesimally small living organisms or that they are autocatalytic substances producing disease and manufactured in certain injured and stimulated host cells. Gordon (in Jordan and Burrows) makes two pertinent points: 1. the distinction between animate and inanimate matter may prove pointless as regards the viruses, and 2. it is unnecessary and undesirable to assume that all viruses are of essentially the same nature. Both he and Rivers suggest that the very small viruses may be inanimate and the larger ones micromicrobes.

Antibiotics block bacterial metabolism and prevent the assimilation of amino acids in the synthesis of bacterial protoplasm. The smallest viruses are presumed to be nucleoprotein molecules and are thus unable to carry out metabolism; they therefore resist antibiotic action. Antibiotic action against virus diseases appears clinically proportional to their size. The larger the virus, the more susceptible it is to antibiotic action. The Common Cold Research Unit of Great Britain, however, has suggested that somatic cells produce minute amounts of a protein derivative that tends to negate virus production. This substance, called interferon, may represent a major breakthrough in prophylaxis.

BIBLIOGRAPHY

Carter, C. F., and A. L. Smith. *Microbiology and Pathology*. Sixth Edition. St. Louis: C. V. Mosby Company, 1956. Pp. 434–437.

Krueger, W. W., and K. R. Johansson. *Principles of Microbiology*. Philadelphia: W. B. Saunders Company, 1959. Pp. 135–146 and 508–517.

Sarles, W. B., *et al. Microbiology*. Second Edition. New York: Harper and Brothers, 1956. Pp. 398–402.

Stewart, F. S. *Bigger's Handbook of Bacteriology*. Baltimore: Williams & Wilkins Company, 1959. Pp. 473–496.

Chapter XXXII

VIRUS DISEASES

VIRUS DISEASES OF MAN

Of the many virus diseases that affect man, the following list includes the most important known to date:

Variola or smallpox
Vaccinia or cowpox
Alastrim
Varicella or chickenpox
Herpes zoster
Herpes simplex
Molluscum contagiosum
Verruca or warts
Rubeola or measles
Rubella or German measles
Epidemic influenza
Common cold
Parotitis or mumps
Psittacosis or parrot fever
Lymphogranuloma venereum

Rabies or hydrophobia
Anterior poliomyelitis or infantile paralysis
Epidemic encephalitis, including St. Louis encephalitis and Japanese encephalitis Type B
Infectious mononucleosis
Lymphocytic choriomeningitis
Yellow fever
Pappataci fever or sandfly fever
Dengue or breakbone fever
Rift Valley fever
Trachoma
Inclusion blenorrhea
Epidemic keratoconjunctivitis

Coxsackie virus, ECHO virus, and adenovirus infections

Variola or Smallpox and Vaccinia or Cowpox. These two diseases are of historical interest because they were the first for which an effective method of vaccination was devised. In 1798 Jenner reported that a boy inoculated with material taken from a dairymaid infected with cowpox failed to develop smallpox when inoculated with pus obtained from smallpox lesions. Thus Jenner put into practical application his observation that the incidence of smallpox was relatively low among persons who lived in rural districts and came into contact with cows.

Smallpox and cowpox are closely related. An attack of one protects against the other. It is believed by some that vaccinia or cowpox virus is smallpox virus attenuated by passage through cattle.

Smallpox is a generalized, febrile disease characterized by vesicular eruptions which become pustular and crust, often leaving permanent pox marks. The incubation period is 6 to 15 days. Fatality is usually 10–30%. Transmission is by direct or indirect contact. Infectivity is high, starting in the incubation period and lasting well into convalescence. The skin lesions and respiratory discharges contain the virus. Vaccinia in man, produced by smallpox vaccination, is a mild and usually local disease.

1. *The Virus.* The virus passes through most filters, is resistant to low temperatures, glycerin, phenol, and ether, but is heat labile— a temperature of 55° C. or over destroys it. It has been estimated to be 125 to 175 mμ in size. Cytoplasmic inclusions, called Guarnieri or vaccinia bodies, may be found in epithelial cell lesions. In addition, small Paschen or elementary bodies may be demonstrated in vesicular fluid. These are believed to represent the virus.

2. *Pathogenicity.* Various animals, including horses, sheep, and cows, are susceptible to pox diseases.

3. *Laboratory Diagnosis.* Several laboratory procedures are available to differentiate smallpox from other diseases with which it may be confused. Unfortunately, these tests do not distinguish between smallpox, vaccinia, and alastrim. They include:

(a) *Paul Test.* When the scarified cornea of a rabbit is inoculated with material from a suspected lesion, a typical keratitis, in which Guarnieri bodies are demonstrable microscopically, is observed if the lesion was smallpox. The test is positive in about 50% of the cases.

(b) *Buddingh's Test.* Material taken from smallpox lesions is inoculated on the chorio-allantoic membranes of the developing chick embryo. The production of typical pox lesions within 72 hours, smears of which show typical lesions and Paschen bodies, is indicative of smallpox. This test is believed to be more reliable than the Paul test.

(c) *Intradermal Test.* A normal rabbit inoculated intradermally with material from smallpox lesions develops swelling on the second day, proceeding to crust formation and desquamation. An immune animal will develop no such lesion or only a slight response.

4. *Vaccines.* The vaccine most generally employed is that prepared from calf lymph (serum from pustules) obtained by rubbing vaccinia virus into the scarified abdomen of calves. The scarified areas are scraped off 5 days later, using sterile precautions. The pulp is mixed with twice its weight of sterile water and forced through metal sieves. The emulsion of finely divided tissue is mixed with 50% glycerin. Phenol up to 1% or brilliant green dye 1:10,000 is added to reduce bacterial contamination. The pulp is stored at 10° C. Potency tests are made, and cultures are prepared to exclude the presence of contaminants such as Micrococci, Streptococci, and anaerobic bacilli.

More recently introduced are vaccines prepared from virus grown in tissue culture or in the developing chick embryo. These, being bacteria-free, can be injected intradermally and produce immunity without scar formation.

5. *Methods of Vaccinating.* The site to be vaccinated is cleansed with soap and water, wiped with alcohol, and allowed to dry. The methods used for vaccination include:

(a) *Incision or Linear Abrasion Method.* This was first described by Jenner. The skin of the arm is stretched by grasping the underside of the arm. A scratch about one-eighth of an inch is made with the point of a sterile needle. Care should be taken not to draw blood. A drop of vaccine is applied with a wooden applicator and rubbed into the scratch for about 15 seconds and then allowed to dry. The drop of vaccine may be put on the skin first, then the scratch may be made and the vaccine rubbed in with the side of the needle or a sterile toothpick.

(b) *Multiple Pressure Method.* A drop of vaccine is applied. With a sharp, sterile needle held parallel to the skin, the operator presses the side of the needle point firmly and rapidly into the drop about 30 times in about 5 seconds. An area not more than one-eighth of an inch in diameter is covered, and if the skin has not been rubbed too much in cleansing and if the pressure has been applied perpendicular to the needle, no blood is drawn. After the pressures have been made, the remaining vaccine is wiped off the skin carefully with sterile gauze. This method is favored at present because it results in a minimum scar and is apparently less liable to bacterial infection.

(c) *Drill Method.* The epidermis is perforated with a steel drill having a sharp cutting edge 2 mm. in width and a tip which is sterilized by flaming. The flake of epidermis is removed with a single rotary turn and the vaccine is dropped on the circle exposed and rubbed in with a sterile toothpick. The operation should cause no bleeding.

(d) *Intracutaneous Method.* With the use of a hypodermic syringe about 0.1 ml. of material, tissue-cultured vaccine in most cases, is injected intracutaneously. The ordinary glycerinated virus diluted with 1 part of sterile distilled water has been used when success was not obtained with ordinary scarification methods.

6. *Vaccination Reactions.* These are of four types:

(a) *Primary Vaccinia.* This reaction is observed in persons who have never been vaccinated or who have not had smallpox. It may also be elicited in those in whom immunity against smallpox from previous vaccination has disappeared. About the fourth day a papule appears which quickly develops into a vesicle surrounded by a narrow zone of redness. This gradually enlarges and begins to spread about the seventh day, and reaches its height about the tenth to fourteenth day, after which it recedes. The vesicles dry up, and a scab is formed which falls off about the twenty-first day, leaving a scar.

(b) *Vaccinoid or Accelerated Reaction.* Persons who have a partial immunity from a previous vaccination or an attack of smallpox give this reaction, in which all the stages of a primary vaccinia, but milder and having a more rapid course, are observed.

(c) *Immune Reaction.* Persons who have had smallpox or those protected by previous vaccination give a reaction characterized by redness with some elevation in 24 to 48 hours, occasionally 72 hours, with itching and no vesicle formation. No scar or scab results after the quick fading of the redness. The reaction is indicative of complete protection.

(d) *Negative Reaction.* Revaccination is advised when no reaction is observed. Three failures to respond are considered as proof of immunity when the vaccine used is known to be potent.

7. *Immunity.* An attack of smallpox or cowpox confers immunity. Protection for one year or several years is afforded by vaccination with vaccinia virus. Following exposure the individual should be revaccinated. Persons immune to variola have antibodies (neutralizing, complement-fixing, agglutinins, and precipitins) demonstrable in their sera.

Alastrim. This is a disease which resembles a mild attack of smallpox. It is believed to be related to smallpox since it protects against smallpox, and an attack of smallpox protects against alastrim. Vaccination confers protection against both.

Varicella or Chickenpox. This is a rather mild, highly contagious disease characterized by fever and the appearance of vesicles after an incubation period of 14 to 21 days. The disease has been shown to be distinct from smallpox, with which it may be confused clinically in the early stages. An attack of chickenpox does not protect against smallpox, nor does smallpox confer immunity against chickenpox.

Elementary bodies which are agglutinated in convalescent serum are present in vesicle fluid.

Chickenpox is believed by some investigators to be related to herpes zoster, for cases have been reported of children who developed chickenpox on contact with herpes zoster. Complement-fixing antibodies against fluid from herpes zoster and varicella vesicles have been found in the sera of patients with herpes zoster and varicella. Some cross agglutination has been observed. One attack of varicella usually protects against subsequent infection.

Herpes Zoster. This disease, also termed shingles or zona, is an acute dermotropic virus disease characterized by a vesicular dermatitis in which the vesicles follow a nerve trunk. The virus has as yet not been transmitted to any experimental animals with filtered material. Inclusion bodies have been described in cases of herpes zoster. One attack confers a lasting immunity.

Herpes Simplex. This is a mild, dermotropic, acute, eruptive, virus disease of the skin and mucous membranes characterized by vesicles containing serous fluid, with subsequent scab formation but no scars.   Vesicles on the lips and nostrils (cold sores), on the genitalia, or associated with fevers (herpes febrilis, fever sores) are most common. The vesicles usually appear within 24 to 48 hours, and last about 7 to 14 days.   Herpetic infections have been reported following contact, local irritation, exposure to cold, heat, ultraviolet light, and other factors, dependent largely on the individual.   Intranuclear inclusion bodies have been described in the lesions.   Herpes virus is filterable through Berkefeld V and N candles.   It is about 150 mμ in size, and is thermolabile, for a temperature of 50° C. destroys it.   It may be cultivated in tissue cultures and in the developing chick.

When inoculated onto the scarified surface of a rabbit's cornea, a keratitis is produced.   Virulent strains may become neurotropic.   A fatal encephalitis has been produced in rabbits, and the intracerebral injection of herpes virus into mice produces characteristic transmissible encephalitis.

Infection with herpes simplex virus differs from most virus diseases in that the duration of immunity is uncertain or, at best, of short duration.   Neutralization tests have demonstrated that the serum of persons recovering from herpes have neutralizing antibodies. Complement fixation antibodies also have been found.   Antibodies have been noted in the serum of persons who paradoxically develop herpes simplex, perhaps indicating a persistent latent infection.

Molluscum Contagiosum.   This is a children's disease characterized by benign, epithelial nodules formed around the face or on the mucous membrane of the genitalia, and transmitted by contact.   Inclusion bodies, which are believed to contain the elementary bodies, are noted in the cytoplasm.   The disease has been experimentally transmitted to man.

Verruca or Warts.   This disease is characterized by benign papilloma of the skin.   Emulsions made of wart tissue have produced warts in human beings injected intradermally, after an incubation period of 4 weeks.   Warts contain intranuclear inclusion bodies.

Rubeola or Measles.   This is an acute, infectious disease of childhood, characterized by fever, catarrh, coryza, Koplik spots on the buccal mucous membrane, and papular rash which spreads over the surface of the body, with a final desquamation or scaling of the epidermal cells.   Secondary invaders may account for the dangerous sequelae, such as bronchopneumonia, endocarditis, meningitis, etc.   This disease has an incubation period of 10 to 14 days, and one attack usually confers a lasting immunity.   Filtered nasopharyngeal washings and blood of patients at the height of the disease have been injected into monkeys, with the development of measles-like eruptions.

Convalescent serum, pooled adult serum, parental whole blood, and placental extract have been used both therapeutically and prophylactically to afford complete temporary protection or to assure an attenuated form of the disease and resultant immunity in those exposed to infection.

Rubella or German Measles. This disease is similar to measles, from which it may be differentiated by a longer incubation period of 10 to 21 days, a short prodromal period, and absence of Koplik spots. The disease runs a benign course and is conspicuous by a complete absence of complications and sequelae. The etiological agent has not been isolated, but many believe it to be a virus. One attack of the disease confers immunity.

Epidemic Influenza. This disease was believed to be caused by *Hemophilus influenzae* or the Pfeiffer bacillus, until 1933, when a virus was isolated from the throat washings of patients suffering from influenza. The virus has been estimated to be about 70 to 100 mμ in size. It resists freezing for about 2 weeks, and retains its potency in glycerin for the same length of time. When dried and frozen by means of the lyophile apparatus, it remains potent in mouse lungs after 6 weeks in the refrigerator.

Research on influenza control during global epidemic outbreaks of Asian influenza in the late 1950's failed to produce effective chemoprophylactic or chemotherapeutic agents. Russian virologists reported some success with live virus vaccines produced from attenuated Asian influenza strains and administered by inhalation. In the United States killed viral vaccines given by a pararespiratory route were tried, with antigenic variation countered by incorporation with type A virus strains. A broad-spectrum influenza vaccine includes four families of influenza A and two of influenza B. Major antigens of strains formerly prevalent may reappear cyclically. Antibody levels are higher and endure longer when influenza vaccine is given intramuscularly as an emulsion of water in light mineral oil. High antibody levels may persist without significant change for four to five years. Influenza viral vaccine is grown in the extra-embryonic fluid of chick embryos from types A, A', and B virus strains.

Neutralizing antibodies against influenza virus are present in a large proportion of normal individuals, and following an attack of influenza there is a rise in the titer of such antibodies. Complement-fixing antibodies also have been demonstrated. An *in vitro* test is based on *Hirst's phenomenon,* the ability of influenza virus to agglutinate chicken erythrocytes mixed with it. The reaction is prevented by immune serum. An attack of influenza results in only a temporary immunity.

Common Cold. This is an acute, frequently epidemic, highly communicable disease transmitted by droplet infection. It is characterized by catarrhal inflammation of the nose, throat, trachea, nasopharynx, and upper bronchi. Numerous microorganisms, including pneumococci, *Hemophilus influenzae,* and others, have been considered the etiologic agents from time to time.

Filtrates of nasal secretions of persons with colds have produced infection in human volunteers inoculated intranasally. Clinical symptoms similar to those observed in man have been incited in chimpanzees with such filtrates, and it was possible to transmit the infection from ape to ape.

Only three or four of the nine known strains of cold virus indicate a relationship to cold outbreaks sufficiently positive to warrant their inclusion in viral-cold vaccines. The cold virus can be grown on human kidney cells in standard medium enriched with amino acids, vitamins, and minerals, but containing less bicarbonate to lower alkaline content. The cold virus is extremely resistant and stable.

Parotitis or Mumps. This is an acute, communicable disease characterized by inflammation of the parotid or other salivary glands and frequent involvement of the testes and meninges. The disease is transmissible to monkeys. The production of orchitis and parotitis has also been reported in cats. One attack usually confers a lasting immunity. Convalescent serum has been used prophylactically during the incubation period.

Psittacosis or Parrot Fever. This is a communicable disease of parrots transmissible to man through the respiratory tract by direct or indirect contact. It is caused by a virus with an affinity in parrots for the liver and spleen. The intranasal or intramuscular injection of the virus produces the disease in parrots. These animals after recovery are refractory to subsequent infections. The disease is transmissible to mice, guinea pigs, rabbits, and *Macacus rhesus* monkeys.

In man the virus produces a pneumonia, after an incubation period of 10 to 30 days, and the virus may be demonstrated in the sputum and blood. The sputum may be inoculated intraperitoneally into white mice. The development by the animals of an illness which is fatal usually within 5 to 14 days, and occasionally not before 30 days, is diagnostic. Lesions are found in the liver and spleen. Cytoplasmic inclusion and elementary bodies are also demonstrable in infected material. They are small, coccoid forms arranged singly and in pairs, and are believed to be one stage in a developmental cycle.

In monkeys intracerebral inoculation causes a meningo-encephalitis. In parrots virus infection usually subsides following a single injection of aureomycin.

One attack of the disease usually produces an active immunity lasting for some time. Complement-fixing and neutralizing antibodies are demonstrable in the sera of vaccinated and convalescent individuals. Intramuscular injections of fresh, unattenuated virus result in successful immunization.

Lymphogranuloma Venereum. Also known as lymphogranuloma inguinale, this is a human contagious disease acquired by sexual contact and characterized by inflammation of the lymphatic glands of the inguinal region and development of small herpetiform lesions on the external genitalia. Intracerebral inoculation of monkeys and mice with suspensions of gland tissue result in meningo-encephalitic lesions. Cats have also been experimentally infected. The virus is filterable through Chamberland L_3 and Berkefeld V filters. It may be preserved by freezing at $-2°$ to $-3°$ C. for about 10 days, but is destroyed by heating at $60°$ C. for 30 minutes. Elementary bodies and a developmental cycle similar to the psittacosis virus have been observed.

Lymphogranuloma venereum may be diagnosed by a skin test known as the *Frei test,* performed by injecting intracutaneously into the forearm an antigen of heated pus from an infected gland. The reaction reaches its height in about 48 hours, and consists of an inflammatory, infiltrated, dome-shaped area about 0.5 cm. in diameter, having in the center a small area of necrosis surrounded by a red zone.

Lymphogranuloma is reported to produce a toxin similar to bacterial endotoxins, capable of causing production of specific antitoxin. It is also one of the few viruses that appear to be affected by the sulfonamides.

Rabies or Hydrophobia. This is a specific, fatal, virus disease of dogs and other animals, characterized by extreme irritation of the central nervous system and transmitted to man by the bites of infected animals.

1. *The Virus.* The saliva of infected animals contains the virus. In rabies the affected nerve cells contain cytoplasmic inclusion bodies called Negri bodies; their presence indicates rabies. The virus may

be attenuated by drying in KOH, by exposure to 1% phenol, and by temperatures above 45° C. It remains potent in glycerin at 7° C. for some time. Rabies virus has been cultivated in tissue culture and in the chick embryo. Such preparations have been studied for their efficacy as immunizing agents. Mice have been made immune for about 9 months by the intraperitoneal injection of such virus, and dogs have also developed a considerable immunity.

2. *Disease.* In man, after an incubation period, usually of 26 to 70 days, often longer, the first symptoms of difficulty in breathing and inability to swallow are observed, with a rise in temperature. In the dog the disease is manifested by increased aggressiveness, characteristic restlessness, loss of appetite, desire to bite, paroxysms of fury, rapid emaciation, paralysis, and death. Death in man occurs in the majority of cases on the third or fourth day after the appearance of symptoms. The development of rabies following the bite of a rabid animal is dependent upon the point of inoculation, the amount of virus introduced, and the strength of the virus. The period of incubation is shorter and the disease is generally severe when the bite is on the face or in the tips of the fingers, where there is an abundance of nerves.

3. *Laboratory Diagnosis.*

(a) *Spread Method.* Spreads made from the cortex in the region of the fissure of Rolando or in the region corresponding to it in lower animals, from Ammon's horn, and from the gray matter of the cerebellum are examined after staining with Giemsa's method for the presence of Negri bodies, which appear magenta with blue granules.

(b) *Animal Inoculation.* If Negri bodies are not detected in the spreads, small portions of the hippocampus are emulsified in sterile physiologic saline and about 0.125 ml. is inoculated subdurally into guinea pigs or rabbits. Generally, death occurs after 16 days or longer, and Negri bodies may be demonstrated in the brain tissue on autopsy.

A mouse inoculation test has been introduced recently. It consists of intracerebral inoculation of brain material. Some of the mice are examined for Negri bodies in the brain and others are studied for the appearance of characteristic weakness and paralysis of the hind legs, prostration, and death. This method has the advantage of earlier diagnosis.

(c) *Serological Reactions.* A mouse protection test has been devised for measuring the antibodies in serum against rabies virus. Complement fixation technics have thus far yielded unsatisfactory results.

4. *Treatment and Prophylaxis.* Following the bite of an animal all wounds should be immediately cleaned and, if possible, thoroughly cauterized with fuming nitric acid. Various vaccines have been prepared for the prophylactic treatment of rabies following the bite of a rabid animal and for immunization of dogs.

(a) *Pasteur's Method.* This procedure, with some modifications, is still in use today. It is based upon the fact that rabies virus may be intensified by passage through rabbits. Virus taken from the street dog is known as *street virus.* The virulence for the rabbit of this virus, which to begin with produces rabies after an incubation period of from 12 to 14 days, may be enhanced so that the incubation period is reduced to about 6 or 7 days, after which the virus remains constant in virulence and is known as *fixed virus.* (Negri bodies are not demonstrable in fixed virus infections, perhaps because of the shorter incubation period.) The fixed virus is then subjected to various periods of drying to decrease its virulence. The course of treatment consists of 14 to 21 separate inoculations of fixed virus of increasing virulence.

(b) *Semple's Method.* Treatment consists of a course of 14 injections, given daily, of 2 ml. of a vaccine prepared by incubating an 8% emulsion of brains of rabbits injected with rabies fixed virus, in 1% phenol in normal salt solution for 24 hours at 37° C. and then diluting it with an equal volume of saline (thus making a 4% emulsion of the virus). The injections are made subcutaneously, usually over the abdomen.

(c) Chloroform- and ultraviolet-light-treated virus have also been shown to be efficient immunizers.

(d) The most efficient methods of prevention lie in the muzzling of dogs and in the quarantine of incoming animals for observation.

A slight decrease in mortality from rabies has been shown in the statistics from various parts of the world. Cases of paralysis following the administration of vaccines have been reported. The immunity produced by antirabic treatment is variable.

Anterior Poliomyelitis or Infantile Paralysis. This is a disease of the central nervous system in which the nerve cells of the anterior horn of the spinal cord are injured, resulting in flaccid paralysis of the muscles enervated by the damaged nerve cells. The disease is ushered in after an incubation period of 7 to 14 days with symptoms of fever, headache, stiffness of the neck, irritability, and gastrointestinal disturbances. *Abortive poliomyelitis* is a form of the disease in which there is no evidence of muscle involvement. There may be only slight transient weakness, in which case the attack is termed *nonparalytic.*

1. *The Virus.* Poliomyelitis was successfully transmitted to monkeys by Landsteiner and Popper in 1908 by the intraperitoneal injection of spinal cord from a case of poliomyelitis. These investigators were unable to carry the virus from monkey to monkey. Flexner and Lewis, in the same year, inoculated monkeys intracerebrally and found that with this route they could transmit the virus serially from monkey to monkey. Attempts to transmit the disease to other laboratory animals were unsuccessful, until recently when Armstrong reported the transmission of poliomyelitis (Lansing strain)

to Eastern cotton rats by intracerebral injection. The infection of white mice has also been accomplished. The virus is extremely small, and its size is estimated as 8 to 12 mμ. It resists desiccation, freezing, glycerin, and exposure to 5% phenol. It is, however, thermolabile, a temperature of 45° C. being lethal.

2. *Portal of Entry.* The virus has been demonstrated in the nasopharyngeal washings of patients and of healthy carriers. Monkeys have been infected by intranasal instillations of the virus. Certain experiments in which the olfactory nerve has been severed or the nasal mucosa blocked by chemicals have shown that infection is prevented in monkeys so treated. Therefore, some believe that infection with poliomyelitis virus occurs through the respiratory tract. However, the isolation by many investigators of the virus from stools of human beings and monkeys infected with the disease has raised the possibility of a gastrointestinal portal of entry. The virus has also been isolated from sewage and from flies.

3. *Immunity.* Well-nourished individuals seem to be more susceptible to poliomyelitis than are poorly-nourished ones. One attack of the disease confers a strong immunity, second attacks being rare. The serum of convalescents contains antiviral substances which can be demonstrated by neutralization tests in monkeys. A large proportion of adults also have these neutralizing substances, and there does not seem to be much correlation between neutralizing titer and either the development of poliomyelitis or recovery from it.

4. *Prophylaxis.* Various methods have been attempted to prevent poliomyelitis, but at present there is no effective procedure for conferring an immunity of either slight or considerable duration.

(a) *Convalescent Serum.* Although a number of studies have been made on the prophylactic value of convalescent serum or normal adult serum, there is no conclusive evidence that either one is of value.

(b) *Nasal Spray.* A high degree of resistance to poliomyelitis in monkeys was reported by several investigators but questioned by others after spraying the nasal mucosa with a solution of sodium aluminum sulfate (alum), 4% tannic acid, or picric acid. However, trial of sprays in children during two epidemics furnished no evidence that either picric acid alum spray or 1% zinc sulfate reduced the incidence of poliomyelitis.

(c) *Active Immunization.* Salk developed trivalent formalin inactivated vaccine types 1, 2, and 3 in dosages of 1 cc. two shots 7–10 days apart, followed by a booster shot 6–7 months later. Monkey sensitivity tests and a second filtration through Seitz-type filters during production, along with small amounts of penicillin and streptomycin, have rendered the vaccine safe.

Antibody response shows up 7–10 days after the initial injection intramuscularly. The effectiveness of the vaccine is reported to be 70%–76%.

Epidemic Encephalitis. During the past few years a number of infections of the central nervous system have been recognized as caused by viruses. These diseases have been termed *epidemic encephalitis* and include St. Louis encephalitis and Japanese encephalitis Type B. They may be distinguished from encephalitis lethargica or von Economo's disease, for which no accepted etiologic agent has yet been isolated.

Von Economo's disease is a chronic sporadic infection. Following an acute stage, muscle spasticity, mental deterioration, and personality alteration occur. These may become worse and the patient may die, there may be recovery with some disability, or the patient may make a complete recovery. Herpes virus has been isolated in some of these cases, and is believed by a number of workers to be the cause of the disease. In Japan there is a type of encephalitis which is similar to Von Economo's disease. It has been referred to as *Japanese encephalitis Type A*.

1. *St. Louis Encephalitis.* This disease was first recognized in epidemic form in 1933 in St. Louis. Following an incubation period of 4 to 21 days, symptoms of fever and meningeal irritation occurred. The disease had a short course and few sequelae. A virus was recovered by the intracerebral inoculation of monkeys and mice with brain tissue of fatal cases. Neutralizing antibodies against the virus were demonstrated in the sera of patients recovering from the infection and also in the sera of persons living in various parts of the United States. Reservoirs for the disease appear to be birds and a number of mammals. Mosquitoes and probably ticks play a role in transmission.

2. *Japanese Encephalitis Type B.* In Japan there has been recognized a type of encephalitis which differs from Type A in that it is more prevalent in the summer months. The disease is more acute and the incidence is highest among older persons. Although the disease was first differentiated from Type A in 1928 and designated Type B, it was not until 1935 that its etiologic agent was isolated by the intracerebral inoculation of mice. The virus differs from that of St. Louis encephalitis in that it produces a more severe disease in monkeys. It has been shown to be distinct from St. Louis encephalitis virus by neutralization tests, but shows partial immunologic relationship to it by complement fixation. Insects may be involved in transmission.

3. *Postinfection Encephalitis.* Cases of encephalitis following varicella, pertussis, typhus fever, and mumps have been reported. Vaccination against smallpox is occasionally followed by encephalitis

known as postvaccinal encephalitis. The cause of these types of encephalitis is not known.

Infectious mononucleosis. Also called glandular fever and "kissing disease," this benign lymphadenitis is characterized by an increase in the number of mononuclear cells. The etiological factor appears to be an adenovirus transmitted in the saliva.

The disease has a predilection for young adults, and produces slight fever, laryngitis, and slightly swollen lymph glands.

Lymphocytic Choriomeningitis. This is an infection of the central nervous system, beginning with a mild, upper respiratory infection, for which a virus was isolated in 1934 by Armstrong and Lillie in the course of infecting monkeys with material from a patient who died of encephalitis. The virus has also been transmitted to mice. It is present in the brain, spinal fluid, blood, and urine of monkeys that are injected. A virus similar to that of lymphocytic choriomeningitis has been found in normal mice and in monkeys. The virus has been shown to be distinct from the viruses of equine encephalomyelitis, St. Louis encephalitis, and poliomyelitis by means of neutralization and complement fixation tests. Arthropods may play a role in transmission.

Yellow Fever. This disease is transmitted by the female mosquito, *Aedes aegypti*, and is characterized by fever, chills, muscular pain, jaundice resulting from liver damage, severe gastrointestinal symptoms, and vomiting of blood. Filterability of the causal agent, mosquito transmission, incubation period in the mosquito, and infectivity of the patient's blood were definitely established by the American Army Commission consisting of Reed, Carroll, Agramonte, and Lazear at the beginning of this century, following earlier work by Carlos Finlay.

1. *The Virus.* Yellow fever is transmissible to monkeys by intraperitoneal or intradermal injections of the virus. Mice injected intracerebrally develop encephalitis, and the virus is altered from one with primarily viscerotropic properties to one that is neurotropic. Intranuclear inclusions are demonstrable in the liver of infected monkeys and man and in the ganglion cells of mice. The virus has been estimated to be 17 to 28 mμ in size. It resists freezing, drying, and exposure to glycerin and formalin, but it is destroyed by heating at 60° C. for 10 minutes. It has been cultivated in tissue culture.

2. *Immunity.* One attack confers immunity. Neutralizing antibodies have been demonstrated in the sera of a large number of individuals in areas where yellow fever is prevalent.

3. *Prophylaxis*. The subcutaneous injection of vaccine prepared with human immune serum and dried living virus fixed for mice produces a rise in titer within a few weeks. Vaccination with virus attenuated by prolonged cultivation in tissue culture has given some satisfactory results.

4. *Jungle Yellow Fever*. Yellow fever in the absence of *Aedes aegypti* has been observed in certain areas, and termed jungle yellow fever. This form of the disease and the causing virus appear identical with the classical type.

5. *Prevention*. Many areas have been freed from yellow fever by precautions taken to prevent the breeding of mosquitoes.

Pappataci Fever, Sandfly Fever, or Phlebotomus Fever. This is an acute, mild, febrile disease of tropical countries, which is transmitted by sandflies. The disease is characterized by chills, fevers, conjunctivitis, and pains. It has an incubation period of 1 to 6 days. The virus is demonstrable in the blood. One attack confers immunity for at least 4 months and probably a year or more.

Dengue or Breakbone Fever. This is a mild disease of warm climates, characterized by fever, headaches, muscular pains, and often a maculopapular rash. The disease is carried by the *Aedes aegypti* and a related mosquito, and it has been experimentally transmitted to man after an incubation period of 3 to 4 days by the injection of blood and serum filtrates from cases of the disease. The immunity conferred by an attack is of uncertain duration.

Rift Valley Fever. This is primarily a disease of sheep, cattle, and goats, but produces in man an infection with symptoms of malaise, nausea, fever, and severe headache. The disease, which is probably mosquito-borne, has been transmitted experimentally to mice, monkeys, and ferrets. Intranuclear inclusion bodies are demonstrable in the liver. The virus has been estimated to be about 23 to 35 mμ in size, and has been cultivated in tissue culture. One attack confers immunity. Neutralizing and complement-fixing antibodies are found in the serum of convalescents from the disease.

Trachoma. This is an eye disease, probably of virus origin, which is found in unhygienic surroundings and is characterized by granulated eyelids, conjunctivitis, and lacrimal gland injury. Cell inclusions containing elementary bodies are found in the infected conjunctiva. After an incubation period of several days to a month, monkeys have been experimentally infected with material taken from cases of trachoma. However, characteristic inclusion bodies have not been found in the experimental disease.

IMPORTANT VIRUS DISEASES OF MAN

NAME OF DISEASE	PERIOD OF INCUBATION	PORTAL OF ENTRY AND MODE OF TRANSMISSION	SELECTIVE ACTION AND LESIONS	INCLUSION BODIES	SUSCEPTIBLE ANIMALS	IMMUNITY	PROPHYLAXIS
Variola or smallpox	6-15 days	Contact, probably mucous membrane of fauces	Skin and other organs, dermotropic	Cytoplasmic (Guarnieri body) and intranuclear (Paschen or elementary body)	Monkeys, calves, and rabbits	One attack confers immunity	Vaccination with vaccinia virus
Vaccinia or cowpox	3 days	Contact or vaccination	Skin, dermotropic	Cytoplasmic	Cattle, sheep, horses, and rabbits	Immunity of variable duration following vaccination	Vaccination with vaccinia virus
Varicella or chickenpox	14-21 days	Contact, by droplet infection via nasopharyngeal secretions	Skin and mucous membranes, dermotropic	Cytoplasmic and intranuclear reported	Not known	One attack confers immunity	Convalescent serum may be of value
Herpes zoster or shingles	12-15 days	Contact	Skin, nerves, and ganglia, dermotropic	Cytoplasmic and intranuclear reported	Not known	One attack confers immunity	None
Herpes simplex	1-2 days	Contact, fever, exposure to ultraviolet rays, etc.	Skin, dermotropic	Intranuclear	Rabbits, guinea pigs, mice, monkeys	Immunity of short duration, probably none	None
Molluscum contagiosum	Experimentally, 14-25 days	Contact	Skin, dermotropic	Cytoplasmic	None	Little is known	None
Verruca or warts	Experimentally, 4-9 weeks	Contact	Skin, dermotropic	Intranuclear	None	May be local	None
Rubeola or measles	10-14 days	Contact, by droplet infection	Skin and mucous membranes of the respiratory tract, dermotropic	Cytoplasmic reported	Monkeys	One attack confers immunity	Convalescent or normal serum, normal blood, placental extract
Rubella or German measles	10-21 days	Contact, by droplet infection	Skin and mucous membranes	None	None	One attack confers immunity	None
Epidemic influenza	3-8 days	Contact, by droplet infection	Lungs and nasal mucosa	None	Ferrets and mice	Immunity is of short duration	Vaccine may be of value
Common cold	1-5 days	Contact, by droplet infection	Mucous membrane of respiratory tract	None	Chimpanzees	Immunity is of short duration	None

Disease	Incubation period	Mode of transmission	Tissues affected	Inclusion bodies	Susceptible animals	Immunity	Convalescent serum
Parotitis or mumps	5–21 days	Contact, by droplet infection	Salivary glands	Cytoplasmic	Monkeys, and possibly cats	One attack confers immunity	
Psittacosis or parrot fever	10–30 days	Contact with infected parrots	Lungs	Cytoplasmic, elementary bodies	Parrots, mice, rabbits, monkeys, and guinea pigs	One attack confers immunity	Vaccination with active virus intramuscularly
Lymphogranuloma venereum	Several days to about 2 weeks	Sexual contact	Lymphatic system in genital region	Cytoplasmic, elementary bodies	Monkeys, mice, rabbits, cats, and guinea pigs	Little is known	None
Rabies or hydrophobia	Variable, 27–60 days, or longer	Bite of rabid animal	Central nervous system, neurotropic	Cytoplasmic, Negri body	All mammals and some birds	Vaccination confers immunity of short duration	Vaccine of fixed virus and cauterization of wound
Anterior poliomyelitis or infantile paralysis	7–14 days	Contact, by droplet infection through respiratory tract, or possibly gastrointestinal tract	Central nervous system, nerve cells of anterior horns, neurotropic	None	Monkeys, cotton rats, and mice reported	One attack confers immunity	Salk trivalent inactivated viral vaccine
Epidemic encephalitis: St. Louis encephalitis Japanese encephalitis Type B	4–21 days; Not known	Contact, insects	Central nervous system, neurotropic	None	Mice and monkeys	One attack confers immunity	None
Lymphocytic choriomeningitis	Experimentally 4–7 days	Probably by contact, possibly insects	Central nervous system, neurotropic	None	Monkeys, mice, and guinea pigs	One attack confers immunity	None
Yellow fever	4–13 days	Through skin by bite of female *Aedes aegypti* mosquito	Liver and other organs	Intranuclear	Mice and monkeys	One attack confers immunity	Vaccine of virus and immune serum, or tissue-cultured virus
Pappataci or sandfly fever	1–6 days	Through skin by bite of sandfly	No definite lesions described	None	None	Immunity of 4 months to 1 year or more	None
Dengue or breakbone fever	3–6 days	Through skin by bite of *Aedes aegypti* mosquito	Skin and throat	None	Possibly monkeys	Immunity of uncertain duration	None
Trachoma	Not known	Contact, through conjunctiva	Mucous membrane of eye	Cytoplasmic and elementary	Possibly monkeys	Little is known	None

305

NOTE: For discussion of Coxsackie virus, ECHO virus, and adenovirus infections, see p. 309.

Inclusion Blennorrhea. This disease, clinically similar to trachoma, is most prevalent in the newborn. Inclusions are found that are morphologically similar to trachoma.

Epidemic Keratoconjunctivitis. This is a highly infective eye disease recently appearing among West Coast shipyard workers, where it was known as "shipyard disease." There are frequent swelling of the regional lymph nodes, and systemic symptoms.

EXTRAHUMAN VIRUS DISEASES

Viruses have been reported to cause disease in practically every type of living thing—mammals, fowl, birds, amphibians, fish, insects, plants, and perhaps bacteria (see Chapter XXXIII on Bacteriophage). A partial list of these follows:

1. *Sheep.* Agalactia, catarrhal fever, contagious pustular dermatitis, louping ill, Nairobi disease, Rift Valley fever, sheep pox.

2. *Horses or Cattle.* African horse sickness, Borna disease, equine encephalomyelitis types A and B, equine influenza, foot-and-mouth disease, horse pox, malignant catarrh of cattle, periodic ophthalmia of horses, rinderpest (cattle plague), vesicular stomatitis.

3. *Hogs.* Hog cholera, swine influenza, swine pox.

4. *Dogs and Foxes.* Distemper, encephalitis of foxes, pseudorabies, rabies.

5. *Fowl.* Fowl plague, fowl pox (contagious epithelioma), infectious laryngotracheitis, leucemia of chickens, Newcastle disease, Rous sarcoma.

6. *Rabbits.* Infectious fibroma, infectious myxomatosis, infectious papilloma, rabbit pox, spontaneous encephalitis, virus III.

7. *Birds.* Avian diphtheria, canary pox, pigeon pox, psittacosis, ornithosis.

8. *Guinea pigs.* Guinea pig epizoötic, guinea pig paralysis, salivary gland disease.

9. *Mice.* Infectious ectromelia, influenza-like disease of Swiss mice, lymphatic leucemia.

10. *Rats.* Salivary gland disease.

11. *Ferrets.* Epizoötic disease.

12. *Frogs.* Carcinoma.

13. *Fish.* Carp pox, epithelioma of barbus, lymphocytic disease.

14. *Insects.* Polyhedral diseases, sacbrood of honey bees, silkworm jaundice.

15. *Plants.* Mosaic diseases, rosette of wheat, tulip break, etc.

16. *Bacteria.* Bacteriophage(?).

Of these diseases of animals, rabies, psittacosis, equine encephalomyelitis, foot-and-mouth disease, Rift Valley fever, and cowpox are transmissible to man. Some of these have been discussed in the preceding section on virus diseases of man.

Foot-and-Mouth Disease. This is an acute, febrile disease of cattle, sheep, and hogs, which is characterized by vesicular eruptions inside the mouth and around the feet.

1. *Virus*. The causative agent of foot-and-mouth disease is present in the vesicular lymph and in the blood in the early stages. It is 8 to 12 mμ in diameter. It resists alcohol, chloroform, phenol, and glycerin, but is destroyed by alkali. It is destroyed at 37° C. in 24 hours, but remains active in the icebox for several months. Inoculation into guinea pigs, rabbits, dogs, and cats produces the disease in these animals. Inclusion bodies have been observed in lesions after 24 and 48 hours. The virus has been cultivated in tissue culture containing guinea pig embryo tissue and clotted guinea pig plasma.

2. *Transmission*. The highly contagious disease is transmitted by direct or indirect contact with the virus released by rupture of the vesicles. Ingestion of contaminated raw milk is one of the factors involved. When transmitted to man, the disease is usually mild, with transient vesicles on the hands and feet and sometimes gastrointestinal disturbances.

3. *Immunity*. Immunity following an attack usually lasts for a year. Vaccines of living virus, virus in combination with immune serum, or formalinized virus have been used, but they do not protect animals against the disease.

Swine Influenza. This disease of swine has been shown to be caused by the combined action of a filterable virus and the bacterium *Hemophilus influenzae suis*. The injection of virus alone produces a mild disease in swine. The bacterium has no effect on the animals. The combination of the two, however, causes an illness which is similar to the natural disease in swine. Intramuscular injection of virus alone produces an immunity in swine against both the mild virus disease and the severe natural disease, but injection with the bacillus produces no immunity whatsoever. Antibodies against swine influenza virus have been observed in the sera of adults, but were lacking in the sera of children tested. When ferrets are injected with swine influenza virus a disease is produced similar to that caused by human influenza virus. It is believed that swine influenza virus is an adapted strain of human influenza virus causing the pandemic of influenza in man in 1918–1919. By means of complement fixation tests swine influenza virus cannot be differentiated from human influenza virus. Sera of ferrets and mice that have been repeatedly inoculated with human influenza virus contain antibodies against both human and swine virus.

Equine Encephalomyelitis. This is a disease first observed in horses but since found in a wide variety of wild and domestic animals and birds. It is characterized by inflammation of the meninges, with pyrexia, inco-ordination, vertigo, paresis, and motor and sensory paralysis. It is transmitted by mosquitoes and other blood-sucking insects. Three strains of virus causing the disease have been identified by immunological methods—the severe eastern, the less severe western, and the Venezuelan. All three have been known to infect man. Formolized tissue culture and chick embryo virus vaccines have been used with good results in the immunization of horses and laboratory workers. For treatment of this disease an antiserum may be given with some benefit, particularly in the early stages.

Rinderpest or Cattle Plague. This is generally a fatal disease characterized by catarrh, fever, conjunctivitis, diarrhea, and emaciation, and largely spread through contaminated food and water. The virus is present in the blood, secretions, and intestinal contents of infected animals. One attack confers immunity. Chloroform-treated emulsions of spleens and lymph nodes of animals killed in the acute stage of the disease have been used for vaccination of cattle. However, the disease may be spread by such vaccinated animals.

Fowl Pox or Contagious Epithelioma. This is a disease of chickens manifested by wartlike scabs on the combs and head. Elementary bodies were described by von Prowazek. Immunization with vaccine is highly successful.

Canine Distemper. This is a disease of young dogs, foxes, and other animals. It is characterized by catarrhal inflammation of the respiratory tract and diarrhea. It has an incubation period of from 3 to 5 days. The disease may be prevented by immunizing vaccines. Convalescent serum has been used to confer passive immunity, and has therapeutic value also.

Hog Cholera. This disease was shown by Dorset, in 1903, to be due to a filterable virus, complicated by the presence of *Salmonella choleraesuis* (*suipestifer*) as a secondary invader. The virus is present in the urine. One attack of the disease produces a fair immunity, and simultaneous inoculation of hyperimmune antiserum and virus confers a lasting active immunity.

Infectious Myxomatosis of Rabbits. This is a highly fatal, virus disease of domesticated rabbits, characterized by tissue swellings forming gelatinous tumors and a purulent conjunctivitis. The virus is present in discharges from the nose and eyes, and in the blood and serous exudates. The rapid spread of tumors to all parts of the body reminds one of a similar phenomenon in the case of human cancer. It is remarkable that injection of relatively benign fibroma virus into rabbits produces immunity against the cancer-like disease, infectious myxomatosis.

Coxsackie Virus, ECHO Virus, and Adenovirus Infections.

Coxsackie viruses include thirty distinct types which fall into two groups, A and B, of the so-called enteroviruses.

These small viruses characteristically occur in the feces of infected persons, and are epidemiologically similar, being widely distributed throughout the world. They are highly infectious, especially to the young.

Group A viruses have been known to cause poliomyelitis-like lesions in mice and monkeys. In addition there are six immunotypes associated with herpangina, a febrile disease of children characterized by ulceration and visicles of fauces and soft palate.

The group B viruses have five serologic types and apparently are indistinguishable clinically, in mice, and in tissue culture. They are frequently the cause of epidemic pleurodynia with meningitis, and myocarditis. Myocarditis frequently complicates poliomyelitis. The etiological significance of the group B virus in aseptic meningitis has been firmly established by recent and repeated isolation of these viruses from cerebrospinal fluid. All five types of group B viruses may be recovered from cerebrospinal fluid.

Coxsackie virus infections may be identified by isolating the virus from feces collected during the acute phase of the illness. Irradiation appears to increase and prolong the tissue yield of injected Coxsackie virus in mice.

ECHO (*E*nteric *C*ytopathogenic *H*uman Orphan) *viruses* are found in tissue cultures and stools. They produce transitory infections of the alimentary tract. There are twenty distinct types. ECHO viruses types 4, 6, 9, 16, and 18 may be found in the stools of healthy human beings without acute febrile illness. Summer diarrheal disease of infants and very young children may be caused by ECHO viruses. Specific virologic diagnoses are obtained by isolation of viruses from throat or rectal swabs, stools, and cerebrospinal fluid. Some ECHO and exanthema viruses may be present in respiratory-enteric illnesses during winter months. Sites of ECHO virus recovery are the throat, stools, and very frequently cerebrospinal fluid; rarely are they recovered from the blood or mouth.

Adenovirus infection with ten human of eighteen known serotypes may cause a catarrhal inflammation of the mucous membranes or the respiratory and ocular systems. Pathological lesions show follicular enlargement of the submucous and regional lymphoid tissues, the intestinal tract, and the mesenteric lymph nodes.

NOTE: For virus bibliography, see p. 289.

Chapter XXXIII

BACTERIOPHAGE

Twort, in 1915, while studying certain Staphylococcus cultures, observed curious transparent areas or breaks in growth, now called *plaques*. On touching one of the transparent areas with a platinum needle and drawing the needle across the surface of a young Staphylococcus culture, he found the culture seemed to dissolve, and the needle track became clear and transparent within a few hours. In 1917, d'Herelle obtained a similar effect with filtered cultures of the dysentery bacillus and with filtered stools from dysentery patients. He named the lytic agent *bacteriophage* (bacteria-eater).

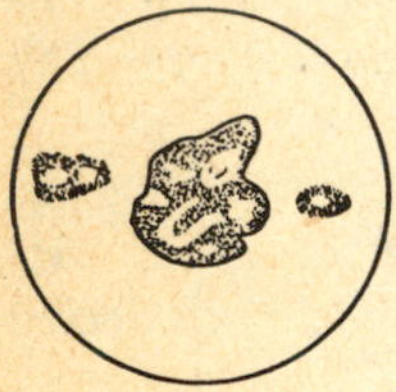

FIG. 70. Bacteriophage plaques in three colonies undergoing lysis.

Definition. Bacteriophage is an ultramicroscopic agent or substance which causes transmissible lysis of bacteria.

1. A small amount of bacteriophage, one part in a billion, added to a young, actively growing broth culture of susceptible bacteria will, in the course of 3 or 4 hours, cause solution of all or nearly all of the cells.

2. A billionth of 1 ml. of the clear solution, unfiltered or filtered through a Berkefeld candle, added to a new, young culture of growing bacteria will again cause the lysis of the cells in a few hours. Thus, the bacteriophage has been transmitted from one culture to another.

Mechanism. After the bacteriophage enters the bacterial cell, swelling of the latter occurs. It is during this period that the phage multiplies. The swollen cell bursts and disappears, liberating phage into the medium. If the lysed culture is then incubated, the few cells which may remain multiply. These are resistant to lysis by the phage used and often are variants (cultural, serological, etc.) of the original bacterial culture.

Properties of Bacteriophage. 1. Is particulate, as evidenced by (a) plaque formation and proportionality of number of plaques to dilution of phage, (b) electromicrographs (see Fig. 71), (c) passage through certain filters and retention by smaller ones, and (d) concentration by ultracentrifugation.

2. Varies in size for different "races" from 8–12 mμ to 50–75 mμ. A crystalline product prepared by Northrop and inseparable from phage activity had a molecular weight of 300,000,000 and was apparently a nucleoprotein.

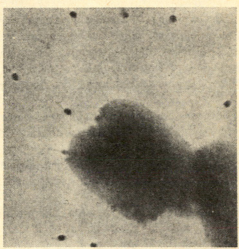

Courtesy, R.C.A.

FIG. 71. Electromicrograph of *Escherichia coli* and bacteriophage particles. Note the tail-like appendages. ($\times$ 67,000.)

3. Can be transferred from one culture to another and so be propagated indefinitely, the lytic potency of the filtrate increasing with each transfer.

4. Propagates only in cultures of living, actively metabolizing bacteria, but not in lifeless media or when only dead bacteria are present.

5. Is active in high dilution (1:100,000,000 or even higher).

6. Passes through the compact clay and infusorial earth filters which generally keep back bacteria.

7. Is inactivated, as a rule, by exposure to temperatures of 72° to 75° C. for 30 minutes and to ultraviolet light.

8. Is inactivated by certain chemicals, but seems to be somewhat more resistant than most bacteria.

9. Exists in different species or "races" of bacteriophage, specific for and associated with particular species or even strains of bacteria. The phage specificity is believed to be associated with the antigenic structure of the bacteria.

10. Is antigenic. Injection of lytic filtrates causes production of inhibitory antibodies specific for the phage inoculated. The bacteriophages are immunologically heterogeneous, but they comprise a number of distinct groups. There may be some correlation between immunological types and other characteristics, such as ability to develop in the absence of calcium, resistance to urea and ultraviolet light, etc.

Nature of Bacteriophage. There are various views concerning the nature of bacteriophage. None has been generally accepted.

1. D'Herelle believes that bacteriophage is a living organism parasitic on bacteria, and that all the phage "races" or types are one species, the different strains developing as an adaptive response to the environment.

2. Many workers—probably the majority—believe that phage is an autocatalytic enzyme or enzyme-like substance. (See Northrop's work above.)

3. Hadley has proposed that phage is a genelike substance representing a phase in the life history of the susceptible bacteria. This view is not widely accepted.

Some have said that phage is a virus that produces disease of bacteria. However, since the nature of viruses is also unsolved, this does not clarify the subject to any degree.

Distribution. Bacteriophage is widely distributed in nature. It is commonly present in the contents of the small and large intestines of man and consequently in feces and sewage. It is particularly abundant in individuals who are convalescing from a bacterial infection.

Uses. 1. *Therapeutic.* D'Herelle believes that bacteriophage is an integral part of the normal immunity mechanism, through direct antagonism for the disease-producing agent, by preparation of bacteria for phagocytosis, and by rendering the bacteria particularly potent as antigens. Bacteriophages against various microorganisms, e.g., those of Staphylococcus infections, cystitis, bacillary dysentery, plague, cholera, etc., have been employed therapeutically. Results on the whole have been disappointing and their use largely discontinued. Some work has shown that phage will not act in the presence of blood, pus, feces, or any colloidal material.

2. *Identification.* Highly specific bacteriophage types, prepared by adaptive and selective cultivation, have been employed as a basis for identifying species and strains of Shigella, Staphylococcus, hemolytic Streptococci, and species whose position in a genus is doubtful. Use of specific phages with organisms isolated in various typhoid fever patients and epidemics has sometimes made possible the tracing of sources and carriers.

BIBLIOGRAPHY

Boyd, W. C. *Fundamentals of Immunology.* Third Edition. New York: Interscience Publishers, 1956.

Dubos, R. J., ed. *Bacterial and Mycotic Infections of Man.* Third Edition. Philadelphia: J. B. Lippincott Company, 1958.

Eberson, Frederick. *The Microbe's Challenge.* Lancaster, Pa.: Jacques Cattell Press, 1941.

Jawetz, E., J. L. Melnick, and E. A. Adelberg. *Review of Medical Microbiology.* Los Altos: Lange Medical Publications, 1962. Pp. 358–366.

Swingle, D. B. *General Bacteriology,* rev. by G. W. Walter. New York: D. Van Nostrand Company, 1947. Pp. 256–268.

Umbreit, W. W. *Modern Microbiology.* San Francisco: W. H. Freeman and Company, 1962. Pp. 460–465.

Chapter XXXIV

PATHOGENIC PROTOZOA

Protozoa are a heterogeneous group of unicellular animal organisms with diversification and highly specialized functions of the protoplasm. Specialized portions of the cell are called organelles. They have a well-defined nucleus, cytoplasm, and cell wall, and often granules and vacuoles. Many pass through a developmental life cycle.

CLASSIFICATION

Subphylum **Plasmodroma**—Movement by flagella or pseudopodia.

CLASS I. **Rhizopoda,** with movement by means of pseudopodia or false feet, and cytoplasm which has no limiting membrane. In this class are included: *Endamoeba histolytica, Endamoeba coli, Endolimax nana, Iodamoeba williamsi,* and *Dientamoeba fragilis.*

CLASS II. **Mastigophora,** with motility by means of whiplike appendages or flagella. They include the hemoflagellates, such as members of the genus Trypanosoma, which invade the bloodstream of man and animals causing a tropical sleeping sickness in the former. Examples of organisms classified among the Mastigophora are: *Leishmania donovani,* the cause of kala azar; *Trypanosoma gambiense, Trypanosoma rhodesiense, Trypanosoma cruzi,* the causes of trypanosomiasis in man; *Trichomonas hominis,* found in the human intestine; *Trichomonas vaginalis,* found in vaginal secretions; and *Giardia lamblia,* present in the small intestine, sometimes causing symptoms.

CLASS III. **Sporozoa,** obligate parasites which are nonmotile and do not engulf or capture food. They are transmitted through intermediate hosts, and may reproduce asexually (schizogony) or sexually (sporogony), and with or without resistant spores. Examples of these organisms are *Plasmodium malariae, Plasmodium vivax, Plasmodium falciparum,* and *Plasmodium ovale,* the causes of malaria.

Subphylum **Ciliophora**—Movement by cilia. Most have two kinds of nucleus, a macronucleus and a micronucleus. Example: *Balantidium coli,* an intestinal parasite of man, monkeys, and pigs.

ENDAMOEBA HISTOLYTICA

This organism is the most frequently encountered of the pathogenic intestinal protozoa. It is found in the feces, intestinal ulcers, and liver abscesses of infected persons.

Morphological Characteristics. It constantly changes its shape and is about 18 to 25μ in size. It exhibits amoeboid movement by means of blunt pseudopodia of hyaline ectoplasm. It contains food vacuoles and ingested red blood corpuscles. *Endamoeba histolytica* reproduces by mitosis and ultimately forms spherical cysts in which form the disease is spread. The resistant cysts are about 5 to 20μ in size.

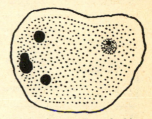

Fig. 72. *Endamoeba histolytica*, showing ingested red blood cells, granular nucleus, ectoplasm, and endoplasm.

Cultural Characteristics. Cultures of *Endamoeba histolytica* are best made on such media as Craig's Locke-serum medium, St. John's medium, or others consisting of a mixture of serum, egg albumen, and other synthetic mixture like Locke's solution.

Pathogenicity. The onset of infection with *Endamoeba histolytica* is more gradual than that due to *Shigella dysenteriae*. The stools contain both mucus and blood. In the disease in man, which varies widely in severity, the *Endamoeba histolytica* makes its way between the epithelial cells into the submucosa of the large intestine, an abscess is formed, and edema of the submucosa is observed. Abscesses may also form in the liver and lungs.

Transmission. The *Endamoeba histolytica* cysts are transmitted through flies, fresh vegetables, water, and carriers. The healthy carriers who pass cysts in their stools, although they have never had any symptoms of the disease, play a great role. Convalescents also may pass cysts in their stools for some time. Active amoebic dysentery patients pass the noninfective ameboid forms.

Prophylaxis and Immunity. The cysts of *Endamoeba histolytica* probably resist the action of chemical disinfectants, such as chlorine or potassium permanganate. Therefore, food and water which may be contaminated with them should be subjected to boiling for at least 30 minutes. Thorough cleansing of hands should be carried out before handling food. Access of flies to food should be prevented.

There is no direct evidence of acquired immunity, but some resistance is believed to be present after recovery.

Diagnosis. The stools are examined for the presence of *Endamoeba histolytica* by direct smear in saline, smear in Lugol's solution, or fixed stained preparations. Trophozoite, precystic, or encysted forms of the protozoan are looked for. *Endamoeba coli* may be confused with *Endamoeba histolytica*, but it is distinguished from the latter as follows:

Endamoeba Histolytica	Endamoeba Coli
Pathogenic for man.	Nonpathogenic for man.
The trophozoite form is 20–35μ in size.	The trophozoite form is 15–30μ in size.
Living trophozoites ingest red blood cells, and degeneration and culture forms contain bacteria. This is diagnostic.	Living trophozoites do not ingest red blood cells, and bacteria and starch grains are the principal inclusions.
The cysts contain 1–4 nuclei, rarely more, and mature cysts contain 4 nuclei.	The cysts contain 1–8 nuclei, rarely more, and mature cysts contain 8 nuclei.

The complement fixation test has been used for diagnosis of infection with *Endamoeba histolytica*. This test has been found highly specific and gives positive results in about 90% of the cases.

Related Protozoa Nonpathogenic to Man. These include: *Endamoeba coli, Endolimax nana, Iodamoeba williamsi*, which occur in the intestine, and *Endamoeba gingivalis*, which occurs in the mouth and was once believed to be the cause of pyorrhea.

TRYPANOSOMA GAMBIENSE

This organism and *Trypanosoma rhodesiense*, with which it may be identical, cause African sleeping sickness.

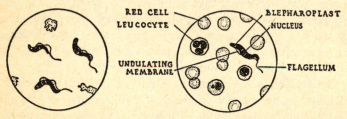

Fig. 73. *Trypanosoma gambiense*, the cause of African sleeping sickness.

Morphological Characteristics. 1. Form—an elongated body, thick in the middle and tapering at each end. It has a long flagellum at the anterior pole, and a nucleus in its interior at about the center. Near the posterior pole is a granule called the blepharoplast, to which the flagellum is attached. 2. Size—from 15 to 30μ in length by 1.5 to 2.5μ in breadth. 3. Motility—very motile, by means of the flagellum and contractions of the body; the undulating membrane also aids in movement.

Pathogenicity. Trypanosomes of African sleeping sickness are transmitted by the bite of the tsetse fly. Irregular fever is observed after the incubation period of 2 weeks to several months. During the first stage, the parasites are found in the blood and also in the lymph. The lymph nodes and spleen become enlarged, often accompanied by anemia, cardiac injury, and edema. Death may occur in this stage, but the disease usually runs through the sleeping sickness stage of somnolence and apathy, when the invaders are found in large numbers in the spinal fluid. Arsenicals and germanin are fairly effective therapy in the earlier stages. Certain wild animals probably act as reservoirs of the disease.

Related Organisms. 1. *Trypanosoma lewisi*—infects the blood of wild rats. 2. *Trypanosoma cruzi*—causes Chagas' disease, a Brazilian form of trypanosomiasis, and lodges in the cells of the voluntary muscles, heart, and other organs. The disease is transmitted by *Triatoma megista*, a reduvid bug. 3. *Trypanosoma brucei*—causes nagana, trypanosomiasis of African animal life, and is transmitted by the tsetse fly. 4. *Trypanosoma equinum*—causes disease known as Mal de Caderas and attacks horses in South America. 5. *Trypanosoma theileri*—causes trypanosomiasis of cattle in South America.

LEISHMANIA DONOVANI

This organism is the cause of kala azar, a parasitic disease of the reticulo-endothelial system, which is believed to be transmitted through the sandfly, *Phlebotomus argentipes*. The disease lasts for some time and is characterized by involvement of the spleen, liver, bone marrow, and lymphatic glands. The organism has been demonstrated in the nasopharyngeal discharges of patients with the disease. Treatment with antimony compounds is effective.

Morphological Characteristics. 1. Form—round or oval body with a sharp outline. 2. Size—very small, about 1 to 3μ in length and 1.5 to 2.5μ in breadth. 3. Staining properties—not readily stained.

Diagnosis. The identification of infection with *Leishmania donovani* includes the following laboratory procedures:

1. *Blood film*—stained with either Leishman's stain or Wright's modification of the Romanowsky stain.

2. *Blood culture*—made on Nicolle's modification of the Novy-MacNeal medium.

3. *Sternal bone marrow, liver, and spleen puncture.*

4. *Aldehyde test*—serum from patients with kala azar gels in a few seconds to 30 minutes when two drops of formalin are added to it, because of the greatly increased amount of serum euglobin present.

Related Organism.   Related to *Leishmania donovani* and similar to it in morphology is *Leishmania tropica*.   This organism is the cause of cutaneous leishmaniasis, which manifests itself by ulcerative cutaneous lesions.   *Leishmania tropica* is found in large numbers in the lesions.   It is believed that this organism, too, is vectored by a member of the genus Phlebotomus.

OTHER FLAGELLATES

In the class Mastigophora are also included: 1. *Giardia lamblia* and 2. *Trichomonas hominis*, which inhabit the intestinal tract and may cause enteritis; 3. *Chilomastix mesnili*, a common saprophytic inhabitant of the intestinal tract; 4. *Trichomonas vaginalis* and 5. *Trichomonas foetus*, which inhabit the vagina of human beings and cattle respectively and may cause vaginitis.

MALARIA (GENUS PLASMODIUM)

Malaria is characterized by recurring fever, chills, and sweats.   It is transmitted by the bite of the anopheline mosquito.

Malarial Parasites.   Four species of the genus Plasmodium cause malaria.   These are:

1. *Plasmodium vivax*—causes *benign tertian malaria*, having a cycle of 48 hours, with fever recurring every third day, counting from the first day of fever.

2. *Plasmodium malariae*—causes *quartan malaria*, having a cycle of 72 hours, with fever recurring every fourth day.

3. *Plasmodium falciparum*—causes *malignant subtertian malaria* or estivo-autumnal malaria, with fever recurring at irregular intervals between the second and third days.

4. *Plasmodium ovale*—causes a *tertian type of malaria*, and is difficult to differentiate from *Plasmodium malariae*.

Life Cycle of Malarial Parasites.   There are two stages in the life cycle of the Plasmodium parasite, one in the female anopheline mosquito—the sexual cycle or *sporogony*, the other in the blood of man—the asexual cycle or *schizogony*.

1. *Sporogony, the sexual cycle in the mosquito*, has for its function the formation of new parasites (sporozoites) from the gametocytes ingested following the biting of a person infected with malaria. The *gametocytes* are taken into the stomach of the mosquito. The *microgametocytes* undergo several nuclear divisions, and the organisms later form 6–8 whiplike processes. These, becoming detached as free-swimming *microgametes*, later penetrate the *macrogamete* to fertilize it. The *oökinete* or zygote (the fertilized female) attaches itself to the stomach wall, and forms the *oöcyst* which may contain 1000–10,000 *sporozoites*. These are liberated into the mosquito's body when the oöcyst reaches maturity and bursts. They then travel to the salivary glands of the insect and are injected when it bites a human being. This stage takes about 12 days at temperatures of 25–30° C.

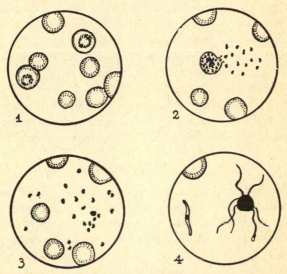

Fig. 74. *Plasmodium malariae*. (1) Ring-shaped trophozoites inside erythrocytes. (2) Schizont with merozoites being set free. (3) Free merozoites in bloodstream ready to enter corpuscles. (4) Male gametocyte undergoing exflagellation with a threadlike microgamete or spermatozoa set free.

2. *Schizogony, the asexual cycle in the blood of man*. During this process, the sporozoites injected through the bite of the mosquito penetrate the red blood corpuscles, where they grow rapidly to form *trophozoites* having the well-known signet ring appearance. The trophozoites divide to form many segments, and are then known as *schizonts*. When the corpuscle ruptures, the segments or *merozoites* are liberated, circulate in the blood, and then invade normal red blood cells to form trophozoites again. Every time periodic generations of the merozoites are liberated, chills and fever symptomatic of malaria occur.

After a few such cycles, some merozoites become differentiated and grow into large, mature, adult, sexual forms or *gametes*. These circulate in the blood stream and either perish or are taken up by the female Anopheles mosquito.

The time consumed by this stage is 24–72 hours and is characteristic for the Plasmodium species involved.

Diagnosis.   The diagnosis of malaria depends upon the demonstration of parasites in the blood.   The blood films are prepared in the usual way, and are stained with Wright's stain or the Leishman stain.   *Plasmodium vivax* will be found in various stages of development in blood corpuscles, which will be enlarged and show Schüffner's dots.   *Plasmodium falciparum* can be identified since it forms only ring forms and crescent-shaped gametocytes in the peripheral circulation, without enlargement of the red blood cells.   In infection with *Plasmodium malariae* the red blood cells are not enlarged, and the typical signet ring appearance of the trophozoites is diagnostic.

Prevention.   Malaria prevention lies in the effective eradication of the Anopheles mosquito by preventing its breeding.   Its access to infected persons should also be prevented by screening of houses. The taking of quinine or atabrine is effective in preventing infection in areas where malaria is prevalent.   The incidence of malaria in the United States has declined almost to zero, however the need for continuing surveillance continues.   The vector persists in many areas, and the possibility of introduction from abroad is ever present.

OTHER SPOROZOA

These include: 1. *Babesia bigemina,* the cause of red water fever in cattle which is transmitted through the tick *Margaropus annulatus* in America; 2. *Babesia bovis,* the cause of European hemoglobinuric fever of cattle; and 3. *Babesia ovis,* the cause of a similar disease in sheep.

BIBLIOGRAPHY

Anderson, W. A.   *Pathology.*   Second Edition.   St. Louis: C. V. Mosby Company, 1953.

Chandler, Asa C.   *Introduction to Parasitology.*   Eighth Edition.   New York: John Wiley and Sons, 1949.

Craig, Charles F.   *Etiology, Diagnosis and Treatment of Amebiasis.*   Baltimore: Williams & Wilkins Company, 1944.

————.   *Laboratory Diagnosis of Protozoan Diseases.*   Second Edition.   Philadelphia: Lea and Febiger, 1948.

Culbertson, J. T.   *Immunity against Animal Parasites.*   New York: Columbia University Press, 1941.

Hagan, W. A.   *Infectious Diseases of Domestic Animals.*   New York: Comstock Publishing Company, 1943.   Pp. 383–482.

Hegner, R. W., and K. A. Stiles.   *College Zoology.*   Sixth Edition.   New York: Macmillan Company, 1951.

Hunter, G. W., and F. R. Hunter.   *College Zoology.*   Philadelphia: W. B. Saunders Company, 1949.

Jawetz, E., J. L. Melnick, and E. A. Adelberg. *Review of Medical Microbiology.* Los Altos: Lange Medical Publications, 1962. Pp. 367–387.

Kudo, R. R. *Protozoology.* Fourth Edition. Springfield, Ill.: Charles C. Thomas, 1954.

Mackie, T. T., *et al. Manual of Tropical Medicine.* Second Edition. Philadelphia: W. B. Saunders Company, 1954.

Napier, L. Everard. *The Principles and Practice of Tropical Medicine.* New York: Macmillan Company, 1946.

Nauss, R. W. *Medical Parasitology and Zoology.* New York: Paul B. Hoeber, 1944.

Russell, Paul F., Luther S. West, and Reginald D. Manwell. *Practical Malariology.* Philadelphia: W. B. Saunders Company, 1946.

Strean, L. P. *Oral Bacterial Infections.* Brooklyn, N. Y.: Dental Publishing Company, 1949. Pp. 144–146.

Strong, R. P. *Stitt's Diagnosis, Prevention, and Treatment of Tropical Diseases.* Philadelphia: The Blakiston Company, 1945. Pp. 1–525.

Swingle, D. B. *General Bacteriology,* rev. by G. W. Walter. New York: D. Van Nostrand Company, 1947.

Umbreit, W. W. *Modern Microbiology.* San Francisco: W. H. Freeman and Company, 1962. Pp. 294, 300–304.

Villee, C. A., W. F. Walker, and F. E. Smith. *General Zoology.* Philadelphia: W. B. Saunders Company, 1958.

Zinsser, H., and J. T. Enders. *Immunity.* New York: Macmillan Company, 1939. Pp. 440–462.

PART III

INFECTION AND IMMUNITY

CHAPTER XXXV

INFECTION

Definitions. *Infection* may be defined as invasion of the tissues of the body by pathogenic organisms, which then multiply, and cause disease. Bacteria range from those which produce disease (*pathogenic*) to those which do not produce disease (*nonpathogenic*). The latter organisms may reside in the body without manifesting any pathogenic properties. However, there are certain organisms which are nonpathogenic but under circumstances may become pathogenic. Thus, *Escherichia coli* is a normal inhabitant of the gastrointestinal tract of man, but it has been known to produce certain infections like peritonitis, colitis, meningitis, and urinary infections.

An *infectious disease* is one that is caused by a microorganism or virus. Such diseases are usually *communicable* to other individuals. A *contagious disease* is an infectious disease which is *readily* spread to other individuals. The type includes most of the infectious diseases like scarlet fever, measles, whooping cough, tuberculosis, and mumps, which are spread by droplet infection. On the other hand, yellow fever, pneumonia, syphilis, herpes simplex, and others, although also infectious diseases, are conveyed only when certain conditions are fulfilled.

Types of Infections. Infections may be classified as:

1. *Primary.* This is the first disease noted in an illness.

2. *Secondary.* When the body has been weakened by the primary infection there is in many instances predisposition to secondary infection with the same or another organism. Example: Measles is in most cases a primary infection, which may be followed by a secondary infection like pneumonia. Scarlet fever may be followed by a secondary otitis media or mastoiditis.

3. *Mixed.* This occurs when the disease is caused by two or more organisms. Thus, a person may have a pneumonia in which two organisms are isolated, or a wound infection with *Micrococcus pyogenes* var. *aureus* and *Clostridium tetani*.

Another classification of infections is based on the distribution of the microorganisms in the host.

1. *Local*, when the symptoms are confined to one area, as in a wound.

2. *Focal*, when the organisms are originally confined to one area, which may serve as a source for further dissemination of organisms or toxic materials to other parts of the body.

3. *Systemic* or *general*, when there is a general invasion and the entire body seems to be affected. Such infections may in turn be classified as:

(a) *Bacteremia*. Microorganisms invade the bloodstream but there is no active multiplication of them. (This term has often been used synonymously with septicemia.)

(b) *Septicemia*. The bloodstream is invaded and there is multiplication of the microorganisms.

(c) *Pyemia*. Pus-producing bacteria repeatedly invade the blood and localize at different points causing new metastatic foci of infection.

(d) *Sapremia*. This is a toxic result of presence in the blood of the products formed by the growth of saprophytes on diseased or injured tissues.

(e) *Toxemia*. Bacteria are localized and produce a toxin which is spread throughout the body and absorbed by the body cells.

Factors Determining Infection. Whether or not infection will result from contact of the host with the microorganism depends on conditions of the contact and the balance of the virulence of the organism and the resistance of the host. $D = \dfrac{V}{R}$, where D = disease, V = virulence and number of pathogens, and R = resistance of the host.

1. *Conditions of Contact*.

(a) *Number of Organisms*. For infection to be possible the number of organisms must be sufficient to overcome the host's local defenses and to gain a foothold for growth.

(b) *Portal of Entry*. This is the route the organism must take in entering the body in order to have any effect. *Corynebacterium diphtheriae* enters through the nasopharynx. Tetanus spores may be swallowed without producing any symptoms in man. However, when introduced into a lacerated wound, tetanus develops. The portal of entry of *Clostridium tetani* is the skin, but it is also necessary that some tissues be destroyed before it can gain a foothold.

2. *Virulence* or disease-producing power of the organism. This is defined as the ability of bacteria to injure the body of the host by

invasion and multiplication or by poisoning. The known contributing factors of this ability are classified below:

(a) *Toxins*—kill or injure tissue cells, may paralyze defenses. There are two kinds of bacterial toxins: (1) *exotoxins* or soluble or true toxins, and (2) *endotoxins*. Their main characteristics are tabulated below.

EXOTOXINS	ENDOTOXINS
Diffuse out of intact bacterial cell.	Do not diffuse out of intact bacterial cell.
Extremely high potency.	Low potency.
Incubation period.	Incubation period.
Good antigen.	Poor antigen.
Thermolabile.	Thermostable.
Selective tissue action.	No characteristic pharmacological action.
Protein in nature.	Protein or glucolipoid in nature.
Destroyed by proteolytic enzymes.	Resist proteolytic enzymes.
Detoxified by formaldehyde.	Not detoxified by formaldehyde.

Among the organisms producing exotoxins are *Corynebacterium diptheriae*, *Clostridium tetani*, and *Clostridium botulinus*.

(b) *Invasiveness*. (1) *Hemolysins**—destroy red blood cells. There are two demonstrable types, one which is filterable, extra-cellular, and antigenic, and the other exhibited in culture on semi-solid media containing blood. Relation between the two is obscure. (2) *Leucocidins*—destroy white blood cells. (3) *Anti-opsonins*—drive away white blood cells. (4) *Coagulases*—accelerate formation of blood clots; found especially in Micrococcus. (5) *Fibrinolysins*—dissolve blood clots; found especially in Streptococcus. (6) *Spreading* or *Duran-Reynals factor*—produces great increase in permeability of host tissues; may be identical with enzyme hyaluronidase.

(c) *Capsules*. The presence of capsules is associated with virulence. They appear to function as a bacterial defense against the phagocytic activity of the white blood cells. (See also p. 17.)

(d) *Mechanical Action*. Multiplication of organisms may result in blocking of capillaries of the heart, lungs, brain, etc.

(e) *Miscellaneous*. Various other factors have been reported. (1) *Necrotoxin* or *necrotizing factor*—kills tissue cells; produced by some Staphylococci. (2) *Hypothermic factor*—lowers body tempera-

---

* Not to be confused with immune hemolysins formed by an animal body in response to injection of red blood cells of another species. See p. 345.

ture; produced by *Shigella dysenteriae*. (3) *Edema-producing factor*—formed by pneumococcus.

3. *Resistance* of the host.

(a) *General Well-being*, including nutrition, fatigue, etc.

(b) *External Defenses*. (1) The unbroken *skin* and *mucous membranes* are generally impervious to particulate material of bacterial size. Bacteria may, however, enter the skin through hair follicles and sweat gland ducts. Clean skin is also actively bactericidal, but not to the normal skin flora. Cilia in the nasopharyngeal tract aid in mechanical removal of bacteria. (2) *Secretions* may destroy bacteria by acidity, e.g., perspiration, gastric juice, vaginal secretions. Tears and saliva protect by mechanical flushing; mucus, by forming a protective covering.

(c) *Nonspecific Internal Defenses*. (1) The *natural bactericidal power* of blood (serum) and other body fluids is quite feeble, but is believed to play some role. (2) Activity of some body cells, e.g., *phagocytosis* (see Chapter XXXVII).

(d) *Specific Resistance Due to Antibodies*. See Chapters XXXVI and XXXVII.

Spread of Infection. Epidemiology is the term given to the distribution of the diseases of mankind. Diseases may be *epidemic*, when there occurs in a limited time among a limited population an unusual number of cases of a communicable disease; *pandemic*, when the population concerned is much larger, that of an entire country, or even a continent; *endemic*, when the disease is prevalent in the usual numbers in a given population and there is no increase in the number of cases.

1. Diseases may be spread by *contact*, in which case isolation of sick persons and those coming in contact with them is indicated. Such diseases include those spread by droplet infection, i.e., pneumonia, colds, scarlet fever, measles, mumps, diphtheria, tuberculosis, whooping cough.

2. Enteric diseases, including typhoid, paratyphoid, amoebic and bacillary dysentery, infant diarrheas, cholera, and food infections, are spread through human *excreta* containing the organisms causing these diseases.

3. *Insect-vectored* diseases are transferred to man through intermediate hosts which harbor infective agents. (a) Mosquitoes—transmit yellow fever, malaria, and dengue. (b) Flies—transmit trypanosomiasis or African sleeping sickness. (c) Fleas—transmit bubonic plague. (d) Lice—transmit typhus fever and infectious

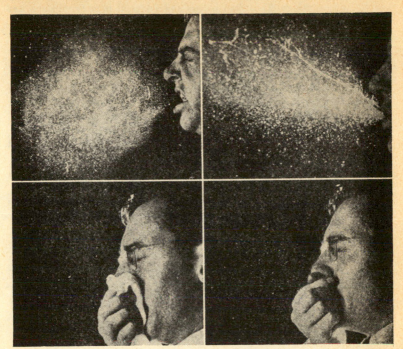

Courtesy, M. W. Jennison and American Assn. for the Advancement of Science.

FIG. 75. Atomization of droplets into the air during sneezing, demonstrated by high-speed photographs. Upper left: violent, unstifled sneeze. Upper right: head cold sneeze. Lower left: sneeze stifled with handkerchief. Lower right: sneeze stifled with hand.

jaundice. (e) Ticks—transmit Rocky Mountain spotted fever and Texas fever of cattle.

BIBLIOGRAPHY

Boyd, W. C. *Fundamentals of Immunology.* New York: Interscience Publishers, 1947. Pp. 369–392.

Burnet, F. M. *The Natural History of Infectious Disease.* Second Edition. Cambridge, England: Cambridge University Press, 1953.

Dubos, R. J. *The Bacterial Cell.* Cambridge, Mass.: Harvard University Press, 1945. Pp. 188–228.

———. *Bacterial and Mycotic Infections of Man.* Third Edition. Philadelphia: J. B. Lippincott Company, 1958.

Faust, E. C. *Animal Agents and Vectors of Human Disease.* Philadelphia: Lea and Febiger, 1955.

Maxcy, K. F. *Preventive Medicine and Public Health.* Eighth Edition. New York: Appleton-Century-Crofts, 1956. Section 1.

Parish, H. J. *Antisera, Toxoids, Vaccines and Tuberculins in Prophylaxis and Treatment.* Third Edition. Baltimore: Williams & Wilkins Company, 1955.

Rivers, T. M., and F. L. Horsfall, eds. *Viral and Rickettsial Infections of Man.* Third Edition. Philadelphia: J. B. Lippincott Company, 1959.

Rosebury, T. *Microorganisms Indigenous to Man.* New York: McGraw-Hill Book Company, 1962. Pp. 310–343.

Sulzberger, M. B., and R. L. Baer. *Office Immunology.* Chicago: Yearbook Publishers, 1947. Pp. 95–129.

Swingle, D. B. *General Bacteriology,* rev. by G. W. Walter. New York: D. Van Nostrand Company, 1947. Pp. 250–256.

Umbreit, W. W. *Modern Microbiology.* San Francisco: W. H. Freeman and Company, 1962. Pp. 398–409.

Winslow, Charles-Edward A. *The Conquest of Epidemic Disease.* Princeton, N. J.: Princeton University Press, 1943.

Zinsser, H., and J. T. Enders. *Immunity.* New York: Macmillan Company, 1939. Pp. 440–721.

CHAPTER XXXVI

TYPES OF IMMUNITY

Immunity may be defined as the ability of the living individual to resist or overcome infection. This state of resistance is indicated either by the failure of the individual to develop the disease upon exposure, or in some cases by the demonstration of specific immune bodies in the blood which are considered effective against the invading organisms.

Natural Immunity. This is a type of immunity with which an individual is born. It enables him to resist infection without first having the disease.

1. *Species.* This immunity is characteristic of a particular species. Example: Dogs are immune to anthrax and tuberculosis.

2. *Racial.* This immunity is characteristic of a particular race within a species. Example: Algerian sheep are immune to anthrax. In the races of man, Negroes are said to be relatively resistant to yellow fever and more susceptible to tuberculosis than whites. The issue, however, is obscured by many other factors, including a possible congenital immunity in the former case and economic status in the latter.

3. *Individual.* This immunity, characteristic of particular individuals, may be largely attributable to acquired immunity due to an earlier, mild, unrecognized attack of the disease.

4. *Congenital.* This immunity, found in the newborn, is due to the passive transfer of antibodies from mother to offspring through the placenta. Thus, infants in the first year of life are resistant to diphtheria and scarlet fever.

Acquired Immunity. This immunity is acquired by the individual during his lifetime.

1. *Active.* This is a relatively lasting immunity due to the development within the individual of antibodies as a result of contact with the microorganisms or their products. The body cells and tissues themselves react to produce the specific immunity.

(a) *Naturally Acquired.* This immunity is attained as a result of an attack of the disease itself. One attack of certain diseases confers lifelong immunity. Examples: diphtheria, whooping cough, typhoid fever, scarlet fever, yellow fever, most virus diseases.

(b) *Artificially Acquired.* This immunity results from a course of immunization with attenuated cultures* (e.g., smallpox vaccination, Pasteur's rabies "treatment"), killed cultures (e.g., formalinized pneumococcus vaccines), sensitized bacterial vaccines, toxin, toxin-antitoxin, toxoid (e.g., diphtheria immunization).

2. *Passive.* This is a short-lived immunity in which the antibodies are produced in another animal whose blood or serum is injected into the person. The body cells of the treated individuals take no part in producing the immunity.

BIBLIOGRAPHY

Burrows, W. *Textbook of Microbiology.* Eighteenth Edition. Philadelphia: W. B. Saunders Company, 1963.

Dubos, R. J. *Bacterial and Mycotic Infections of Man.* Third Edition. Philadelphia: J. B. Lippincott Company, 1958.

Jawetz, E., J. L. Melnick, and E. A. Adelberg. *Review of Medical Microbiology.* Los Altos: Lange Medical Publications, 1962. Pp. 109–111.

Maxcy, K. F. *Preventive Medicine and Public Health.* Eighth Edition. New York: Appleton-Century-Crofts, 1956. Section 1.

Raffel, S. *Immunity, Hypersensitivity, Serology.* New York: Appleton-Century-Crofts, 1953.

St. Whitlock, O., and F. N. Furness, eds. *Natural Resistance to Infections.* Annals of the New York Academy of Sciences, Vol. 66, p. 233. New York: 1956.

Smith, D. T., *et al. Zinsser's Textbook of Bacteriology.* Twelfth Edition. New York: Appleton-Century-Crofts, 1960.

Walter, W. G., and R. H. McBee. *General Microbiology.* New York: D. Van Nostrand Company, 1958.

Wilson, G. S., and A. A. Miles. *Topley and Wilson's Principles of Bacteriology and Immunity.* 2 vols. Fourth Edition. Baltimore: Williams & Wilkins Company, 1955.

---

* A vaccine is currently defined as a preparation of killed or attenuated infective agent used to produce active, artificial immunity. *Stock* vaccines are those prepared from stock cultures kept in the laboratory. *Autogenus vaccines* are those prepared from the patient's own infection. *Sensitized bacterial vaccines* are vaccines mixed with a corresponding amount of serum from an immune animal. Commercially prepared for most bacteria of which plain vaccines are available, the sensitized vaccines produce immunity more rapidly and with fewer undesirable reactions.

CHAPTER XXXVII

IMMUNOLOGICAL REACTIONS

Parenteral introduction of foreign proteins, including microorganisms, into the animal body results in the production of specific protective substances. These are termed antigens and antibodies respectively.

An *antigen* is any substance which stimulates the production of specific antibodies. Antigens are protein in nature, and practically all proteins, except the incomplete ones such as gelatin, are antigenic. Specificity of the antigen is determined by its chemical composition.

An *antibody* is formed by the animal body in response to the presence of antigen with which it combines in a specific, antagonistic manner. The antibodies are closely associated, and may be identical, with serum globulin. They may be separated from other serum constituents by dilution with distilled water or more often by salting out.

When an antibody is tied to the appropriate fluorescent dye and then allowed to combine with the virus or bacterium for which the antibody is specific, the combined particle will glow under the improved ultraviolet microscope. Labeled antibodies are now available to detect antigens for rabies, psittacosis, poliomyelitis, measles, chicken pox, mumps, and influenza.

The five main types of antibodies, as determined by their action, and the leading organisms stimulating their production are listed below.

1. *Antitoxin—Corynebacterium diphtheriae, Streptococcus scarlatinae, Shigella dysenteriae, Clostridium tetani, Clostridium perfringens, Clostridium feseri.*

2. *Agglutinin—Diplococcus pneumoniae, Salmonella typhosa, Salmonella paratyphi, Salmonella schottmuelleri.*

3. *Precipitin—Diplococcus pneumoniae, Bacillus anthracis.*

4. *Opsonin*—Most microorganisms, *Neisseria intracellularis, Diplococcus pneumoniae, Salmonella typhosa.*

5. *Cytolysin—Treponema pallidum, Hemophilus pertussis, Neisseria gonorrheae, Neisseria intracellularis.*

Ehrlich believed that the different demonstrable antibodies were separate and distinct substances. Zinsser and others, however, have proposed a "unitarian" hypothesis—that the various antibodies stimulated by a single antigen are essentially identical regardless of the consequences of the antigen-antibody union.   These are variable and depend on the nature of the antigen and the conditions of the reaction.

Haptenes, or *partial antigens*, described by Landsteiner, are relatively simple substances unable to stimulate antibody production when injected by themselves but determining immunological specificity when combined with antigenic protein.   The specific antibodies produced react with the haptene alone or with the combined protein-haptene, both *in vivo* and *in vitro*.   Not necessarily proteins, the haptenes are exemplified by the polysaccharide Specific Soluble Substances (S.S.S.) found in the pneumococcus capsule.

A widely distributed *heterophile antigen*, described by Forssmann, stimulates production of hemolysin against sheep's red blood cells when injected into rabbits.   The antigen has been found in organs of guinea pigs, horses, dogs, cats, mice, fowl, and tortoises, and in some bacteria.

TOXINS AND ANTITOXINS

Ehrlich regarded toxin-antitoxin interaction as purely chemical and equivalent to the neutralization of acids by alkalies.   Thus, the union was postulated to take place according to the law of multiple proportions.   Therefore, if one part of antitoxin neutralizes one part of toxin, 500 parts of antitoxin should neutralize 500 parts of toxin. Danysz, however, observed that when toxin is added to antitoxin in fractions after some lapse of time between additions, a mixture which is nontoxic when toxin and antitoxin are added at once, becomes toxic. This is the *Danysz phenomenon*.   Bordet and Landsteiner believe that the toxin-antitoxin reaction is an adsorption phenomenon, and there is strong evidence to substantiate this view.

Toxoid.   It has been observed that toxin, when kept for a long time, deteriorates, until it is no longer toxic.   Exposure of toxin to 0.4% formalin at a temperature of 37° to 40° C. for about a month produces a product (*toxoid* or *anatoxin*) which is innocuous but is at the same time antigenic and capable of producing a high titer of antibody.   Such preparations have been made from the toxins of *Corynebacterium diphtheriae* and *Clostridium tetani* and are used in immunization against diphtheria and tetanus, respectively.   *Alum-precipitated toxoid* is preferred by many because it is largely protein-free and because the antigenic stimulus is operative for a longer period owing to the slow liberation of toxoid.

Standardization of Toxin and Antitoxin.

1. *Diphtheria Toxin and Antitoxin.*

(a) *M.L.D.* or *minimum lethal dose* of toxin is the smallest amount which kills an average guinea pig of 250 gm. at the end of 4 days after subcutaneous injection. The instability of the M.L.D. makes it unsuitable for the standardization of antitoxin.

(b) *L+ dose* is the smallest amount of toxin which, when mixed with one unit of antitoxin, will kill the average guinea pig of 250 gm. at the end of the fourth day. This is the unit used for the standardization of antitoxin.

(c) *Lo dose* is the largest amount of toxin which can be added to one unit of antitoxin to completely neutralize it so that no symptoms are produced in the animal injected.

(d) *Lf* or *flocculation unit* is the amount of toxin or toxoid that will combine with a proportionate amount of antitoxin in the *in vitro* Ramon flocculation test. (See footnote, p. 218.)

(e) *Lr dose* is the largest amount of toxin which when mixed with one unit of antitoxin produces only slight redness when injected intracutaneously into a guinea pig.

(f) A *unit of diphtheria antitoxin* is that amount of antitoxin which, when added to an L+ dose of toxin, will neutralize this dose so that the life of the average 250-gm. guinea pig will be prolonged up to the end of the fourth day. It was originally defined as the amount of antitoxin which would neutralize 100 minimum lethal doses of toxin.

2. *Tetanus Toxin and Antitoxin.*

(a) Tetanus toxin is constant in potency and is used for the standardization of tetanus antitoxin.

(b) One M.L.D. of tetanus toxin is the smallest amount of tetanus toxin that will kill a 250-gm. guinea pig on the fourth day in the presence of ¼ unit of standard antitoxin furnished by the National Institute of Health in Washington.

(c) One unit of tetanus antitoxin is that amount of antitoxin which will protect a 350-gm. guinea pig against 1000 fatal doses of standard toxin furnished by the National Institute of Health.

3. *Scarlatinal Streptococcus Toxin and Antitoxin.* Since there is no laboratory animal susceptible to this toxin, the toxin and antitoxin are standardized in children.

(a) The unit of scarlatinal Streptococcus toxin is the *Skin Test Dose (S.T.D.)*, or the smallest amount of toxin which produces an erythema at least 1 cm. in diameter in a susceptible child in 48 hours after intracutaneous injection.

(b) The unit of scarlatinal Streptococcus antitoxin is 10 times the smallest amount of antitoxin which will neutralize five skin test doses of toxin so that no reaction as great as 1 cm. is observed in 48 hours in susceptible children.

Tables of Toxins, Antitoxins, and Antisera. In the following tables are summarized certain important points concerning various toxins, antitoxins, and antisera.

IMPORTANT TOXINS *

Product	Production	Test Animal	Unit of Potency	Use
Diphtheria toxin	Veal broth	250-gm. guinea pig	One M.L.D. is the smallest amount of diphtheria toxin that will kill a 250-gm. guinea pig on the 4th day. One L+ dose of diphtheria toxin is the smallest amount which when injected with N.I.H. ** standard unit of antitoxin will cause death of a 250-gm. guinea pig on the 4th day.	Schick test. Standardization of diphtheria antitoxin. Immunization of horses. Production of toxin-antitoxin. Production of toxoid.
Erysipelas Streptococcus toxin	Special bouillon	Man	One S.T.D. of erysipelas Streptococcus toxin is the smallest amount which will cause an erythema at least 1 cm. in diameter when injected intradermally in a susceptible person.	Standardization of erysipelas Streptococcus antitoxin. Immunization of horses.
Meningococcus toxin	Special liquid media	Man	One S.T.D. of meningococcus toxin is the smallest amount that will give an erythema at least 1 cm. in diameter in a susceptible person when injected intradermally.	Standardization of meningococcus antitoxin. Immunization of horses. Skin test for susceptibility.
Perfringens (C. perfringens) toxin	Special bouillon	Pigeon	One test dose of perfringens (C. perfringens) toxin is the smallest amount that will kill a 350-gm. pigeon in 24 hours in the presence of one standard unit of perfringens antitoxin.	Standardization of perfringens (C. perfringens) antitoxin. Immunization of horses.
Puerperal septicemia Streptococcus toxin	Special bouillon	Man	One S.T.D. is the smallest amount which will cause an erythema at least 1 cm. in diameter when injected intradermally in a susceptible person.	Standardization of puerperal septicemia antistreptococcic serum (antitoxin). Immunization of horses.
Scarlet fever toxin	1. Veal bouillon for skin testing and human immunization 2. Special broth for immunizing horses	Man	One S.T.D. of scarlet fever toxin is the smallest amount which will give an erythema at least 1 cm. in diameter in 48 hours when injected intradermally in a susceptible person.	Dick test. Active immunization of susceptibles. Standardization of scarlet fever antitoxin. Immunization of horses.
Tetanus toxin	2% glucose broth	250-gm. guinea pig	One M.L.D. is the smallest amount of tetanus toxin that will kill a 250-gm. guinea pig on the 4th day in the presence of 1/4 unit of N.I.H. standard antitoxin.	Standardization of tetanus antitoxin. Immunization of horses. Tetanus toxoid.

* After Parke, Davis & Company.
** National Institute of Health.

ANTITOXINS AND ANTISERA*

Product	Production	Test Animal	Unit of Potency
Anti-anthrax serum	Native antiserum from the horse	*In vitro* test	Potency proved by agglutination tests. Each lot must agglutinate *B. anthracis* in 1:6400 dilution.
Antidysenteric serum (polyvalent)	Native antiserum from the horse	1. *In vitro* test 2. Mouse	1. The finished antiserum must agglutinate Hiss-Y, Flexner, and Shiga strains of *S. dysenteriae*, and compare favorably with N.I.H.** standard antiserum. 2. In addition, each lot may be standardized by mouse protection tests as required by the British Ministry of Health.
Antimeningococcic serum	Native antiserum from the horse	*In vitro* test	The finished antiserum must agglutinate all four Gordon types of *Neisseria intracellularis* and compare favorably with N.I.H. standard antiserum.
Antipneumococcic serum Types I and II (Felton)	Euglobulin fraction of antiserum from the horse	White mouse	One unit of Felton's antipneumococcic serum is the smallest amount which will protect a white mouse against one million lethal doses of pneumococci.
Antistreptococcic serum (polyvalent)	Refined, concentrated antiserum from the horse	*In vitro* test	Potency proved by agglutination tests with various strains of hemolytic and non-hemolytic Streptococci.
Diphtheria antitoxin	Refined, concentrated antiserum from the horse	250-gm. guinea pig	One unit of diphtheria antitoxin is the least amount which will protect a 250-gm. guinea pig from one L+ dose of diphtheria toxin for at least 4 days.
Erysipelas Streptococcus antitoxin	Refined, concentrated antiserum from the horse	Man	One unit of erysipelas Streptococcus antitoxin is that amount of antitoxin which will completely neutralize one S.T.D. of erysipelas Streptococcus toxin.
Meningococcus antitoxin	Native antiserum from the horse	Man	One unit of meningococcus antitoxin is 10 times the amount that neutralizes one S.T.D. of meningococcus toxin.
Perfringens antitoxin	Refined, concentrated antiserum from the horse	Pigeon	One unit of perfringens antitoxin is the amount which will protect a 350-gm. pigeon against one test dose of perfringens toxin for 24 hours.
Puerperal septicemia antistreptococcic serum (antitoxin)	Refined, concentrated antiserum from the horse	Man	Standardized by antitoxin content. One unit of puerperal septicemia antistreptococcic serum is that amount which completely neutralizes one S.T.D. of puerperal septicemia Streptococcus toxin.
Scarlet fever antitoxin	Refined, concentrated antiserum from the horse	Man	One unit of scarlet fever antitoxin (N.I.H.) is the least amount which completely neutralizes 50 S.T.D.'s of scarlet fever toxin. The original neutralizing unit of the Scarlet Fever Committee is that amount of antitoxin which completely neutralizes one S.T.D. of scarlet fever toxin.
Tetanus antitoxin	Refined, concentrated antiserum from the horse	350-gm. guinea pig	One unit of tetanus antitoxin is 10 times the least amount which will protect a 350-gm. guinea pig from one L+ dose of tetanus toxin for at least 4 days.

* After Parke, Davis & Company.
** National Institute of Health.

AGGLUTINOGENS AND AGGLUTININS

Certain bacteria, foremost of which is *Salmonella typhosa*, when inoculated into animals cause production of antibodies in the serum which clump a suspension of the specific organism when the serum is mixed with it. Such antibodies are called *agglutinins* or *receptors*. They are also present in the serum of patients with typhoid fever, and the clumping of typhoid bacilli by immune serum has been used by Widal in his agglutination test for diagnosis.

Nature of Agglutinins. 1. Agglutinins are heat resistant, but are destroyed by temperatures between 60° and 70° C. 2. They are destroyed by alkalies. 3. Exposure to heat and acids causes agglutinins to lose their clumping power, but they can still combine with the agglutinogen. 4. Agglutinins are specific for a particular organism, but related organisms will agglutinate them at a low titer. 5. Agglutinins persist for a long period of time in sera dried *in vacuo* and stored. 6. They do not kill bacteria, and both living and dead bacteria can be agglutinated.

Mechanism of Agglutination. 1. Bordet showed that agglutination occurs upon the mixture of immune serum and a suspension of bacteria only when salts are present. The bacteria combine with the agglutinins in the absence of salts, but for visible agglutination or clumping to occur, the presence of salts is necessary. 2. Bacteria carry a negative charge. The electrical charge is also negative in the usual menstruum. Therefore the bacteria having the same charge as the medium repel one another, and in so doing stand apart in a free suspension. Specific immune serum, however, carries a positive charge. Therefore, when it is mixed with a corresponding negatively charged suspension of bacteria, the unlike charged particles attract one another, resulting in the clumping or agglutination of the bacteria. 3. Bacteria may be agglutinated in the absence of antibodies by the addition of acid to a suspension (acid agglutination). 4. Some bacteria, like Streptococci, undergo spontaneous clumping.

Prozone or Proagglutinoid Phenomenon. The agglutinating titer of a serum is the highest dilution at which agglutination takes place. It has been observed that in some instances low dilutions of serum cause poor agglutination or even fail to agglutinate bacteria, when higher dilutions will produce the phenomenon. This is called the prozone or proagglutinoid phenomenon and has been attributed to the presence of *agglutinoids* which prevent clumping.

H and O Agglutinins. There are two types of bacterial agglutinins: 1. The flagella or motile H (hauch) agglutinins are specific for the flagella antigens. 2. The somatic or body agglutinins O (ohne hauch) are found within the body cell of flagellated organisms or of nonmotile organisms without flagella. Thus, motile bacteria stimulate the production of two agglutinins, one formed against the flagellar antigen (H) and the other against the body cell (O).

The H agglutinins are highly specific and form large clumps and more of a floccular type of agglutination. The O agglutinins are not as specific and produce a granular type of agglutination. Recently, a Vi or virulence antigen has been described for *Salmonella typhosa* and some paratyphoid bacilli. This antigen is believed to be related to the virulence of the bacillus.

H antigens may be destroyed by treatment of the bacteria with warm (37° C.) alcohol. O antigens may be inactivated by treatment with formalin or phenol, which cause the flagella to stiffen out and the bodies of the bacilli to be held apart.

1. *Preparation of H Antigens.*

(a) Grow actively motile, smooth strain of the organism on 2% beef heart agar, pH 7.0–7.2, for 18 to 20 hours at 37° C.

(b) Wash off growth with a small amount of 0.85% saline solution containing 2% formalin or 5% chloroform.

(c) Place in icebox until sterile.

(d) Using 0.85% saline with 2% formalin in it, adjust the turbidity of the suspension to 10 times that of McFarland's Standard No. 3.

(e) For the test, dilute the antigen with 9 parts of 0.85% saline.

2. *Preparation of O Antigens.*

(a) Wash off growth of smooth culture of the organism after incubation on 2% beef heart agar, pH 7.0–7.2, at 37° C. for 18 to 20 hours, with 0.5% phenol in 0.85% saline solution.

(b) Add to saline suspension 33.3% of 95% alcohol, mix, and incubate at 35° to 37° C. overnight.

(c) Decant supernatant fluid and adjust the turbidity to 10 times that of the McFarland Standard No. 3, reducing the alcohol content to 30% by this adjustment.

(d) For the test, dilute the antigen with 9 parts of 0.85% saline.

Technic of Agglutination Tests.

1. *Macroscopic Tube Test.* Various dilutions of serum are mixed with an equal volume (0.5 ml.) of the suspension of bacteria to be agglutinated. The tubes are incubated at 37° C. The reading is made in a few hours when the positive tubes show the flaking of the

organisms at the bottom of the tube or clumps suspended throughout the liquid.

2. *Widal or Microscopic Agglutination Test.*   The technic of this test requires the mixture of dried blood containing agglutinins and a suspension of motile *Salmonella typhosa.*   In a hanging drop preparation, it is observed that in the presence of specific immune serum the typhoid bacilli lose their motility and stick together in clumps.   The Widal test depends upon the specific power of the blood serum of a patient with typhoid fever to agglutinate *Salmonella typhosa.*

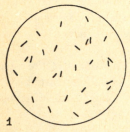 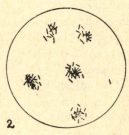

FIG. 76.   Widal agglutination test.   (1) Normal arrangement of *Salmonella typhosa* in hanging drop preparation.   (2) *Salmonella typhosa* agglutinated in the presence of typhoid immune serum.

(a) Place a drop of the patient's blood on a cover-slip and add a number of drops of physiologic salt solution until a cherry-red color is obtained.   This color indicates a 1:10 dilution.

(b) Mix and transfer five drops to a second cover-slip and add five drops of saline, giving a 1:20 dilution.

(c) From this dilution make a 1:40, and then a 1:50, dilution.

(d) To each cover-slip add one drop of a young broth culture or saline suspension of *Salmonella typhosa* and mix.   This results in serum dilutions ranging from 1:20 to 1:100.

(e) Invert the cover-slip over the depression in a hollow-ground slide and examine the preparations under the microscope.   Compare them with a hanging drop made of the broth culture as a control.

(f) The test is positive if complete agglutination occurs within 1 hour in dilutions of 1:40 to 1:80.

(g) The original slide test is often modified to a macroscopic test using both O and H antigens.   It has recently been suggested that a significant titer should be at least 1:100 for O antigen and 1:200 for H antigen.

3. *Modification of Widal Test Using Serum.*   The Widal test may be performed using serum instead of dried blood.   The serum has the advantage that dilutions can be made more accurately.   A 1:10 dilution of the blood is designated as a cherry-red color.   However, since the amount of hemoglobin in blood varies, the arbitrary accept-

ance of a cherry-red color as a 1:10 dilution is erroneous to some extent. Dried blood, however, has the advantage over serum that it is more easily obtained and can be readily transported.

(a) A throttled pipette with a long stem is used, and on this a mark is made about 1 cm. from the extremity.

(b) Saline solution drawn up to the mark of the pipette is expressed onto each of seven clean slides.

(c) One volume of the serum to be tested is taken up in the pipette and discharged on a separate clean slide.

(d) A second volume is mixed on the slide on which a volume of normal saline has been placed. The serum on this slide is thus diluted 1:2.

(e) One volume of this dilution is drawn up into a pipette and discharged onto and thoroughly mixed on the next slide containing the normal saline solution. This represents a 1:4 dilution of the serum.

(f) The process is repeated through the series of slides, with the exception of the seventh.

(g) One volume of the last dilution of serum must be discarded.

(h) The slides contain one volume of a dilution of serum from 1:2 to 1:64. One slide contains only normal saline as a control.

(i) To each of the slides is added an amount of a young motile culture of *Eberthella typhosa* equal to the volume of serum present.

(j) The slides are then studied for evidences of agglutination.

4. *Production of Agglutinating Sera.* Sera of high agglutinating titers may be produced in rabbits by injecting them intraperitoneally or subcutaneously with 1 ml. of killed bacteria grown in bouillon. The initial injection is followed by another ml. 4 days later, and four or five subsequent injections made at intervals of 3 to 5 days. Small amounts of blood are obtained by bleeding the rabbits from the veins. These are tested for agglutinins, and when serum of a sufficiently high titer is obtained, it is employed in various agglutination tests.

5. *Microscopic Slide Agglutination Test.* The agglutination test is used in identifying bacteria which may be confused with other similar organisms.

(a) Bacteria from suspected colonies are mixed on a glass slide with separate drops of potent antisera against *Salmonella typhosa, Salmonella paratyphi, Salmonella schottmuelleri, Shigella dysenteriae, and Salmonella enteritidis.* Various dilutions of the sera are made before the test is started.

(b) A control is prepared by mixing some of the unknown organisms with a drop of diluted normal serum.

(c) With a needle a small portion of the suspected colony is suspended in each of the drops of serum. The needle must be flamed after the preparation of each suspension to avoid mixing the sera.

(d) Observe the slide with the naked eye or a hand lens for specific flocculation which may occur within a few minutes. Gentle warming may hasten the reaction, but a positive agglutination test is of presumptive value only.

6. *Rapid Agglutination Test.* Recently a method of testing for agglutinins by using a concentrated antigen has been devised. When a quantity of serum from an infected animal or man is mixed with a quantity of the known antigen, clumping of the organisms in the antigen occurs within a few minutes. This phenomenon is useful in testing for Malta fever in man, contagious abortion in cattle, and bacillary diarrhea in chickens.

Agglutinin Absorption Test. Certain organisms of a particular group of bacilli exhibit agglutination to some low degree in agglutinating serum prepared against one member of that group. To identify an organism which is related immunologically to other bacteria, it is necessary to perform an agglutinin absorption test in which each organism removes from the serum the agglutinins specific for itself. Thus, in an outbreak of food infection where an organism A has been isolated and is believed to be related to either B or C, organisms which are antigenically related to it, the procedure for identifying it is as follows:

1. Inoculate agar slants with a culture of each of A, B, and C and incubate for 24 to 48 hours.

2. Wash off the slant growth with a small quantity of sterile physiologic salt solution and place suspensions in 15 ml. centrifuge tubes.

3. Centrifuge the tubes at high speed for 20 minutes or more. The bacteria should be tightly packed in the bottom of the tubes.

4. Pour off the supernatant fluid and drain thoroughly.

5. Prepare 1:10 dilutions of known antisera specific for organisms B and C.

6. Into two sets of three tubes each, place 10 ml. of each diluted serum and add suspensions of organisms A, B, and C, respectively.

7. Mix and incubate in water bath at 45° C. for 4 hours, and in the refrigerator overnight.

8. Centrifuge the tubes until the bacteria are well packed in the bottom of the tube and the fluid is clear.

9. Pipette off clear supernatant, representing the absorbed sera.

10. Use this supernatant for agglutination tests with suspensions of organisms A, B, and C, respectively.

11. If the unknown organism A has completely absorbed all the agglutinins in serum B, it will cause no agglutination in the test with B serum. Therefore, the suggestion is that it is identical with B. If, however, no agglutination is observed with serum C, the organism A has absorbed all the agglutinins from serum C, and therefore may be similar to C.

12. The result of the agglutinin absorption test is not considered as absolute proof of the identity of two organisms until *reciprocal* agglutination tests are performed. In this case, if antiserum prepared against organism A agglutinates organism C, and organism C absorbs serum A, and the absorbed serum fails to agglutinate organism C, then organism A is identical with organism C. The same test should be done with organism B, if the first absorption tests show that organism A may be identical with B.

Practical Applications of Agglutination Tests.

1. Agglutinin absorption tests help to identify and differentiate culturally similar microorganisms. This is useful during outbreaks of food infection, in pneumococcus typing (see p. 203), and for classification studies.

2. The test is of value in the diagnosis of such diseases as typhoid fever, paratyphoid fever, typhus fever (Weil-Felix reaction), undulant fever, and tularemia.

3. On the basis of the agglutination test, the blood of human beings has been found to fall into four classes, O, A, B, and AB. These are based on the presence of iso-agglutinins in sera which cause the agglutination of the red blood corpuscles of one man by the serum of another. The agglutination tests on blood are used for the typing of human blood for transfusions. Since blood groupings are inherited according to Mendelian law, the test has also been of value in paternity disputes.

PRECIPITINOGENS AND PRECIPITINS

Precipitins are antibodies formed in response to the injection of soluble antigens, which, when mixed with the antigen, aggregate the molecules with the formation of a precipitate. Precipitation does not occur in the absence of electrolytes.

Applications of the Precipitation Test. 1. Because of its high specificity the test is useful in the detection of any other meat which may be substituted and distributed as beef. 2. The test has wide application in law in the detection of stains made by human blood. The stained material is extracted and the solutions are tested with various antisera prepared in rabbits by the injection of blood of cows, horses, dogs, and man. 3. The test is used in bacteriological investigations in the differentiation of types of pneumococci, and also in the detection of anthrax infection of animal tissues (Ascoli thermoprecipitin test). 4. The Kahn precipitation and the Kline microscopic precipitation tests are used in the diagnosis of syphilis. (See p. 274.)

Technic of the Test (Pneumococcus Typing). For a precipitation test it is necessary to have a serum of a high titer, and one whose degree of specificity is known. The test is performed as follows in typing pneumococci:

1. Into each of a series of small agglutination tubes place 0.5 ml. of properly diluted antipneumococcic serum and float over it 0.5 ml. of clear supernatant fluid taken from the centrifuged peritoneal washings of a mouse injected with sputum or a suspension containing pneumococci.

2. Usually a precipitation reaction occurs immediately in the tube containing the homologous immune serum, but no precipitation occurs with the heterologous serum.

3. If the supernatant fluid is added carefully so that it does not mix with the serum, a positive test will be indicated by a ring of white precipitate at the point of contact.

4. When the fluid and serum are mixed, a positive test is indicated by a precipitate which is visible throughout the mixture. If no immediate reaction occurs, place the tubes in a water bath at 37° C., and observe after incubation for 15 minutes, 30 minutes, and 60 minutes.

CYTOLYSINS AND COMPLEMENT

The cytolysins, also called *amboceptors* or *sensitizers*, are antibodies which lyse or dissolve bacteria or red blood cells. The visible reaction of lysis does not occur unless there is present a normal non-specific constituent of serum, called *alexin* or *complement*, which also combines with the cell.

Complement, normally present in the blood of all animals, deteriorates rapidly. It is thermolabile and is inactivated by a temperature of 56° C. for one-half hour. In order for complement to act, the cells must already have been sensitized by the amboceptor or sensitizer. Complement does not combine with antigen in the absence of amboceptor, but antigen and amboceptor will unite regardless of the presence of complement.

Mechanism. Ehrlich believed that complement acts upon antigen only indirectly through the amboceptor, which functions as a bridge between the first two. The Bordet view, held by most investigators, is that the union is a specific adsorption, the sensitized antigen being rendered susceptible to the action of complement.

Pfeiffer Phenomenon. It was noticed by Pfeiffer that living cholera vibrios, injected into the peritoneal cavity of guinea pigs that had been rendered immune, underwent certain changes when the peritoneal exudate was examined at various intervals. The organisms were observed to become (a) nonmotile, (b) swollen, (c) coarsely granular, (d) indistinct in outline, and (e) finally dissolved completely. This phenomenon of lysis was attributed by Pfeiffer to the presence of *bacteriolysin*, a specific substance which was present only in immune serum and could be transferred to normal animals by injection of immune serum.

The process was later shown by Bordet to be the result of the activity of two substances—one, the thermostable bacteriolysin which acted only in the presence of the other, the thermolabile complement, a constituent of normal serum. If the serum containing complement was inactivated by heating at 56° C. for one-half hour, the immune serum lost its ability to lyse the bacteria. However, if serum containing complement was added to such an inactivated serum, the bacteria were dissolved.

Bordet-Gengou Phenomenon. It was noted by Bordet and Gengou in 1901 that when inactivated specific immune serum and a culture of *Pasteurella pestis* were mixed together with complement, the complement was bound or fixed in destroying the bacilli. This was visibly shown by using a hemolytic system consisting of a suspension of washed rabbit red blood cells and antirabbit hemolysin. If complement had been free it would have been detected by the hemolytic system, and lysis of the red blood cells would have occurred, for hemolysin can act only in the presence of complement. However, in this case no hemolysis occurred. When normal serum was used instead of immune serum it was noted that hemolysis of the red blood cells occurred. Since specific amboceptor was not present in the normal serum to sensitize the cells, the complement had not been bound, but was left free to take part in the lysis of the red blood cells by the hemolysin of the hemolytic system.

The reactions observed may be represented as follows:

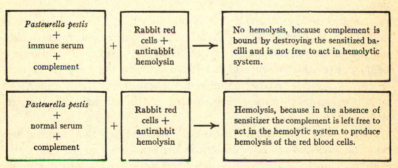

Wassermann or Complement Fixation Test. The hemolytic system employed by Bordet and Gengou to give a visible reaction for the union of antigen and antibody has been utilized in similar tests for the diagnosis of such diseases as glanders, syphilis, gonococcic and meningococcic infections, tuberculosis, and whooping cough. Wassermann perfected a complement fixation test for the diagnosis of syphilis. This test requires a bacteriolytic system and a hemolytic system, which gives an index as to whether the complement has been bound by the bacteriolytic system.

1. The *bacteriolytic system* consists of:

(a) The *antigen*, which is an alcoholic extract of ether-extracted powdered beef heart containing 0.4% cholesterol.   This antigen is nonspecific.   The Wassermann test, therefore, though it is a disease-specific test, is not immunologically specific, for a nonspecific antigen distinct from the causative agent of the disease (*Treponema pallidum*) is used.

(b) The *patient's serum*, which should be clear and not hemolyzed.   It is inactivated by heating at 56° C. for one-half hour on the day of the test to destroy the varying amounts of complement present.

(c) Standard, titrated *complement* obtained by using pooled serum of two or more guinea pigs which have been bled within 24 hours.

2. The *hemolytic system* consists of:

(a) Red-cell amboceptor or *hemolysin*, which is present in the serum of rabbits that have been immunized against sheep red blood cells.   In some cases human red blood cells are used.

(b) *Blood cell suspension*, which is a 5% suspension of sheep red blood cells or, if human red blood cells are used to prepare the hemolysin, a 5% suspension of human red blood cells.

The object of the test is to determine whether the complement will be bound in the bacteriolytic system.   In syphilis the specific cytolysins are increased.   These have an affinity for the fixing reagent (antigen), and the complement is fixed by the combination of these two.

A *positive* Wassermann test is one in which the complement is bound by the bacteriolytic system so that none is free to cause hemolysis of the red blood cells in the hemolytic system.   Therefore, no hemolysis, or the presence of red blood cells, in the tube indicates a ++++ positive Wassermann test.   There are various grades of hemolysis ranging from +++, which is still strongly positive, to ++, and +.

A *negative* Wassermann test is one in which the patient's serum has no antibodies and the complement is left free to act in the hemolytic system.   Therefore, hemolysis of the red blood cells will be observed, and the tube will have a uniform pink appearance.

In performing the Wassermann test it is necessary to determine certain units in standardizing the various reagents.

1. *Antibody unit* is the smallest amount of serum that with two units of a homologous fixing reagent gives complete fixation of complement under standard conditions.

2. *Unit of fixation* is the smallest amount of fixing reagent that gives complete fixation of complement with two units of homologous immune serum.

3. *Anticomplementary dose* is the smallest amount of fixing reagent that inhibits hemolysis.

4. *Minimum hemolytic dose of complement* is the smallest amount of complement which will completely hemolyze 0.1 ml. of a 5% suspension of erythrocytes which have been sensitized with an excess amount of hemolysin. Note: *Sensitization* of the red blood cells is accomplished by exposing one part of a 5% suspension of the erythrocytes to one part of the standard dilution of hemolysin. This is so standardized that 0.2 ml. of such a mixture will contain 0.1 ml. of the 5% suspension and two hemolytic units of hemolysin.

5. *Hemolytic unit* is the smallest amount of hemolysin which will give complete hemolysis of 0.1 ml. of a 5% suspension of erythrocytes with an excess of complement at the end of 45 minutes in a water bath at 37° C.

OPSONINS AND PHAGOCYTOSIS

One of the mechanisms whereby the body cells get rid of certain bacteria and foreign material is by ingestion of these substances. Any cell which destroys microorganisms by enveloping and absorbing them is called a *phagocyte*. The intracellular digestive process whereby the leucocytes and certain fixed cells eat up or phagocytize the various invading bacteria is known as *phagocytosis*.

Opsonins. Metchnikoff* observed that the process of phagocytosis occurs more readily in the presence of immune serum than with normal serum. To the substance responsible for the enhancing of the tendency of leucocytes to engulf bacteria, Wright gave the name *opsonin*, and Neufeld called it *bacteriotropin*. That opsonins are necessary for the process of phagocytosis is shown by the observation that when bacteria or white blood corpuscles are washed free of serum, the absorption and ingestion of invading bacteria do not take place.

Opsonins are also present in normal serum. Following infection with certain microorganisms the amount of opsonins is increased. Opsonins exhibit characteristic antibody specificity. Their activity appears to be the result of two components—one thermostable, and one thermolabile, present in normal serum, and resembling complement in many ways.

Generally, there is a positive chemotactic influence exerted between the phagocytes and the bacteria. The leucocytes engulf the bacteria and ingest them. Bacteria so phagocytized first become swollen, then coarsely granular, finally lose their outline, and then disappear entirely within the cytoplasm of the phagocytes.

Opsonic Index. The opsonic action of a serum is measured by determining its opsonic index, which is the number of bacteria pha-

---

* Metchnikoff's insistence on the importance of phagocytosis as the sole basis of immunity is incorporated in his *cellular theory* of immunity. This was opposed by Ehrlich's *humoral theory*, emphasizing the importance of chemical substances (antibodies) dissolved in the bloodstream.

gocytized by the unknown serum of a patient divided by the number destroyed by the normal serum (control).

Phagocytic Cells. Phagocytosis is produced by the microphages or wandering phagocytes, and the macrophages, and also the cells of the reticulo-endothelial system.

1. The *microphages* include the leucocytes. Following an infection of the skin with Streptococci there is a migration of polymorphonuclear leucocytes to the site of infection, and these ingest the bacteria. The leucocytes degenerate, become cloudy, swollen, and fatty, and disintegrate. These degenerated phagocytes plus debris, blood serum, and the digested bacteria constitute the substance commonly known as *pus*.

2. The *macrophages* include the large mononuclear leucocytes. In infection with *Mycobacterium tuberculosis* these cells surround the bacteria. The polymorphonuclear leucocytes do not appear in the later stages of the disease, but may be of some importance in the beginning of the disease in disposing of bacteria.

3. The *reticulo-endothelial system*, including the endothelial cells lining the capillaries and the liver, the spleen, the bone marrow, and the lymph sinuses, play some role in phagocytosis. It is claimed that these tissues are responsible for the production of humoral antibodies, and also phagocytize bacteria.

OTHER ANTIBODIES

It has been suggested that there are at least two other types of antibody actions. 1. *Ablastins*—reproduction-inhibiting antibodies which prevent multiplication of the invading organism. 2. *Neutralizing antibodies*—which render the infectious agent, generally a filterable virus, noninfective, when mixed and incubated with it.

MECHANISM OF ANTIGEN-ANTIBODY REACTION

The antigen-antibody reaction takes place in two stages. The first involves union of the elements, and the second, the consequences of that union which appear as agglutination, etc. Historically there have been two major theories proposed to explain the mechanism of the reaction.

Ehrlich's Side-Chain or Receptor Theory.

1. Interaction of antigen and antibody is a chemical phenomenon.

2. Body cells obtain nutriment through localized cell substances called receptors or side-chains, which have combining affinities with food and other substances.

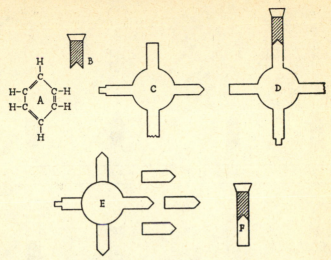

Fig. 77.   Ehrlich's explanation of formation of antitoxin.   *A*. Benzene ring with side chains.   *B*. Toxin molecule consisting of a haptophore group which serves to unite with antitoxin or body cell, and toxophore group which produces the toxic effects.   *C*. Body cell with different types of receptors.   *D*. Toxin molecule combines with body cell through suitable receptors, limiting the activity of the cell.   *E*. Body cell is stimulated to overproduction of the specific receptors which are shed into the bloodstream; these **constitute antitoxin.**   F. Antitoxin combines with toxin through haptophore group **to neutralize it.**

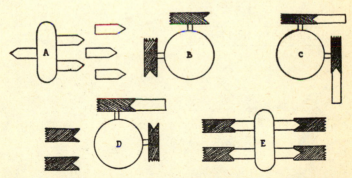

Fig. 78.   Ehrlich's theory for formation of agglutinins.   *A*. Bacterial cell with attached and free agglutinogens.   *B*. Body cell with receptors with which agglutinogen can combine.   The receptors consist of a haptophore portion for combining with the agglutinogen, and a zymophore portion which causes the clumping.   *C*. Agglutinogen in contact with body cell through a suitable receptor prevents proper cell function.   *D*. The body cell overproduces specific receptors as agglutinins, and sheds them into the bloodstream.   *E*. Overproduced agglutinins unite with body cells, causing agglutination.

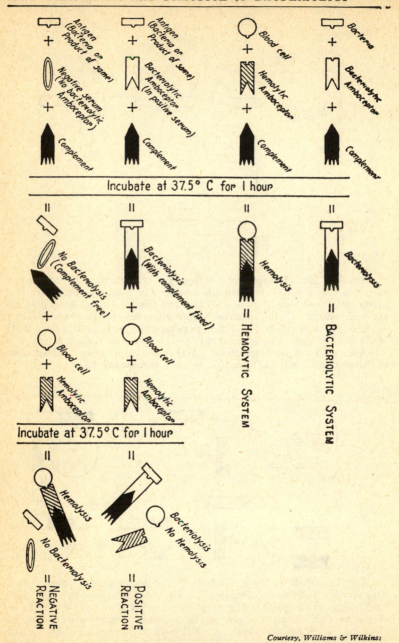

Incubate at 37.5° C for 1 hour

Incubate at 37.5° C for 1 hour

FIG. 79. Diagrammatic representation, according to Ehrlich's theory, of the Wassermann complement fixation test. (After Kelser, *Manual of Veterinary Bacteriology*.)

3. When bacteria or other foreign cells enter the body, the combining affinity of certain body cells may by chance be satisfied by bacterial or other substances.

4. By their union with toxin or other bacterial substance, the receptors are rendered useless for their normal physiological function.

5. The receptors are cast off and the body cell tends to regenerate the lost part and usually tends to overcompensate.

6. The receptors formed in excess of body needs are discharged into the bloodstream.

7. These free receptors are the antibodies.

8. Receptors are of varying degrees of complexity. Antitoxins are receptors of the first order, agglutinins and precipitins, second order, and lytic reactions, third order.

9. Similar representations were made of antigen and complement. For example, toxin was considered to have two functional parts—a haptophore which unites with the receptor, and a toxophore which exerts the poisonous effect. Toxoid was considered toxin in which the toxophore was destroyed or inactivated.

Although many immunological phenomena can be explained neatly by Ehrlich's theory, newer experimental evidence has often failed to confirm it and necessitated modifications of the original concept.

Bordet's Adsorption Theory.

1. Antigen and antibody solutions may be regarded as colloidal systems.

2. Union of the elements is an adsorption phenomenon, physical in nature rather than chemical.

This theory has been found satisfactory in many respects, its biggest failure being the inability to account for specificity.

Modern Concept. The modern concept of antigen-antibody reaction is an outgrowth of both of the previous theories.

1. The reacting substances are regarded as colloids.

2. Union of the elements is a highly specific adsorption phenomenon of surface chemistry, partly physical and partly chemical in nature.

3. The second stage of the reaction is an extension of the process of union and is therefore also specific.

BIBLIOGRAPHY

Boyd, W. *Fundamentals of Immunology*. Third Edition. New York: Interscience Publishers, 1956.

Burrows, W. *Textbook of Microbiology*. Seventeenth Edition. Philadelphia: W. B. Saunders Company, 1959.

Carpenter, P. L. *Immunology and Serology.* Philadelphia: W. B. Saunders Company, 1956.

Cushing, J. E., and D. H. Campbell. *Principles of Immunology.* New York: McGraw-Hill Book Company, 1957.

Hare, R. *Bacteriology and Immunity.* New York: Longmans, Green and Company, 1956.

Jawetz, E., J. L. Melnick, and E. A. Adelberg. *Review of Medical Microbiology.* Los Altos: Lange Medical Publications, 1962. Pp. 112–122.

Komer, J. A., and L. Tuft. *Clinical Immunology, Biotherapy and Chemotherapy.* Philadelphia: W. B. Saunders Company, 1941.

Landsteiner, K. *The Specificity of Serological Reactions.* Springfield, Ill.: Charles C. Thomas, 1936.

Marrack, J. R. *The Chemistry of Antibodies and Antigens.* Med. Res. Council, Spec. Rept. Ser. No. 230. London: His Majesty's Stationery Office, 1939.

Raffel, S. *Immunity, Hypersensitivity, Serology.* New York: Appleton-Century-Crofts, 1953.

Rivers, T. M., and F. L. Horsfall, eds. *Viral and Rickettsial Infections of Man.* Third Edition. Philadelphia: J. B. Lippincott Company, 1959.

Rosebury, T. *Peace or Pestilence.* New York: McGraw-Hill Book Company, 1949.

Sherwood, N. P. *Immunology.* Third Edition. St. Louis: C. V. Mosby Company, 1951.

Smith, D. T., *et al. Zinsser's Textbook of Bacteriology.* Eleventh Edition. New York: Appleton-Century-Crofts, 1957.

Society of American Bacteriologists. *Manual of Methods for Pure Culture Study of Bacteria.* Geneva, N. Y.: Biotechnical Publications, 1949. Chap. VII.

Strean, L. P. *Oral Bacterial Infections.* Brooklyn, N. Y.: Dental Publishing Company, 1949. Pp. 160–171.

Top, F. H. *Communicable Diseases.* Third Edition. St. Louis: C. V. Mosby Company, 1955.

Topley, W. W. C. *An Outline of Immunity.* Baltimore: William Wood and Company, 1935.

Wilson, G. S., and A. A. Miles. *Topley and Wilson's Principles of Bacteriology and Immunity.* 2 vols. Fourth Edition. Baltimore: Williams & Wilkins Company, 1955.

Young, G. G. *Witton's Microbiology.* Third Edition. New York: McGraw-Hill Book Company, 1961. Pp. 210–219.

Zinsser, H., and J. T. Enders. *Immunity.* New York: Macmillian Company, 1939.

Chapter XXXVIII

ALLERGY

Considerable confusion and disagreement exist over the definitions of terms used to describe various conditions of hypersensitivity. The most widely accepted uses, which will be employed in this text, are given below.

Allergy is a condition of unusual or exaggerated susceptibility to a specific substance which is harmless in similar amounts for the majority of members of the same species. The *allergen* (exciting substance) may or may not be protein and may or may not be antigenic.

Anaphylaxis is a particular kind of allergy, in which the allergen is an antigenic protein. The acute, severe, and often fatal type of anaphylaxis, in which sense the term is most frequently used, is not common in man and is usually reserved for the condition of sensitization experimentally produced in laboratory animals. It has been observed that animals which do not react to the first injection (*sensitizing dose*) of a harmless protein will react violently to a second injection (*toxic dose*) of the same protein after a lapse of several weeks.

Symptoms. Susceptibility to allergic reactions varies markedly in different individuals, as do the symptoms and pathological changes. The reactions are manifested mainly by dilation of the blood vessels and affectation of tissues with well-developed smooth musculature. Although symptoms of anaphylaxis vary in different animals, common to most are abrupt fall in blood pressure, leucopenia, reduction in blood complement and fibrinogen, capillary dilation, and general irritation of smooth muscles. In man, the tendency is not toward general shock, but rather to local symptoms particularly in the respiratory and gastrointestinal tracts and the skin.

Shwartzman Reaction. Shwartzman observed that a severe, hemorrhagic, necrotic reaction is produced in rabbits at the site of an intradermal injection with filtrates of cultures of *Salmonella typhosa* or certain other organisms, when this is followed within 24 hours by an intravenous injection of the filtrate. The reaction appears within 24 hours after the second injection.

Arthus Phenomenon. This is a peculiar local effect following repeated subcutaneous injections of horse serum into a rabbit. The earlier injections are without local effect. As the animal becomes immune, however, the site of inoculation becomes transiently swollen and edematous. Further injections produce a firm, indurated area with local necrosis. The local response can be elicited at any site and is not necessarily confined to the area of the earlier injections.

The phenomenon is demonstrable in rabbits and to a limited extent in guinea pigs, but not at all in other experimental animals; it occurs in man, sometimes during the course of the rabies prophylactic. It appears to be dependent on the union within the tissues of circulating precipitin and its specific antigen.

Serum Sickness. Intramuscular or subcutaneous injection of horse serum in certain individuals results in a skin eruption or urticaria, fever, edema, enlargement of the lymph nodes with pain and tenderness, or pain in the joints. Occasionally there is death due to extreme dyspnea. Reactions may be prevented by preliminary skin testing to ascertain sensitivity.

Bacterial Allergy. Hypersensitiveness to microorganisms or their products may be produced experimentally in animals and human beings or may occur spontaneously during the course of a disease. This may be used as a diagnostic aid. The tuberculin reaction in tuberculosis, the mallein reaction in glanders, the reaction to abortin in infectious abortion, the reaction to luetin in syphilis are considered examples of bacterial allergy.

It is possible that allergic reactions may be responsible for many disease symptoms. Von Pirquet has suggested that the exanthemata (skin eruptions) of some acute infectious diseases, such as scarlet fever, are anaphylactic reactions.

Allergy to Pollen, Hair, Etc. The reactions are often localized in the upper respiratory tract, with coryza, sneezing, and sometimes asthma. The condition is familiarly termed hay fever.

Food Allergy. Certain individuals exhibit hypersensitiveness to certain foods such as strawberries or oysters. The symptoms are generally gastrointestinal but may include urticaria and others.

Drug Allergy. Drug idiosyncrasies are usually characterized by skin reactions as itching and rash and in some cases by body pain, edema, etc.

Inheritance. Although most allergies are probably acquired, in some cases there appears to be an inherited general tendency to become sensitized. Inheritance of a specific sensitization does not occur.

This inherited sensitiveness in human beings has been termed *atopy* by Coca.

Prausnitz-Küstner Reaction. This is the passive transfer of local hypersensitivity in man by the intradermal injection of serum of the allergic person.

Tests for Sensitivity. Hypersensitivity to a specific substance may be determined by application of the material to scarification made on the skin (cutaneous method or scratch test), intracutaneous (intradermal) injection, or patch test on the surface of the skin or mucous membrane. A positive reaction consists of a raised, urticarial area or wheal.

Desensitization. In some cases, persons may be desensitized by repeated small injections of the specific protein at short intervals. It is believed that a strong reaction is avoided because only small amounts of the allergen reach the sensitive tissues and an increasing resistance is built up.

Antianaphylaxis is a state of desensitization in which the allergic reaction is not obtained in response to the injection of the allergen. It is believed due to the presence in the circulating blood of sufficient free antibodies to unite with all the antigen injected so that it cannot reach the sensitized tissues in which the reaction takes place.

Mechanism. The phenomenon of allergy is essentially immunological. It is believed to be an antigen-antibody reaction occurring within the tissue cells and producing inflammation. Immune (not sensitive) animals presumably possess sufficient circulating antibodies to combine with and immobilize the antigen before it reaches the fixed tissue cells. To explain the inflammatory reaction, the anaphylotoxin theory proposes partial digestion of the antigen with the liberation of toxic substances. A second theory postulates the disturbance of the colloidal equilibrium.

BIBLIOGRAPHY

Cooke, Robert A. *Allergy in Theory and Practice*. Philadelphia: W. B. Saunders Company, 1947.

Gershenfeld, Louis. *Bacteriology and Allied Subjects*. Easton, Pa.: Mack Publishing Company, 1945. Pp. 451–455, 509–513.

Todd, J. C., *et al*. *Clinical Diagnosis by Laboratory Methods*. Twelfth Edition. Philadelphia: W. B. Saunders Company, 1953.

Urbach, Erich, and Philip M. Gottlieb. *Allergy*. Second Edition. New York: Grune & Stratton, 1946.

Vaughan, W. T. *Practice of Allergy*. Third Edition. St. Louis: C. V. Mosby Company, 1954.

Chapter XXXIX

PREVENTION OF THE SPREAD OF COMMUNICABLE DISEASES

(Points for Nurses and Attendants)

The most effective means for control, prevention, and ultimate eradication of communicable diseases are: (a) isolation (separation of the sick from the well, and also of contacts), (b) disinfection of materials containing microorganisms, as in diphtheria, scarlet fever, tuberculosis, typhoid fever, plague, whooping cough, and cholera, (c) preventive inoculation, and (d) education.

Isolation. Whenever a case of an infectious disease occurs, the most important act is to isolate the patient from other individuals in order to avoid spread. All those persons in contact with the patient before the disease was recognized should also be kept separately and watched for the appearance of symptoms. In this way any possible outbreak may be confined to a small group before it begins to spread.

Disinfection. All materials, such as sputum, excreta, and clothes worn by the patient, should be properly disinfected. The use of a 5% solution of carbolic acid or Lysol is efficient. For further information see Chapter VI. The proper disposal of the belongings of the patient and anything with which he may come in contact is imperative. In infectious disease hospitals the danger of cross infection from one ward to another should be considered. The hands of persons caring for the sick should always be kept scrupulously clean and should be washed each time they come near the patient. Gowns worn by nurses should be changed frequently, especially when going from one ward to another. To avoid the necessity of having too many things to disinfect, rooms should be freed of all unnecessary articles of furniture, hangings, and carpets. Care should be taken to avoid handling any food, utensils, or clothes of the patient, without first taking the proper precautions to prevent infection.

Preventive Inoculation. Persons working in infectious disease hospitals or entering areas where certain diseases are prevalent should be immunized against such infections whenever possible. Thus, individuals traveling to parts of the world where typhoid fever is common should be vaccinated against this disease. There are also effective means of immunizing against diphtheria, scarlet fever, plague, cholera, smallpox, yellow fever, typhus, and tetanus.

Education. The greatest advance in curbing the spread of communicable diseases lies in making known to the layman the dangers that confront him and the means whereby he can avoid certain diseases. Mothers should be taught the value of immunization against diphtheria, the desirability of keeping children at home when they appear to have only a "cold," and the hazards of exposing children unnecessarily by allowing them to play with other children who may have infectious diseases, in order that they may more quickly get over the so-called inevitable diseases of childhood, such as measles and whooping cough. Only by revealing the facts and by getting the earnest cooperation of the people concerned can persons engaged in public health work hope to achieve their goal of lowering the incidence of certain common infectious diseases for which there are known methods of prevention and therapy.

BIBLIOGRAPHY

Anderson, Gaylord W., and Margaret G. Arnstein. *Communicable Disease Control.* Third Edition. New York: Macmillan Company, 1953.

Boyd, Mark F. *Preventive Medicine.* Seventh Edition. Philadelphia: W. B. Saunders Company, 1945.

Broadhurst, Jean, and Leila I. Given. *Microbiology Applied to Nursing.* Fifth Edition. Philadelphia: J. B. Lippincott Company, 1945.

Calder, R. M. *Bacteriology for Nurses.* Third Edition. Philadelphia: W. B. Saunders Company, 1943.

Dubos, R. J. *Bacterial and Mycotic Infections of Man.* Third Edition. Philadelphia: J. B. Lippincott Company, 1958.

Marshall, M. S. *Applied Medical Bacteriology.* Philadelphia: Lea and Febiger, 1947. Pp. 1–340.

Morse, Mary Elizabeth, and Martin Frobisher, Jr. *Bacteriology for Nurses.* Seventh Edition. Philadelphia: W. B. Saunders Company, 1946.

Mustard, Harry Stoll. *Introduction to Public Health.* Third Edition. New York: Macmillan Company, 1953.

Netter, Erwin, and D. R. Edgeworth. *Medical Bacteriology for Nurses.* Philadelphia: F. A. Davis Company, 1957.

Parish, H. J. *Antisera, Toxoids, Vaccines and Tuberculins in Prophylaxis and Treatment.* Third Edition. Baltimore: Williams & Wilkins Company, 1955.

Sinclair, Charles George. *Microbiology for Nurses.* Sixth Edition. Philadelphia: F. A. Davis Company, 1945.

Smith, A. L. *Carter's Microbiology and Pathology.* Seventh Edition. St. Louis: C. V. Mosby Company, 1960.

Top, F. H. *Communicable Diseases.* Third Edition. St. Louis: C. V. Mosby Company, 1955.

Young, G. G. *Witton's Microbiology.* Third Edition. New York: McGraw-Hill Book Company, 1961. Pp. 231–238.

GLOSSARY

Abacterial. Free from bacteria.

Abbé condenser. An attachment to a microscope consisting of a mirror and a series of wide-angled achromatic lenses placed beneath the stage to give strong illumination.

Abscess. A localized collection of pus in a cavity formed by the disintegration of tissues.

Acetobacter. A genus of aerobic bacilli which secure their energy by the oxidation of alcohol to acetic acid.

Acid-fast. The property of not being readily decolorized by acids or other means when stained.

Actinomyces. A genus of parasitic fungi characterized by a radiating arrangement of the mycelium and by small spores.

Acute. Having a short, severe course.

Adenovirus. A virus causing mild infection of the upper respiratory tract.

Aerobe. An organism that grows best in the presence of free oxygen.

Aerogens. Gas-producing bacteria.

Afebrile. Without symptoms of fever.

Agar-agar. A gelatinous substance prepared from Ceylon moss and used as a solidifying agent in culture media. It dissolves in boiling water and solidifies at about 38° C.

Agglutination. The phenomenon of clumping of the cells distributed in a fluid, observed when certain bacterial cultures are treated with serum containing an agglutinin specific for the particular organism.

Agglutinin. An antibody found in an immune serum which when added to a suspension of its homologous microorganism causes the organisms to adhere to one another, forming clumps.

Agglutinogen. The agglutinable substance or antigen which stimulates the formation of agglutinins when injected into the animal body.

Agglutinoid. Heat-killed agglutinin which has lost its property of causing agglutination although it can still unite with its corresponding agglutinogen.

Agranulocytosis. Complete or nearly complete absence of granular leucocytes (granulocytes) from the bone marrow and the blood.

Alastrim. A specific, contagious, virus-caused, eruptive fever, resembling a mild smallpox and probably related to that disease.

Alexin. *See* Complement.

Allergen. Any substance capable of inducing or exciting a condition of allergy or hypersensitiveness.

Allergy. A condition of unusual or exaggerated specific susceptibility to a substance which is harmless in similar amounts for the majority of members of the same species. Examples: asthma, hay fever, hives.

Alum. Aluminum and potassium sulfate, $AlK(SO_4)_2 + 12 H_2O$.

Amboceptor. *See* Cytolysin.

Amebiasis. The state of being infected with amoebae.

Amino acid. An organic acid in which one or more hydrogen atoms have been replaced by the amino group NH_2. Amino acids are the building stones of the protein molecule.

Amphitrichous. Having one or a tuft of flagella at each end.

Ampule. A small glass container capable of being sealed so as to preserve its contents in a sterile condition.

Anabolism. The constructive process by which simple substances are converted by living cells into more complex compounds.

359

Anaerobe. An organism which can grow without either air or free oxygen.

Anaphylactic shock. A violent attack of symptoms produced by a second injection of a foreign protein in a sensitive animal.

Anaphylactogen. A substance which is capable of causing anaphylaxis.

Anaphylaxis. Allergy in which the allergen is an antigenic protein. *See also* p. 353.

Anatoxin. *See* Toxoid.

Andrade's indicator. A solution of acid fuchsin, 0.5 gm. in 100 ml. of water, which is decolorized to a yellow color by sodium hydroxide.

Anemia. A condition in which the blood is deficient either in quantity or quality; the latter may consist of a decrease in the amount of hemoglobin or in the number of red blood corpuscles.

Anode. The positive pole of a galvanic battery or other electric source.

Anorexia. Lack or loss of appetite for food.

Anthrax. A fatal infectious disease of cattle and sheep, transmissible to man, caused by *Bacillus anthracis*.

Antianaphylaxis. A state of desensitization in which allergic phenomena do not occur, believed due to free antibodies in the blood preventing the antigen from reaching the sensitized tissues.

Antibiotic. Literally, growth-inhibiting; usually denotes bacteriostatic substance produced by microorganisms.

Antibody. The substance produced in the animal body against an antigen and exerting a specific antagonistic influence on the substance which stimulated its formation.

Anticomplementary. The property of reducing or destroying the power of complement.

Antiformin. A strongly alkaline sodium hypochlorite solution, used as a disinfectant. Because it does not dissolve acid-fast organisms, it is used to isolate them.

Antigen. A substance that stimulates formation of antibodies in the animal organism under suitable conditions.

Antiseptic. A substance that prevents or inhibits the growth of microorganisms without necessarily destroying them.

Antiserum. A serum containing specific antibodies, obtained from the blood of an animal that has received repeated doses of an antigen or that has recovered from an infection.

Antitoxin. An antibody found in the blood serum and in other body fluids which is incited by injections of corresponding toxin and is specifically antagonistic to it.

Arborescent. A branching, treelike growth.

Arnold sterilizer. An apparatus used for sterilizing by actively streaming steam.

Arthritis. Inflammation of a joint.

Arthus phenomenon. A transient, swollen, edematous, local reaction occurring after repeated subcutaneous injections of horse serum into a rabbit, earlier injections having been without local effect. Still further injections produce a firm indurated area with local necrosis.

Ascitic fluid. Serous fluid which accumulates in the peritoneal cavity in cases of dropsy of that area; sometimes used to fortify artificial culture media.

Ascospore. *See* p. 96.

Ascus. *See* p. 96.

Asepsis. Absence of septic matter or freedom from microorganisms.

Aspergillus. A genus of fungi characterized by rounded conidiophores thickly set with chains of colored conidia.

Asporogenous. Not producing spores.

Asthma. A condition marked by recurrent attacks of paroxysmal dyspnea, with wheezing, cough, and a sense of constriction, due to spasmodic contraction of the bronchi.

Atabrine. An acridine preparation used for the treatment of malaria.

Athlete's foot. Ringworm of the feet and toes caused by infection with various members of the genus Trichophyton and other fungi

Atopy. A term devised by Coca to designate the inherited general tendency to become sensitized which may occur in human beings.

Atrichous. Without flagella.

Attenuation. The process of reducing or weakening virulence of a microorganism.

Autoclave. An apparatus for sterilizing by steam under pressure.

Autohemagglutinins. Specific substances in blood serum which agglutinate the cells of the same blood as that in which they are contained.

Autolysin. A substance present in an organism which is capable of causing the disintegration of the cells or tissues of that organism.

Autolysis. Disintegration of an organism by its own enzymes.

Autopsy. Post-mortem examination of a body.

Autosensitization. Sensitization towards one's own tissues; can be effected in an animal by injecting its own lens or placenta protein.

Autotrophic. Capable of living on inorganic matter.

Auxochrome. A chemical group which furnishes salt-forming properties and is responsible for transferring the color of a dye to a substance upon which it acts.

Azotobacter. A genus of bacteria able to fix free nitrogen.

Babes-Ernst bodies. Metachromatic granules.

Bacillus. A rod-shaped bacterium; a genus of rod-shaped bacteria of the family Bacteriaceae.

Bacteremia. A state in which microorganisms invade the bloodstream but do not multiply. This term has been used synonymously with septicemia.

Bacteria. Minute, one-celled, microscopic, plantlike organisms which multiply by fission and lack chlorophyll.

Bacterial synergism. Production of reactions by two or more species of bacteria growing together, which are not produced by either of the organisms growing alone.

Bactericide. Any agent that destroys bacteria.

Bactericidin. A substance present in the body fluids which kills bacteria.

Bacteriolysin. An antibody capable of destroying or dissolving specific bacterial cells which stimulated its formation.

Bacteriolysis. Destruction or dissolution of bacteria inside or outside the animal body.

Bacteriophage. An ultramicroscopic agent which produces a transmissible dissolution of specific bacterial cells and is regarded by some as a living agent and by others as an enzyme.

Bacteriostasis. Prevention of the growth of bacteria, such as results from the addition of certain dyes to culture media.

Beef extract. A watery extract of the soluble constituents of beef, evaporated to semisolid consistency at low temperature in vacuum.

Binary fission. A form of asexual reproduction involving simple cell division of the cytoplasm and nucleus, if present, into two equal parts.

Biogenesis. The origin of living things from things already living.

Bipolar. At both poles or ends of a cell.

Bordet-Gengou phenomenon. *See* Complement fixation.

Botulism. A type of food poisoning caused by the toxin of *Clostridium botulinum* in improperly canned or preserved foods.

Breed count. A microscopic method for counting bacteria on a dried milk film.

Bronchopneumonia. Inflammation of the lungs which usually begins in the terminal bronchioles that become clogged with a mucopurulent exudate.

Brownian movement. Dancing motion of minute particles suspended in a liquid.

Bubo. Inflammatory swelling of a lymphatic gland, particularly in the axilla or groin.

Buccal. Pertaining to the cheek.

Budding. A form of asexual reproduction in which the body divides into two unequal parts, the larger part being the parent and the smaller one the bud.

Buffer. Any substance in a fluid which tends to lessen the change in hydrogen ion concentration upon addition of relatively large amounts of acid or alkali.

Cachexia. A marked state of general ill health and malnutrition.

Camp fever. A popular name for typhus fever.

Cancer. A malignant tumor, made up chiefly of epithelial cells.

Capsule. A broad, colorless, mucoid or gelatinous layer which surrounds some bacteria.

Carboxydomonas. An autotrophic genus of bacteria that obtain their energy by the oxidation of carbon monoxide to carbon dioxide.

Carcinoma. Cancer.

Cardiac. Pertaining to the heart.

Carrier. An individual who harbors specific organisms of a disease in his body without manifesting symptoms but who serves as a means of conveying infection.

Catabolism. The process of destruction or breakdown of tissues and cells from complex to simpler compounds.

Catalase. An enzyme which decomposes hydrogen peroxide, liberating free oxygen.

Catalyst. A substance that accelerates a chemical or physical reaction without itself being destroyed or changed.

Catarrh. Inflammation of the mucous membranes, especially of the nose and throat, with a discharge.

Cathode. The negative electrode or pole of a galvanic cell.

Cellulose. A complex carbohydrate forming the skeleton of most plant cells.

Centigrade thermometer. A thermometer with 100 degrees between the freezing point and the boiling point of water, the former being at zero and the latter at 100 degrees. To convert from Centigrade to Fahrenheit scale:

$$(°C. \times \tfrac{9}{5}) + 32 = °F.$$

Centrifuge. A machine for separating the more solid constituents of a fluid by rotation.

Chancre. The primary lesion in syphilis.

Chancroid. A soft, nonsyphilitic, venereal sore caused by *Hemophilus ducreyi*.

Chemotaxis. Phenomenon shown by living organisms of moving toward (positive) or away from (negative) certain other cells or substances which exert a chemical influence; also chemotropism.

Chemotherapy. The prevention or treatment of diseases or inhibition of their parasitic causes by chemical disinfection.

Chickenpox. Varicella; a rather mild, highly contagious virus disease characterized by fever and the appearance of vesicles.

Chlamydospore. *See* p. 96.

Cholera. An acute infectious disease caused by *Vibrio comma* and characterized by its epidemicity, copious watery discharges, cramps, "rice-water" stool, prostration, and suppression of urine.

Cholera infantum. A common and often fatal noncontagious diarrhea of young children, prevailing in the summer months.

Cholera nostra. Acute gastroenteritis, with diarrhea, cramps, and vomiting, occurring in summer and autumn, and usually caused by improper food.

Chromogenic. Producing a pigment.

Chromophore. Any chemical group whose presence gives a specific color to a compound and which unites with certain other groups (auxochromes) to form dyes.

Chronic. Having a slow, comparatively mild course, continued for a long period.

Clostridium. A genus of bacteria which are anaerobic or microaerophilic and which form spores.

Coagulant. An agent that causes coagulation or clotting.

Coccus. An organism which is round or spherical in shape.

Cohesion. The force which unites the particles of a body.

Colitis. Inflammation of the colon.

Colony. A group of bacteria, on a solid medium, usually derived from the multiplication of a single organism and visible to the naked eye.

Colorimeter. An instrument for measuring color.

Colostrum. The fluid secreted by the mammary gland a few days before and after parturition.

Columella. *See* p. 96.

Comma bacillus. *Vibrio comma,* the cause of Asiatic cholera.

Commensalism. Living together of two species, one of which is benefited by the association while the other is apparently neither benefited nor harmed.

Communicable. Readily transferred from one person to another.

Comparator block. A simple colorimeter consisting of a block of wood with holes in which to place the test tubes to be compared and transverse holes through which to view the colors.

Complement. A thermolabile, normal blood constituent which reacts with sensitized cells to produce visible lysis; also called alexin.

Complement fixation. When antigen unites with its specific antibody, complement, if present, is taken into the combination and becomes inactive or fixed. Its absence or presence as free active complement can be shown by adding sensitized blood cells, or blood cells and hemolytic amboceptor, to the mixture. If free complement is present, hemolysis occurs; if not, no hemolysis is observed. *See also* pp. 344–347.

Complementoid. Complement that has lost its activity by heating but still retains its binding property with amboceptors.

Complication. A disease or diseases concurrent with another infection.

Conidia. *See* p. 96.

Conidiophore. *See* p. 96.

Conjunctivitis. Inflammation of the conjunctiva.

Convalescence. The stage of recovery following an attack of a disease.

Convalescent carrier. A person who carries the infecting organism in his body during the period of convalescence from the infection.

Convalescent serum. Serum of a patient who has recovered from an infectious disease and containing specific antibodies against the cause of the disease.

Corynebacterium. A genus of Corynebacteriaceae consisting of slender, Gram-positive, nonmotile, rod-shaped organisms with a tendency to form club-shaped and pointed forms and to show uneven staining.

Coryza. An acute catarrhal condition of the nasal mucous membrane, attended by a discharge from the nostrils.

Cover-slip. A thin glass plate which covers a mounted object to be studied microscopically.

Cowpox. Vaccinia; a virus disease related to and milder than smallpox, which confers immunity against smallpox. Vaccinia virus is used in vaccination against human smallpox.

Coxiella. A genus of Rickettsieae; pleomorphic rods or coccoid forms; the cause of Q fever, a typhus-like, mild disease.

Coxsackie virus. Several immunotypes of type A and one of type B have caused epidemics resembling poliomyelitis in children.

Crateriform. Depressed or hollowed, used to describe liquefaction of gelatin by certain bacteria.

Culture. A growth of microorganisms.

Culture medium. Any substance or preparation suitable for and used for the growth and cultivation of microorganisms.

Cyanosis. Bluish discoloration of the skin, caused by insufficient oxygenation of the blood.

Cystitis. Inflammation of the bladder.

Cytochrome. A respiratory pigment, widely distributed in plant and animal cells, that is easily oxidized and reduced under suitable conditions.

Cytolysin. An antibody which produces dissolution of cells; also called amboceptor or sensitizer.

Cytolysis. The process of dissolution or destruction of cells.

Danysz phenomenon. Decrease of the neutralizing effect of an antitoxin when a toxin is added to it in divided portions instead of all at once.

Decay. Decomposition of organic compounds under aerobic conditions to form amino acids which are then further broken down.

Denitrifying bacteria. Bacteria capable of reducing nitrates to nitrites, nitrogen, or gaseous ammonia.

Dermatitis. Inflammation of the skin.

Dermatosis. Any skin disease.

Dermomycosis. Any skin disease caused by a fungus.

Dermotropic. Having a selective affinity for the skin.

Desensitization. The process of rendering a person insensitive to a specific protein by giving small injections of the foreign protein to which he is sensitive over a period of time until he no longer reacts to it.

Desiccator. A closed vessel for apparatus or chemicals that are to be dried and kept free from moisture; usually contains a dehydrating agent.

Desoxyribonucleic acid (DNA). One of two types of nucleic acids; found only in nuclei, where its presence in genes indicates a central role in the genetic mechanism.

Desquamation. The shedding of epithelial elements, chiefly of the skin, in scales or sheets.

Deviation of complement. The Neisser-Wechsberg phenomenon of the failure of lysis to occur when considerable excess of amboceptor is added to normal serum (complement) and antigen. The complement unites with unbound amboceptor rather than with amboceptor which has united with antigen.

Dialysis. The separation of crystalloids from colloids by their different diffusibility through a semipermeable membrane.

Dick test. A test for susceptibility to scarlet fever by the development of local redness following intradermal injection of toxin of the causative Streptococcus.

Diphtheria. An acute infectious disease of the mucous membranes of the upper respiratory tract, characterized by patches of pseudomembrane and caused by *Corynebacterium diphtheriae.*

Disease. A definite state of ill health having a characteristic train of symptoms, involving the whole body or any of its parts, and with the etiology, pathology, and prognosis either known or unknown.

Droplet infection. Infection by means of small droplets of sputum and nasal discharges which have been thrown into the air during talking, coughing, or sneezing, and which may remain viable for hours or days. Such diseases as measles, chickenpox, parotitis, vaccinia, and diphtheria are spread in this way.

Dye. A material used for staining or coloring, consisting of benzene rings with chromophore and auxochrome groups.

Dysentery. A term given to a number of disorders marked by inflammation of the intestines, especially of the colon, and attended by pain in the abdomen and frequent stools containing blood and mucous.

Dyspnea. Difficult or labored breathing.

ECHO virus. Causes mild infections in the intestinal tract, or may be present as harmless saprophytes; twenty distinct types.

Eczema. An inflammatory skin disease with vesiculation, infiltration, watery discharge, and development of scales and crusts.

Edema. The presence of abnormally large amounts of fluid in the intercellular tissue spaces of the body.

Electrolysis. Chemical decomposition produced by passing a direct current of electricity through the compound.

Elementary bodies. Minute intracellular bodies found in lesions produced by some viruses, e.g., in vaccinia, and which may be the infectious virus particles.

Empyema. Accumulation of pus in a cavity of the body, especially the chest.

Encephalitis. Inflammation of the brain.

Endemic. Said of a disease, prevalent in the usual numbers in a given population.

Endocarditis. Inflammation of the endocardium or epithelial lining membrane of the heart.

Endoenzyme. Any intracellular enzyme retained in the cell and not excreted into the surrounding medium.

Endotoxin. Any toxin retained within the bacterial cell and liberated only after the death or disintegration of the organism.

Enteric. Pertaining to the intestines.

Enteritis. Inflammation of the intestine, chiefly of the small intestine.

Enterotoxin. A toxin specific for the cells of the intestinal mucosa, giving rise to symptoms of food poisoning.

Enzyme. An organic catalyst produced by a living cell.

Epidemic. A regional outbreak of an infectious disease which attacks many persons and spreads rapidly.

Epidemiology. The study of disease outbreaks.

Epizoötic. Any disease of animals which spreads rapidly and is widely diffused.

Erysipelas. An acute, febrile, contagious disease caused by a strain of *Streptococcus pyogenes* and marked by chills, fever, and intense local redness of the skin and mucous membranes.

Erythema. Redness of the skin.

Erythrocyte. A red blood corpuscle.

Etiology. Disease causation.

Exanthem. Any eruptive disease, e.g., variola, vaccinia, scarlet fever.

Exotoxin. A toxin which is excreted by the cell into the surrounding medium and may be separated by filtration.

Exudate. Any substance which is thrown out from the tissues and becomes deposited in or upon them.

Facultative aerobe. A microorganism which is fundamentally an anaerobe but can grow in the presence of free oxygen.

Facultative anaerobe. A microorganism which is fundamentally an aerobe but can grow in the absence of free oxygen.

Fahrenheit thermometer. A thermometer in which the space between the freezing point and the boiling point of water is divided into 180 degrees, 32 being the freezing point and 212 the boiling point. To convert from Fahrenheit to Centigrade scale:

$$\tfrac{5}{9} \, (^{\circ}F. - 32) = {^{\circ}}C.$$

Family. A group of related genera.

Fermentation. Incomplete oxidation of carbohydrates and carbohydrate-like compounds by microorganisms.

Filamentous. Composed of long threadlike structures.

Filiform. Having a uniform growth along the line of inoculation in streak or stab cultures.

Filtrate. A liquid which has passed through a filter.

Fimbriate. Fringed colony growth.

Fishing. The transfer of a bacterial culture from a plate to a fresh medium by touching the colony with a wire and then inoculating the medium.

Flagella. Whiplike processes for motility.

Flocculent. Containing small adherent masses of bacteria floating in or on a liquid.

Fluctuating variations. Temporary changes which often occur in morphology or physiology of a microorganism, usually because of environmental conditions.

Focus of infection. The chief center of a morbid process.

Fomite. Any substance other than food that may harbor or transmit a disease. Examples: bedding, clothing, dishes.

Formaldehyde. HCHO; a disinfectant gas with a pungent odor.

Formalin. A 40% solution of gaseous formaldehyde.

Frost count. A microscopic colony count for determining the number of bacteria in milk.

Fungi. A subphylum of plants which do not contain chlorophyll, including bacteria, yeasts, and molds.

Fusiform. Radish-shaped liquefaction of gelatin.

Gamete. A sexual cell; a mature germ cell, as an unfertilized egg or mature sperm cell.

Gangrene. Anemic necrosis of tissue combined, usually, with invasion by saprophytic organisms.

Gastroenteritis. Inflammation of the stomach and intestines with symptoms similar to enteritis and dysentery, often caused by the enteric group of bacteria, *Salmonella paratyphi, Salmonella schottmuelleri,* etc.

Genus. A group of one or more related species.

Germ. A microbe or bacillus.

Germicide. Synonymous with disinfectant.

Gingival. Pertaining to the gum or gums.

Glanders. A febrile acute or chronic disease of horses, transmissible to man; caused by *Malleomyces mallei.*

Gonidium. A minute intracellular body which serves as a means of asexual reproduction.

Gonorrhea. A contagious, catarrhal, venereal disease of the genital mucous membranes, caused by *Neisseria gonorrheae.*

Gram. A weight in the metric system equal to about 15 grains troy.

Gram-negative bacteria. Bacteria which lose the initial stain of the Gram stain, are decolorized, and take the color of the final stain.

Gram-positive bacteria. Bacteria which take the initial stain of the Gram stain and are not decolorized, so that they appear purple.

Guarnieri bodies. Cytoplasmic inclusion bodies often found in epithelial cell lesions in variola and vaccinia.

H agglutinin. Agglutinin specific against the (H) antigen found within the flagella of the bacterium.

Haptene. A partial antigen, unable to stimulate antibody production when injected by itself but determining immunological specificity when combined with antigenic protein and reacting with the antibody so produced.

Haptophore. Term used by Ehrlich in his explanation of immunological reactions to denote the specific portion of the toxin or agglutinin molecule through which combination with antitoxin or agglutinogen occurs.

Hay fever. Allergic acute nasal catarrh and conjunctivitis.

Hemagglutinin. A specific substance in blood serum which causes agglutination of red blood corpuscles.

Hematuria. The discharge of urine in which blood is present.

Hemoglobin. Oxygen-carrying red pigment of red blood corpuscles.

Hemolysin. An antibody which, in the presence of complement, is capable of bringing about the dissolution of foreign red blood corpuscles so that their hemoglobin is liberated.

Hemolysis. The dissolution or destruction of red blood corpuscles with liberation of the hemoglobin.

Hemorrhage. Severe bleeding.

Hemotoxin. A cytotoxin capable of destroying blood cells.

Hepatitis. Inflammation of the liver.

Herpes simplex. A mild, acute, eruptive, vesicular virus disease of the skin and mucous membrane.

Herpes zoster. Shingles; an acute virus disease characterized by a vesicular dermatitis which follows a nerve trunk.

Heterologous serum. Serum derived from another species.

Heterotrophic. Living on organic matter.

Homologous serum. Serum derived from the same species.

Hydrocarbon. Any compound of hydrogen and carbon.

Hydrogen ion concentration. The degree of concentration of hydrogen ions in a solution used to indicate the reaction of that solution and expressed as pH, which is the logarithm of the reciprocal of the hydrogen ion concentration.

Hygienic Laboratory coefficient. Phenol coefficient as determined by the U. S. Hygienic Laboratory method.

Hyperemia. Excess of blood in any part of the body.

Hyperglycemia. Excess of sugar in the blood.

Hypertonic. A solution which is of a concentration greater than isotonic concentration.

Hypha. *See* p. 95.

Hypotonic. A solution which is of less than isotonic concentration.

Idiosyncrasy. Individual or peculiar susceptibility to some drug, protein, or other agent.

Immunity. The ability of living individuals to resist or overcome infections.

Immunization. The process of conferring immunity on an individual.

Inclusion bodies. Round, oval, or irregular-shaped bodies occurring in the cytoplasm (cytoplasmic) or nucleus (intranuclear) of cells of the body in some virus diseases, such as rabies, variola, vaccinia, etc.

Incubation period. The period between the time infection occurs and the appearance of the first symptoms.

Incubator. An apparatus for maintaining a constant and suitable temperature for the growth and development of a bacterial culture or other materials.

Indicator. A substance, usually a weak organic acid or base, which changes color when the reaction of a solution changes.

Indirect stain. A stain which does not dye the organism itself but colors the background so that the organism becomes visible by contrast, e.g., nigrosin, India ink.

Induration. The process of hardening; a hardened spot.

Infection. Invasion of body tissues by pathogenic organisms which multiply and cause disease.

Infundibuliform. Funnel-shaped or inverted-cone-shaped liquefaction of gelatin.

Intrathecal. In the spinal column.

In utero. Within the uterus.

In vitro. In the test tube, outside of the animal body.

In vivo. In the living body.

Involution forms. Abnormal shapes taken by bacteria differing from those of the original culture.

Isohemagglutinins. Specific substances in blood serum which agglutinate the blood cells of other members of the same species.

Isotonic. A solution equal in concentration to another solution with respect to a certain solute.

Jail fever. Typhus fever.

Jaundice. Excess of bile pigments in blood, skin, and mucous membranes with a resulting yellow appearance of the individual.

Johne's disease. A chronic enteritis of cattle, caused by *Mycobacterium paratuberculosis.*

Kala azar. A fatal, infectious, protozoan disease of the reticulo-endothelial system, caused by *Leishmania donovani* and believed to be transmitted by the sandfly.

Keratitis. Inflammation of the cornea.

Koplik spots. Small bluish-white spots surrounded by a reddish areola on the mucous membrane, found on the cheeks and lips during the prodromal stage of measles.

Lag phase. The early period of slow multiplication following a bacterial inoculation into a culture medium.

Laryngitis.　Inflammation of the larynx, a condition attended by dryness and soreness of the throat, hoarseness, and cough.

Leishman-Donovan bodies.　Small round or oval bodies found in the spleen and liver of patients suffering from kala azar.

Leucocyte.　A white blood corpuscle.

Locomotor ataxia.　Tabes dorsalis; degeneration of the dorsal columns of the spinal cord and of the sensory nerve trunks, marked by intense pain, inco-ordination, loss of reflexes, etc.

Logarithmic period.　The stage of maximum growth of a bacterial culture at which, if the logarithms of their numbers are plotted against the time, a straight line will be formed.

Lophotrichous.　Having a tuft of flagella at one end.

Lues.　Syphilis.

Luminescence.　The property of giving off light without manifesting an increase in heat energy.

Lysin.　An antibody which, in the presence of complement, dissolves or disintegrates cells.

Macrogamete.　The female form of the malarial parasite which, fertilized by a microgamete in the mosquito, becomes a zygote and goes through a cycle of development.

Macrogametocyte.　The female form of the malarial parasite which, when transferred from man to the mosquito, becomes a macrogamete.

Macrophage.　Large wandering phagocyte, so designated by Metchnikoff.

Macula.　An unraised discolored spot on the skin.

Malaise.　Any indisposition, discomfort, or feeling of ill health.

Malaria.　A protozoan disease caused by members of the genus Plasmodium and transmitted by the bite of the anopheline mosquito; characterized by recurring fever, chills, and sweats.

Malta fever.　*See* Undulant fever.

Mastitis.　Inflammation of the mammary gland.

Measles.　Rubeola; an acute, infectious virus disease characterized by fever, catarrh, coryza, Koplik spots on buccal mucous membrane, and a papular rash.

Meningitis.　Inflammation of the meninges.

Mesophilic bacteria.　Bacteria that grow best at moderate temperatures.

Metabolism.　The sum of all the physical and chemical processes by which the tissues are formed and maintained and energy is made available for use by the body.

Metachromatic granules.　Deeply staining masses, irregular in size and number, which are seen in the protoplasm of various bacteria, e.g., *Corynebacterium diphtheriae*.

Metastasis.　The transfer of disease from one organ or part to another not directly connected with it.

Microaerophilic bacteria.　Bacteria which grow at a low oxygen tension.

Microgamete.　The conjugating male element of the Plasmodium of malaria which fertilizes the macrogamete in the mosquito.

Microgametocyte.　The male form of the malarial parasite which is transferred from man to the mosquito.

Micromicron.　The millionth part of a micron or 10^{-10} cm., designated by the symbol $\mu\mu$.

Micron or micromillimeter.　One-millionth part of a meter, or one-thousandth part of a millimeter, or 1/25,000 of an inch, designated by the Greek letter μ.

Microphages.　Small, actively motile, polynuclear leucocytes which cause phagocytosis of the bacteria of acute infections.

Microsporum.　A genus of parasitic fungi which cause skin diseases such as ringworm and barber's itch.

Milligram.　One-thousandth part of a gram.

Millimicron.　One-thousandth part of a micron, or one-millionth part of a millimeter, or 10^{-7} cm., designated by mμ.

Mixed culture.　Growth of two or more organisms in the same medium.

Monilia. A genus of parasitic fungi which is most frequently encountered in thrush and is characterized by its fermentation of sugar with the production of gas.

Monotrichous. Having a single polar flagellum.

Mordant. A substance used to fix a stain or dye.

Moribund. In a dying state.

Morphology. The science of the form and structure of organized beings.

Much's granules. Non-acid-fast, Gram-positive granules of unknown significance observed in tuberculous lesions in which acid-fast bacilli could not be found.

Mucopurulent. Containing both mucus and pus.

Mucous membrane. A membrane secreting mucus and lining the cavities of the body which connect with the outside air, such as the respiratory, digestive, and genito-urinary tracts.

Mucus. The viscid watery secretion of the mucous glands, composed of water, mucin, inorganic salts, epithelial cells, leucocytes, and granular matter.

Mutation. Transmissible variation that is apt to be permanent, seemingly arising suddenly and spontaneously.

Mycelium. *See* p. 95.

Mycology. The science and study of fungi.

Mycosis. Any disease caused by a fungus.

Myocarditis. Inflammation of the muscular walls of the heart, or myocardium.

Napiform. Turnip-shaped liquefaction of gelatin.

Necrosis. Death of a circumscribed portion of tissue.

Negri bodies. Diagnostic cytoplasmic inclusion bodies found in the affected nerve cells in rabies.

Neisser-Wechsberg phenomenon. *See* Deviation of complement.

Nephritis. Inflammation of the kidney.

Neufeld reaction. Method of typing the pneumococcus based on the fact that the capsule swells in the presence of type-specific immune serum.

Neurotropic. Having an affinity for nervous tissue.

Nitrification. Oxidation of ammonia to nitrites and nitrates accomplished in soil by autotrophic bacteria.

Nitrobacter. A genus of bacteria that can oxidize nitrites to nitrates.

Nitrogen-fixation. The union of free atmospheric nitrogen with other elements to form chemical compounds, such as ammonia and nitrates or amino groups.

Nitrosomonas. A genus of bacteria that can secure their energy by oxidizing ammonia to nitrites.

Nonspecific immunity. Increase of antibodies or production of immunity resulting from the injection of some nonspecific antigen.

O agglutinin. Agglutinin specific for the (O) antigen found within the bacterial cell.

Obligate aerobe. An organism which must have free oxygen for its growth.

Obligate anaerobe. An organism which can live and grow only in an environment with no or minimal amounts of free oxygen.

Oöcyst. An encysted oöspore.

Oöspore. A zygote formed by the conjugation of two sexually differentiated elements.

Opsonic index. The power of the blood of an individual to phagocytize any particular microorganism compared with the power of normal blood.

Opsonin. An antibody in the blood which renders microorganisms or red blood cells more liable to phagocytosis.

Orchitis. Inflammation of a testis.

Osmosis. Passage through a membrane; when two solutions of unequal density are separated by a membrane which selectively prevents the passage of solute particles but is permeable to the solvent, pure solvent passes from the lesser to the greater concentration.

Otitis. Inflammation of the ear.

Pandemic. An epidemic involving a large population.

Papilloma. An epithelial tumor in which the cells cover finger-like processes or ridges of stroma (matrix of an organ), e.g., warts.

Papule. A small, circumscribed, solid elevation of the skin.

Paralysis. Loss of motion or sensation in a living part of the body.

Parasite. A plant or animal which lives upon or within another living organism at whose expense it grows without giving anything in return.

Parenteral. Not through the alimentary canal.

Parotitis. Inflammation of the parotid gland; mumps.

Paschen bodies. Small elementary bodies demonstrable in the vesicular fluid in variola and vaccinia.

Passive immunity. Immunity produced by injection into an individual of serum containing antibodies formed in another.

Pasteurization. The process of heating every particle of milk or milk product, maintaining the temperature for a suitable time, and then rapidly cooling, the temperatures and times being sufficient to destroy all pathogenic organisms and to reduce the bacterial count about 90%.

Pathogens. Disease-producing microorganisms.

Pellicle. A thin membranous film on the surface of a liquid formed as a result of bacterial growth.

Penicillin. A highly effective, chemotherapeutic, nontoxic antibiotic produced by the mold *Penicillium notatum*.

Penicillium. A genus of molds which is characterized by the development of fruiting organs resembling a broom.

Peptonization. Enzymic hydrolysis of milk protein transforming the milk to a clear fluid.

Pericarditis. Inflammation of the pericardium, the membranous sac which contains the heart.

Peritrichous. Having flagella surrounding the entire organism.

Per os. By the mouth.

Pertussis. Whooping cough.

Petechia. A small spot formed by the effusion of blood.

Pfeiffer's phenomenon. When cholera vibrios are injected into the peritoneal cavity of a guinea pig that has been immunized against cholera, it is found on removing a portion of the peritoneal contents from time to time that the vibrios lose their motility, become swollen, disintegrate, and pass into solution.

pH. *See* Hydrogen ion concentration.

Phagocyte. Any cell that destroys microorganisms or other cells by enveloping or absorbing them, e.g., endothelial cells, leucocytes.

Phagocytosis. The process whereby microorganisms and other cells and substances are engulfed by phagocytes.

Phenol. Carbolic acid, a colorless crystalline compound, C_6H_5OH, obtained by the distillation of coal tar and having strong antiseptic and disinfectant properties.

Phenol coefficient. A number indicating the relative efficiency of disinfectants. It is the quotient obtained by dividing the highest dilution of a disinfectant which kills a test organism in a fixed time by the highest dilution of phenol showing the same results.

Photodynamic sensitization. Increased toxicity to bacteria of ultraviolet rays of low intensity and of visible light, resulting from the addition of certain dyes.

Photogens. Bacteria which produce phosphorescence or emit light.

Phototaxis. The movement of organisms under the influence of light.

Plague. An acute, febrile, and highly fatal epidemic disease, caused by *Pasteurella pestis* and transmitted to man by fleas from rats.

Plasmochin. A synthetic quinoline compound used in malaria therapy.

Plasmolysis. Contraction or shrinking of the protoplasm of a cell because of loss of water by osmotic action.

Plasmoptysis. Bursting of protoplasm through the cell wall of a cell as a result of the imbibition of water by osmotic action.

Plating. Cultivation of microorganisms in Petri dishes containing a solid nutrient medium.

Pleomorphism. The assumption of various distinct forms by a single organism or species.

Pneumonia. Inflammation of the lungs, most often of pneumococcic origin.

Polar. Situated at the end or pole of a cell.

Poliomyelitis. A virus disease in which there is inflammation of the gray substance of the spinal cord; commonly called infantile paralysis.

Polyvalent. A term used to designate a stock vaccine made up of many strains of the same organism or different organisms.

Portal of entry. The route a microorganism must take in entering the body in order to have any effect.

Prausnitz-Küstner reaction. Passive transfer of local hypersensitivity in man by the intradermal injection of serum of an allergic person.

Precipitin. An antibody formed in response to the injection of soluble antigens and having the power to cause a precipitation or flocculation of the antigen when mixed with the latter.

Precipitinogen. Any substance which stimulates the formation of precipitin in the bloodstream.

Predisposing cause. Anything which renders a person liable to an attack of disease without actually producing it.

Proagglutinoid or prozone phenomenon. This is noted when a lower dilution of a serum gives no agglutination although agglutination occurs at higher dilutions.

Prodromal. Premonitory, indicating the approach of a disease.

Prognosis. A forecast of the probable result of an attack of disease; the likelihood of recovery.

Prophylaxis. The prevention of disease; preventive treatment.

Protozoa. Unicellular, nucleated, animal organisms with diversification and specialized functions of the protoplasm.

Pseudopodia. "False feet," temporary protrusions of ectoplasm serving for purposes of locomotion.

Psychrophilic bacteria. Cold-loving bacteria whose optimum temperature for growth is 15°-20° C. or below.

Puerperal sepsis. Sepsis occurring after childbirth, usually due to Staphylococcus or Streptococcus.

Punctate. Resembling or marked with points or dots.

Punctiform colonies. Pin-point colonies, measuring less than 1 mm. in diameter.

Pure culture. Specific bacterial growth of only one type of organism.

Purpura. A disease characterized by the formation of purple patches on the skin and mucous membranes, due to subcutaneous extravasation of blood.

Purulent. Containing pus.

Pus. A liquid inflammation product made up of dead and living leucocytes, digested bacteria, tissue debris, lymph, fibrin, and serum.

Pustule. A small elevation of the skin filled with pus or lymph.

Putrefaction. The decomposition of animal or vegetable matter in the absence of oxygen and characterized by the formation of amino acids, mercaptans, indol, skatol, and hydrogen sulfide, with an accompanying unpleasant odor.

Pyemia. A general septicemia in which secondary foci of suppuration occur and multiple abscesses are formed.

Pyogenic. Pus-producing.

Pyrexia. Fever.

Quarantine. Isolation of infected persons and of possible carriers for a period of time to prevent disease transmission.

Quinine. An alkaloid of the cinchona bark which is used as specific malaria treatment.

Rabies. A fatal virus disease of man and animals characterized by extreme irritation of the central nervous system.

Receptor. A term used in Ehrlich's side-chain theory of immunity to indicate that part of the cell molecule which can combine with the haptophore group of toxins or other antigens and may be attached to the cell or remain free in the bloodstream, e.g., antitoxin, agglutinin, precipitin, opsonin.

Relapsing fever. Any one of a group of acute infectious diseases caused by various species of Borrelia, and marked by alternating periods of fever and absence of fever.

Remission. A diminution or abatement of the symptoms of a disease.

Resistance. The ability of the individual to ward off infection.

Rhinitis. Inflammation of the mucous membrane of the nose.

Rhizobium. A genus of bacteria that live symbiotically with leguminous plants and fix free nitrogen.

Rhizoid. Irregular branched or rootlike growth.

Ribonucleic acid (RNA). One of two types of nucleic acids; found in both nucleus and cytoplasm and involved in protein synthesis.

Rickettsiae. Rickettsia bodies; Gram-negative, nonmotile, intracellular parasites whose exact nature is unknown but which are generally considered intermediate between bacteria and viruses.

Rubeola. Measles.

Saccate. Sac-shaped liquefaction.

Saccharose. Cane sugar.

Sapremia. Intoxication due to presence in the blood of the products formed by the growth of saprophytes on diseased or injured tissues.

Saprophytes. Organisms which live on dead organic matter.

Sarcina. A genus of Micrococcaceae characterized by occurrence of cell division in three planes, forming regular packets.

Satellite phenomenon. Growth of larger colonies of *Hemophilus influenzae* in the region of Staphylococci and other bacteria, because of their synthesis of coenzyme.

Scarlet fever. An acute, contagious, exanthematous disease caused by a strain of *Streptococcus pyogenes*.

Schick test. A test for susceptibility to diphtheria by the development of an area of inflammation following intracutaneous injection of diphtheria toxin.

Schizogony. Asexual reproduction by segmentation.

Schizomycetes. The plant class to which all bacteria belong.

Schüffner's dots. Coarse red granules seen in parasitized erythrocytes in tertian malaria on staining with polychrome methylene blue.

Schultz-Charlton phenomenon. When scarlet fever convalescent serum or antitoxin is injected intracutaneously in scarlet fever patients, there is a blanching of the rash at the site of injection.

Sensitizer. *See* Cytolysin.

Sepsis. A state of infection.

Septate. Divided by a septum into various compartments.

Septicemia. Invasion and multiplication of organisms in the bloodstream.

Serum. The clear liquid which separates in the clotting of blood from the clot and the corpuscles.

Serum sickness. A form of hypersensitive reaction following the injection of foreign serum and marked by urticarial rashes, edema, fever, joint pains, or swelling and pain in the lymph nodes.

Shwartzman reaction. A severe, hemorrhagic, necrotic reaction produced in rabbits at the site of an intradermal injection with filtrate of cultures of *Salmonella typhosa* or certain other organisms, when this is followed within 24 hours by intravenous injection of the filtrate.

Slant. Solid media allowed to harden in test tubes set at a slant to increase the surface for the growth of colonies.

Slime layer. Carbohydrate jelly layer believed to surround cell wall of all bacteria.

Smallpox. Variola; a generalized febrile virus disease characterized by vesicular eruptions which become pustular and crust, often leaving permanent pox marks.

Somatic. Pertaining to the body.

Specific Soluble Substance. S.S.S.; a complex, highly type-specific, poly-saccharide haptene extracted from the capsular material of various bacteria, e.g., pneumococci.

Spontaneous generation. The theory, now discarded, which postulates that living organisms generate from nonliving matter.

Sporangia. *See* p. 96.

Sporogony. Sexual reproduction by spores.

Sporozoite. A spore formed after fertilization; it is this form which is transmitted from the mosquito to man in malaria.

Stab cultures. Cultures in which the organisms are inoculated far down into the solid butt of the medium to allow for possible anaerobic growth.

Stain. Any dye, reagent, or other material used in coloring tissues or organisms for microscopic study.

Stationary phase. The stage in the growth of a bacterial culture at which multiplication of organisms gradually decreases so that as many are formed as die.

Sterilization. The process of freeing completely from all living microorganisms.

Stomata. Openings in surfaces of leaves of plants for passage of oxygen, carbon dioxide, and water vapor.

Stomatitis. Inflammation of the mouth, usually attended by pain and salivation and often by fetid breath.

Strangles. An infectious disease of horses caused by *Streptococcus equi* and characterized by a mucopurulent inflammation of the respiratory mucous membrane.

Stratiform. Liquefying to the walls of the tube at the top and then proceeding downward horizontally.

Streak. Inoculation of slants or plates with a streak or a direct line movement across the surface of culture media.

Subcutaneous. Beneath the skin.

Subterminal. Situated near but not at the end or extremity.

Sulfur bacteria. Bacteria which obtain their energy from the oxidation of sulfur-containing gases to sulfuric acid.

Suppuration. Pus formation.

Symbiosis. The living together or close association of two dissimilar organisms with mutual benefit.

T.A.B. Abbreviation for a mixed vaccine containing organisms of *Salmonella typhosa*, *Salmonella paratyphi*, and *Salmonella schottmuelleri* (typhoid, paratyphoid A, and paratyphoid B).

Tetanus. Lockjaw; an acute infectious disease caused by the toxin of *Clostridium tetani*, characterized by spasm of the voluntary muscles.

Therapy. The treatment of disease.

Thermal death point. The temperature which destroys all the bacteria present within a given time.

Thermal death time. The length of time required to kill all the organisms in a given substance at a given temperature.

Thermolabile. Easily altered or decomposed by heat.

Thermophilic bacteria. Bacteria which grow best at a temperature of 50°–55° C.

Thermostable. Not easily affected by moderate heat.

Thiobacillus. A genus of bacteria that obtain their energy from the oxidation of sulfides, thiosulfates, or sulfur, forming sulfur, persulfates, sulfuric acid, and sulfates.

Tick. A blood-sucking arachnid parasite.

Titration. Volumetric determination against standard solutions of known strength.

Toxemia. A general intoxication due to the absorption of bacterial products, usually toxins, formed at a local source of infection.

Toxins. Poisonous substances produced by bacteria. When exotoxins are injected into animals in carefully graded doses, they incite the formation of specific substances called antitoxins which nullify the action of the toxin.

Toxoid. Anatoxin; toxin treated with formalin, etc., so that it is no longer toxic but is still capable of stimulating formation of and uniting with antitoxin.

Transfer of cultures. Transplanting viable bacteria from an old medium to a fresh new one.

Transplant. A portion of a culture of bacteria which has been transferred from an old pure culture to a fresh new medium.

Trauma. A wound or injury.

Trench mouth. *See* Vincent's angina.

Trichophyton. A genus of fungi consisting of flat, branched filaments and chains of spores and causing ringworm and athlete's foot.

Tubercle. A nodule characteristic of tuberculosis and containing numerous bacilli; made up of small spherical cells which contain giant cells and are surrounded by a layer of spindle-shaped connective tissue cells, known as epitheloid cells.

Tuberculin. A filterable substance produced in the growth of *Mycobacterium tuberculosis* in culture media. When it is injected intracutaneously in persons who have been exposed to the tuberculosis bacillus or its products, a reaction is produced in 24 to 48 hours consisting of infiltration and hyperemia.

Tuberculosis. An infectious disease caused by *Mycobacterium tuberculosis* and characterized by the formation of tubercles in the tissues.

Tularemia. A disease of rodents, resembling plague, which may be transmitted to man; caused by *Pasteurella tularensis*.

Twort-d'Herelle phenomenon. *See* Bacteriophage.

Undulant fever. Malta fever; an infectious disease of man characterized by continued pyrexia, often with remissions, joint pains, skin rashes, night sweats; caused by *Brucella abortus* and *Brucella melitensis* which cause primarily contagious abortion in cattle and goats, respectively.

Univalent vaccine. A vaccine containing only one variety of organism.

Urticaria. A skin disease characterized by the sudden appearance of a rash which rarely lasts longer than 2 days but may be chronic. It is characterized by smooth, elevated patches which are usually whiter than the surrounding skin and attended by severe itching.

Vaccination. Usually, protective inoculation against smallpox by inoculation with vaccinia virus, but also any act of protective inoculation with a virus or bacteria.

Vaccine. A preparation of killed or attenuated infective agent used to produce active artificial immunity.

Vaccinia. Cowpox.

Varicella. Chickenpox.

Variola. Smallpox.

Vector. A carrier of viruses, bacteria, or protozoa; usually refers to the animal host that carries the pathogenic organism from one human host to another. Example: Anopheles mosquito is the vector of *Plasmodium malariae*.

Venereal. Due to or propagated by sexual intercourse.

Vesicle. A small sac or blister containing liquid.

Viable. Living.

Vi antigen. An antigen found in strains of *Eberthella typhosa* and some paratyphoid bacilli and believed associated with virulence.

Vincent's angina. Trench mouth; a contagious infection of the mouth associated with the presence of *Borrelia vincenti* and the fusiform bacillus and characterized by a false membrane in the tonsillar region.

Virulence. Disease-producing ability.

Virus. *See* p. 285.

Viscerotropic. Having a selective action on the visceral organs, including the liver and spleen.

Viscid. Referring to growth which follows the needle when touched and withdrawn, or to a sediment which on shaking rises as a swirl.

Von Economo's disease. Japanese encephalitis type A; a chronic, sporadic, encephalitic infection of unknown etiology.

Wassermann test. A complement fixation test for syphilis. *See* p. 345.

Weil-Felix reaction. Agglutination of certain Proteus strains by the serum of persons with typhus fever, for which it is used as a diagnostic test.

Weil's disease. Infectious jaundice; an acute infectious disease characterized by jaundice, fever, muscular pain, and enlargement of the liver and spleen, caused by the spirochete *Leptospira icterohaemorrhagiae*.

Wheal. A white or pinkish elevation or ridge on the skin.

Whoop. The convulsive inspiration of whooping cough.

Whooping cough. Pertussis; an infectious disease caused by *Hemophilus pertussis* and characterized by catarrh of the respiratory tract and peculiar paroxysms of cough, ending in a prolonged crowing or whooping respiration.

Widal test. An agglutination test for the diagnosis of typhoid fever. *See* p. 242.

Yaws. A tropical, ulcerative, nonvenereal disease similar to syphilis and caused by *Treponema pertenue*.

Yellow fever. A virus disease characterized by chills, fever, muscular pain, and jaundice, transmitted by the female *Aedes aegypti* mosquito.

Zygospore. A spore formed by the conjugation of two cells which are morphologically identical and do not show sexual differentiation.

Zygote. An organism formed by the union of two germ cells.

Zymogens. Microorganisms which cause fermentation of carbohydrates.

SUPPLEMENT ON
BACTERIAL PHYSIOLOGY

Accounts in medical history clearly show that the "everything-but-the-kitchen-sink" method* for the treatment of infectious disease, though often fruitful, is an illogical approach and an admission of a lack of understanding. Thus, scientists doggedly try the run of the reagent shelves until, by chance (at the 606th try as in the case of Ehrlich) or perhaps never, an effective agent is found.

In recent years a trend in the direction of metabolic studies of the organism in question has been used as a guide in the search for effective natural or synthetic therapeutic agents. That such an approach can be fruitful is demonstrated by recent progress in the use of antibiotic and chemotherapeutic agents in the treatment of disease.

In addition to medical applications, the study of bacterial metabolism (or more properly, microbial metabolism) has greatly advanced our knowledge of the chemistry of living things in general. Apparently, the metabolic pathways in *all* organisms are basically similar. Unicellular organisms are considerably easier to work with than other organisms, and the data so obtained may be applicable to the latter. We shall see in this chapter that it is this *comparative biochemical approach* (so termed by van Niel) which, at the present stage of our knowledge, offers the greatest prospects for success.

NUTRITIONAL REQUIREMENTS OF MICROORGANISMS

It has been noted elsewhere (Chapter II) that bacteria may be grouped as autotrophs or heterotrophs on a nutritional basis. A more detailed account of these groups is given below.

Autotrophs. These are the least exacting of organisms since they can multiply in an inorganic environment and can synthesize cell material from CO_2 and an inorganic nitrogen source. Strict autotrophs (e.g., *Thiobacillus*) are actually *inhibited* by organic substances.

---

* That is, trial and error screening of potential chemotherapeutic agents without consideration of what will most likely be effective in light of the physiology of the organism in question.

1. *Photosynthetic Autotrophs.* Energy is obtained from photochemical utilization of light energy, i.e.,

$$CO_2 + 2 H_2A \xrightarrow{\text{light}} CH_2O + H_2O + 2 A + 112 \text{ Calories.}$$

H_2A represents an H donor; examples, members of sub-order *Rhodobacteriineae.*

2. *Chemosynthetic Autotrophs.* Energy is obtained from oxidation of an inorganic substrate; e.g.,

Thiobacillus thioxidans—$2 S + 3 O_2 + 2 H_2O \longrightarrow 2 H_2SO_4 +$ 141.8 Calories;

Nitrobacter—$HNO_2 + O \longrightarrow HNO_3 + 21.6$ Calories;

Nitrosomonas—$NH_3 + 3 O \longrightarrow HNO_2 + H_2O + 79$ Calories.

Heterotrophs. These are more exacting and vary greatly in their requirements for growth.

1. Require an inorganic nitrogen source and a simple carbohydrate; e.g., *Escherichia coli.*

2. Require certain amino acids (RCHCOOH) and a simple car-
$$\underset{\overset{|}{NH_2}}{}$$
bohydrate; e.g., *Salmonella typhosa.*

3. Require a simple carbohydrate, certain amino acids, and "growth factors"*; e.g., *Lactobacillus casei.*

4. Obligate intercellular parasites (have not yet been cultured on artificial media); e.g., *Treponema pallidans.*

5. Obligate intracellular parasites (little prospect of culturing on artificial media since intimately dependent upon respiration of host cell for essential processes); e.g., *Rickettsia, Virales.*

It should be noted that this is an artificial classification and an oversimplification, since in nature one finds all shades of the physiological spectrum between strict autotrophism and obligate intracellular parasitism.

ENZYMES

Definition. Enzymes may be defined as complex organic catalysts which are produced by living cells but are capable of action independent of them. Enzymes affect the rate of chemical reactions, thus allowing reactions in living organisms to proceed at considerable velocities at physiological temperatures. In the absence of the en-

---

* Growth factors are vitamins or vitamin-like substances which are required in extremely small quantities for growth.

zyme, most of the chemical changes in living tissues would proceed too slowly to be measurable.

Chemical Properties.   The following properties of enzymes point to the fact that enzymes are proteinaceous in nature.
1. Heat lability.
2. Optimum pH for action.
3. Inactivation by extreme acid or alkali.
4. Inactivation by protein precipitants.

Classification.
1. On the basis of site of action.
   (a) Intracellular enzymes.   Act within cell which produces them.
   (b) Extracellular enzymes.   Act outside of cell which produces them.
2. On the basis of reaction catalyzed.   An enzyme may be named by adding the suffix *ase* to the name of the substrate upon which the enzyme acts.   Hence, "urease" (urea + ase) is the name given to the enzyme which acts upon urea.   This convention has not been rigidly adhered to.   Below is a brief classification of enzymes.   There is little agreement among biochemists as to what type of classification is most satisfactory.

ENZYME	TYPICAL SUBSTRATE	END PRODUCTS
I. Hydrolases (catalyze hydrolytic cleavage of substrate)		
A. Esterases		
Lipase	fats and oils	glycerol + fatty acid
B. Carbohydrases		
Maltase	maltose	glucose
C. Proteases		
Pepsin	proteins	peptones
II. Desmolases (catalyze nonhydrolytic cleavage of substrate)		
A. Oxidases		
Ascorbic acid oxidase	ascorbic acid	dehydroascorbic acid
B. Dehydrogenases		
Lactic acid dehydrogenase	lactic acid	pyruvic acid
C. Mutases		
Phosphoglyceromutase	3-phosphoglyceric acid	2-phosphoglyceric acid
D. Transphosphorylases		
Hexokinase	glucose + ATP	glucose-6-phosphate + ADP

Specificity. The specificity of enzymes is one of their most striking properties. Although there are many degrees of specificity, we can readily distinguish certain types.

1. *Steriochemical Specificity.* Certain enzymes attack only one of the optical forms of a compound.

2. *Linkage Specificity.* In the substrate A–B, the linkage between A and B determines whether or not the enzyme will act.

3. *Linkage and Component Specificity.*

(a) In substrate A–B, the enzyme acts only when either A or B and the linkage are of a certain structure.

(b) In substrate A–B, specificity is determined by both A and B and the nature of the linkage between the two.

Nature of Enzyme Action. Although the mechanisms of enzyme action are poorly understood at present, the accompanying figure shows diagrammatically how enzyme action may occur. Direct evidence for an enzyme-substrate complex has been obtained for two enzymes— catalase and peroxidase.

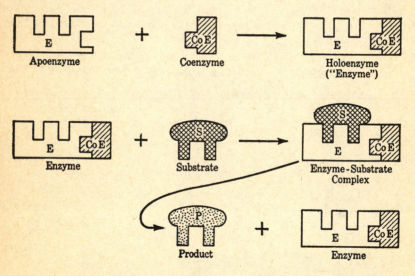

This is usually written for brevity in the following manner:

$$E + S \rightleftharpoons ES \longrightarrow E + P.$$

The union between enzyme and substrate is very specific and probably takes place at definite points on the enzyme surface. Activation of the substrate is the essential function of an enzyme. The enzyme, in some way, subjects the substrate to "internal strain," thereby increasing chemical reactivity.

Activators, Coenzymes, and Prosthetic Groups.

1. *Activators.* Accessory substances which are part of the system activating the substrate and are required before this activation can occur; e.g., Mn^{++} in arginase.

2. *Coenzymes.* Relatively loosely bound accessory substances which play a part in the enzyme catalyzed reaction, but not in the activation of the substrate; e.g.,

$$\text{Lactic acid} \xrightarrow[\text{DPN}]{\overset{\text{lactic}}{\text{dehydrogenase}}} \text{Pyruvic acid}$$

DPN is the coenzyme.

3. *Prosthetic Groups.* Relatively tightly fixed nonprotein parts of certain enzyme molecules. Their functions are similar to those of coenzymes; e.g., hemoglobin—heme is the prosthetic group attached to the protein globin.

These accessory substances are heat-stable and more or less dialyzable. Note that there is no sharp distinction between substrates, coenzymes, and prosthetic groups. As Baldwin has stated, the difference is "one of degree rather than kind."

The Function of Vitamins in Enzymatic Action.

Many of the more fastidious organisms require certain vitamins preformed in the medium. Less fastidious ones can synthesize from simple substances all the needed vitamins. Vitamins appear to act as constitutive parts or precursors of parts of coenzymes. Coenzymes act in a cyclic manner. This explains why very small amounts of vitamins are sufficient for growth. Although the enzymatic functions of all the known vitamins have not yet been elucidated, the functions of some are fairly well known. A list of some of the better known coenzymes is given on p. 382.*

---

* The following abbreviations will be used:

OAA	= oxalacetic acid	FMN	= flavinmononucleotide
DPN	= diphosphopyridine nucleotide	FAD	= flavinadeninedinucleotide
TPN	= triphosphopyridine nucleotide	Co A	= coenzyme A
ADP	= adenosine diphosphate	PABA	= para-aminobenzoic acid
ATP	= adenosine triphosphate		

VITAMIN	COENZYME	REACTION TYPE
(a) Thiamin	cocarboxylase (thiaminpyrophosphate)	decarboxylation of alpha-keto acids
(b) Pyridoxine	cotransaminase (pyridoxal phosphate)	transfer of —NH_2 from amino acid to alpha-keto acid
	codecarboxylase	dicarboxylic amino acid → monocarboxylic acid + CO_2
(c) Riboflavin	FMN FAD	link pyridine nucleotides with cytochromes
(d) Nicotinamide	pyridine nucleotides (DPN, TPN)	link substrates with flavoprotein or other substrates
(e) Pantothenic acid	Co A	reactions involving acetate transfer

Other vitamins whose enzymatic functions are less clear include biotin, folic acid, vitamin B_{12}, and PABA.

Co A and ATP are extremely important since they function in a manner which allows the high energy of the bond to be conserved, thus providing energy for diverse reactions.

BIOLOGICAL OXIDATIONS

If glucose is burned in a calorimeter, approximately 674 Calories of heat per gram mol. are set free. Living organisms can capture only a part of this energy when glucose is metabolized. One of the most interesting and important problems in biochemistry involves the study of the mechanisms whereby this energy is captured and transferred in a usable form. A large portion of energy is conserved by a stepwise degradation of the substrate by enzymatic action and a stepwise transfer of electrons over the electron transport system to oxygen. As will be shown in a later section on glycolysis, energy is needed for certain of the reactions; whereas, at other points, energy in the form of "energy-rich" phosphate bonds ($\sim$ P) is generated at the substrate level. Most of the energy, however, is conserved in $\sim$ P which are generated during electron transport. To illustrate the complex of reactions concerned, let us consider the oxidation of a typical substrate, XH_2.

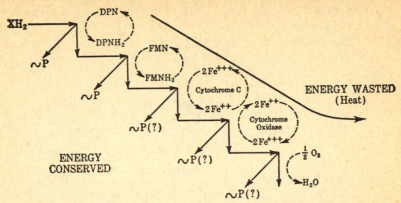

Although this is the path of electron transport in most organisms, certain facultative anaerobes (e.g., *Salmonella typhosa*) lack some of the components of the cytochrome system. Strict anaerobes (*Clostridium*) do not contain cytochromes at all. Therefore, the cytochrome complement of a bacterial cell corresponds roughly to its respiratory character. Although aerobic organisms contain catalase (see Glossary) and the *Clostridia* do not, the sensitivity of the latter to oxygen cannot be attributed solely to the toxicity of H_2O_2. In fact, one cannot demonstrate the production of H_2O_2 under aeration, and added catalase is ineffective in promoting aerobic growth. At any rate, anaerobic life is maintained through coupled oxidation reductions which may involve the flavoproteins.

THE METABOLISM OF CARBOHYDRATES

Kluvyer's Principles.

1. Any biochemical process may be considered as a series of independent reactions.
2. Each step is a chemically intelligible and simple reaction.
3. The common property of all such reactions is electron transport.
4. Thermodynamically, each step is spontaneous approaching the level where all reactions are reversible.

The applicability of the principles to carbohydrate metabolism may be seen below in the scheme for glycolysis.

Glycolysis. The anaerobic phase of carbohydrate metabolism.

1. Definition. The breakdown of carbohydrate to pyruvic acid.
2. If glucose is the initial metabolite, the over-all equation is

$$\text{glucose} + 2 \text{ ADP} \longrightarrow 2 \text{ pyruvate} + 2 \text{ ATP.}$$

This gives a net gain in utilizable energy to the cell of approximately 24 Calories.

3. The Embden-Meyerhof scheme.

glycogen

$\updownarrow$ phosphorylase

glucose-1-phosphate

$\updownarrow$ phosphoglucomutase

glucose-6-phosphate $\xrightleftharpoons[\text{hexokinase}]{\text{ATP}}$ glucose

$\updownarrow$ isomerase

fructose-6-phosphate

ATP $\updownarrow$ phosphofructokinase

fructose-1,6-diphosphate

$\updownarrow$ aldolase

3-phosphoglyceraldehyde $\xrightleftharpoons[\text{isomerase}]{\text{triosephosphate}}$ dihydroxyacetone phosphate

DPN $\updownarrow$ triosephosphate dehydrogenase

1,3-diphosphoglyceric acid

ADP $\updownarrow$ phosphoglycerokinase

3-phosphoglyceric acid

$\updownarrow$ phosphoglyceromutase

2-phosphoglyceric acid

$\updownarrow$ enolase

phosphoenolpyruvic acid

ADP $\updownarrow$ pyruvic kinase

pyruvic acid

Note that glycolysis is a reversible operation. The over-all efficiency of glycolysis from glycogen $\longrightarrow$ lactate is 60%. If glucose is the initial metabolite, the efficiency is 64%.

This scheme was originally worked out in yeast and in muscle. However, the over-all pattern appears to be universal. Although all the enzymes concerned have not been found in all bacteria, in the best known bacterium, *E. coli*, the presence of all the enzymes except phosphofructokinase has been demonstrated.

The Citric Acid Cycle.*   The aerobic phase of carbohydrate metabolism.

The series of reactions leading from pyruvic acid to carbon dioxide and water comprises the aerobic phase of the metabolism of carbohydrates.   The path is cyclic and involves basically a series of reactions in which OAA (enol form) is condensed with "active acetate" (acetyl-Co A) to form citric acid, which is then degraded via decarboxylations and dehydrogenations.   Obviously, more than one turn of the "wheel" is necessary for its complete assimilation.   The cycle as currently conceived is as follows:

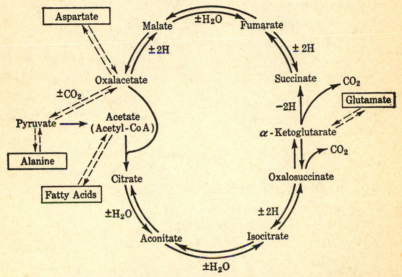

Side reactions (indicated by dash lines) show the interrelation of amino acid and fat metabolism with the citric acid cycle.

At each point on the cycle where hydrogen transfer is involved energy is conserved as $\sim$P.   It is estimated that $16 \sim$ P are generated for each mol. of pyruvate metabolized.   The over-all efficiency of the aerobic phase is 67%.   Note that aerobic metabolism is *not* much more effective than anaerobic metabolism from the point of view of efficiency of trapping the energy available from the reactions concerned.

It appears that both the anaerobic and the aerobic paths occur in all tissues, but all steps of the citric acid cycle have yet to be undisputedly demonstrated in bacteria.

---

* Also termed "Krebs' cycle," "tricarboxylic acid cycle," "cyclophorase."

METABOLISM OF NITROGENOUS COMPOUNDS

Proteolytic Action. Any student who has worked with microorganisms in the laboratory is well aware that many of them exert powerful proteolytic action. The liquefaction of gelatin and the curdling of milk are two well-known cases. Amino acids obtained from the degradation of proteins are, in turn, built up into protein of the organism.

Amino Acids. Amino acids may undergo the following degradation reactions.

1. Deamination (removal of —NH$_2$).

(a) Oxidative deamination.

$$RCHNH_2COOH + O \longrightarrow RCOCOOH + NH_3;$$

e.g., *E. coli* (facultative anaerobe).

(b) Reductive deamination.

$$RCHNH_2COOH + 2 H \longrightarrow RCH_2COOH + NH_3;$$

e.g., *Mycobacterium phlei* (strict aerobe).

(c) "Stickland" reaction.

$$RCHNH_2COOH + H_2O + XCHNH_2COOH \longrightarrow$$
$$RCOCOOH + 2 NH_3 + XCH_2COOH;$$

e.g., *Clostridium sporogenes* (strict anaerobe).

2. Transamination (transfer of —NH$_2$).

(a) Glutamic-oxalacetic transaminase.

$$
\begin{array}{cccc}
\text{COOH} & \text{COOH} & \text{COOH} & \text{COOH} \\
| & | & | & | \\
\text{CH}_2 + & \text{CH}_2 \rightleftharpoons & \text{CH}_2 + & \text{CH}_2 \\
| & | & | & | \\
\text{CH}_2 & \text{C=O} & \text{CH}_2 & \text{CHNH}_2 \\
| & | & | & | \\
\text{CHNH}_2 & \text{COOH} & \text{C=O} & \text{COOH} \\
| & & | & \\
\text{COOH} & & \text{COOH} & \\
\text{glutamic} & \text{OAA} & \text{alpha-ketoglutaric} & \text{aspartic;}
\end{array}
$$

e.g., *Streptococcus faecalis.*

(b) Glutamic-pyruvic transaminase.

glutamic + pyruvic $\rightleftharpoons$ alpha-ketoglutaric + alanine.

3. Decarboxylation (removal of terminal COOH).

$$RCHNH_2—COOH \longrightarrow RCH_2NH_2 + CO_2;$$

e.g., *E. coli.* Distribution among bacteria is haphazard.

PYRUVIC ACID: THE HUB OF METABOLIC DIVERSITY

Pyruvic acid occupies a key position at the "crossroads" of many reactions. We would therefore expect to find considerable variation in end products produced among various groups of organisms. The following diagram of some typical fermentative pathways affirms that pyruvic acid is indeed the unifying intermediate amid metabolic diversity.

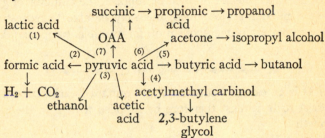

(1) *Lactobacillus.*
(2) *Coli-aerogenes.*
(3) *Coli-aerogenes, Clostridium.*
(4) *Aerobacter aerogenes.*
(5) *Clostridium acetobutylicum.*
(6) *Clostridium butylicum.*
(7) *Propionobacteria.*

ELUCIDATION OF METABOLIC PATHWAYS

Growth Experiments. The determination of the growth rate of an organism when it is supplied various substances in a defined medium may often yield information about possible intermediates in a series of reactions. Thus, if an organism can utilize nitrate as the sole nitrogen source, and it is thought that the NO_3 is reduced to NH_3 during assimilation, the organism should be able to grow on NO_2 and NH_3 as well, provided that these substances are permeable and are not toxic in the concentrations employed. If the organism in question can then grow, the postulation of the pathway is supported but not proved.

Simultaneous Adaptation (Stanier). This technique involves manometric measurement of gas exchange by washed "resting cells" grown on various substrates. If the suspected reaction path is $A \longrightarrow B \longrightarrow X \longrightarrow C \longrightarrow D$, and X is an intermediate, growth on substance A will simultaneously adapt the organism to produce enzymes which will attack B, X, C, and D.

Biochemical Mutants. Recently, mutant strains of microorganisms deficient in the ability to perform some step in an essential reaction have been used with great success in the determination of metabolic paths. The procedure used is as follows.

1. Treat stock culture with some mutagenic agent.

2. Select for mutant strains (these will not grow on the "minimal" medium as did the parent strain).

3. See what substances must be supplied the mutants for growth. If certain mutants are obtained, one may be able to reconstruct as follows the pattern of metabolism of, for example, compound A:

$$A \longrightarrow | \xrightarrow{\text{I}} B \longrightarrow | \xrightarrow{\text{II}} C \longrightarrow | \xrightarrow{\text{III}} D$$

If D is an essential substance and mutant I lacks the ability to convert A to B, it must therefore be supplied with B in the medium for growth.

Mutant II cannot convert B ⟶ C and therefore must be supplied with C, etc. This technic has been applied with great success using *Neurospora* (the common bread mold) and *E. coli*.

BIBLIOGRAPHY

Baldwin, E. *Dynamic Aspects of Biochemistry*. Third Edition. Cambridge: Cambridge University Press, 1957.

Boyer, P. D., H. Lardy, and K. Myback, eds. *The Enzymes*, Vol. I. Second Edition. New York: Academic Press, 1959.

Dixon, M., and E. C. Webb. *Enzymes*. New York: Academic Press, 1958.

Edsall, J. T. *Enzymes and Enzyme Systems: Their State in Nature*. Cambridge, Mass.: Harvard University Press, 1951.

Lardy, H. A., ed. *Respiratory Enzymes*. Minneapolis, Minn.: Burgess Publishing Company, 1949.

Neilands, J. B., and P. K. Stumpf. *Outlones of Enzyme Chemistry*. Second Edition. New York: John Wiley and Sons, 1958.

Oginsky, E. L., and W. W. Umbreit. *An Introduction to Bacterial Physiology*. Second Edition. San Francisco: W. H. Freeman and Company, 1959.

Reiner, J. M. *Behavior of Enzyme Systems: An Analysis of Kinetics and Mechanism*. Mineapolis, Minn.: Burgess Publishing Company, 1959.

Sumner, J. B., and G. F. Somers. *Chemistry and Methods of Enzymes*. Third Edition. New York: Academic Press, 1953.

Thimann, K. V. *The Life of Bacteria: Their Growth, Metabolism, and Relationships*. New York: Macmillan Company, 1955.

SAMPLE EXAMINATION QUESTIONS OF THE NATIONAL BOARD OF MEDICAL EXAMINERS

BACTERIOLOGY

Group 1

Answer any five questions:

1. If direct examination of the sputum from a case of suspected pulmonary tuberculosis fails to reveal acid-fast organisms, what further bacteriological methods should be employed?

2. What are the factors that determine success or failure in the production of an artificial active immunity to (a) typhoid fever, (b) influenza, (c) smallpox, (d) tetanus, and (e) scarlet fever?

3. Give the life cycle of any pathogenic helminth transmitted to man by the bite of an insect.

4. How would you demonstrate: (a) specific bactericidal powers; (b) virus-neutralizing antibodies; and (c) a staphylococcus antitoxin in a serum?

5. Name three members of the group of spore-bearing anaerobic bacteria, and also the diseases for which they may be responsible. List some characteristics of the respective species by means of which they may be identified in the laboratory.

6. Name and describe two organisms that may infect man by means of contaminated water or milk.

7. Are hypersensitive states becoming more common? Give the reasons for your opinion.

Group 2

Answer any five questions:

1. What materials and procedures are used for the development of an artificial, active immunity against (a) typhoid fever, (b) smallpox, and (c) influenza?

2. Name two Gram-negative bacteria, other than cocci, which may be responsible for acute meningitis. How can these be identified in the laboratory?

3. From what locations in the body may the etiological agents of Salmonella infections in man be recovered? Name two selective or differential media used in the isolation or identification of members of this group. List the essential ingredients that give the media their special value.

---

* Reprinted by permission of the National Board of Medical Examiners. If the reader is in doubt about the answer to any question, he should use the Index and consult the relevant discussion in this Outline. Since no single source can supply data on every detail, it may be necessary to utilize in addition the standard textbooks listed on page ix. One of the most comprehensive sources of technical information is *Bergey's Manual of Determinative Bacteriology* (Seventh Edition. Baltimore: Williams & Wilkins Company, 1957).

4. In infestation or infection with the following parasites, indicate those instances in which a serological or a skin test is of practical use. What is the antigen used in any of these tests? How would you interpret such results as may be obtained? (a) *Echinococcus granulosus;* (b) *Enterobius vermicularis;* (c) *Wuchereria bancrofti;* (d) *Trichinella spiralis;* (e) *Plasmodium falciparum;* (f) *Ancylostoma duodenale.*

5. By what means may the virus of yellow fever be identified in the laboratory?

6. How would you identify members of the four major groups of pathogenic cocci by their microscopic and cultural characteristics?

7. Define the following: (a) streptomycin; (b) heterophile antibody; (c) Rh and Hr factors; (d) intermediate host; and (e) brucellergin.

Group 3

Answer any five questions:

1. What bacteriological, serological, or immunological methods will indicate present or past infection with a beta hemolytic streptococcus? For each method describe the results that are accepted as being indicative of such an infection.

2. State some characteristics of the hypersensitivities due to (a) a pollen, (b) a drug, (c) a food. In each case describe how you would satisfy yourself that the suspected agent was the cause. How might recurrence of the reaction be prevented?

3. Name four intestinal helminths of man that occur in the United States. Describe the life cycle and the methods of control of one of these.

4. List at least two properties which the following pairs of organisms have in common: (a) *Hemophilus influenzae* and *Neisseria intracellularis;* (b) the organism of Lymphogranuloma venereum and *Hemophilus ducreyi;* (c) pneumococcus Type II and Friedländer's bacillus Type B; (d) *Clostridium tetani* and *Corynebacterium diphtheriae;* (e) *Treponema pallidum* and *Treponema microdentium;* (f) *Eberthella typhosa* and *Shigella paradysenteriae;* (g) *Bacillus subtilis* and *Bacillus anthracis;* (h) the viruses of smallpox and of vaccinia.

5. For what reasons would you suggest caution in the use of chemotherapeutic and antibiotic agents?

6. Describe the laboratory diagnosis of a viral disease and the means by which an artificial active immunity may be produced to this disease.

7. How would you determine the responsible agent among the several possible etiological agents of: (a) bacillary dysentery; (b) pneumococcal pneumonia; (c) botulism; (d) influenza; (e) bacterial meningitis?

Group 4

Answer any five questions:

1. Name five organisms commonly found on normal mucous membranes of man. By what pathways may these organisms reach other sites wherein they may induce disease?

2. If an epidemic of (a) smallpox and (b) diphtheria were threatening a community, what bacteriological and immunological procedures should be adopted to aid in bringing the epidemics under control?

3. In what circumstances may quantitative serological tests for syphilis be of value?
4. Distinguish between (a) *Leishmania donovani*, (b) *Leishmania tropica*, and (c) *Leishmania brasiliensis*.
5. Select any four of the following organisms and name the technique which you would use in typing each. List the specific materials needed and the basis upon which the differentiation is made: (a) *Hemophilus influenzae;* (b) influenza virus. (c) pneumococcus; (d) beta hemolytic streptococcus; (e) *Eberthella typhosa.*
6. What distinctive characteristics serve to identify two agents, one a bacterium, the other a fungus, which may produce granulomatous lesions in man?
7 Describe the methods of establishing the diagnosis of epidemic typhus fever by (a) the isolation and identification of the parasite and by (b) serological procedures.

Group 5

Answer any five questions:

1. (a) What is the importance of determining the "type" of an organism? (b) What method is used for typing the following organisms: 1. pneumococcus; 2. *Eberthella typhosa;* 3. *Hemophilus influenzae;* 4. influenza virus; 5. *Beta streptococcus hemolyticus?*
2. Name the important animal reservoirs of the agents responsible for the following diseases and indicate the usual means by which man becomes infected: (a) undulant fever; (b) salmonellosis; (c) bubonic plague; (d) psittacosis; (e) tularemia; (f) Rocky Mountain spotted fever.
3. Depict graphically the usual antibody response of an individual to each of three subcutaneous injections of typhoid vaccine at weekly intervals, and the response to a single injection three years later.
4. List, and justify, the precautions that should be observed in the use of (a) diphtheria toxoid, (b) tetanus antitoxin, (c) smallpox vaccine, (d) ragweed antigen, (e) influenza virus vaccine.
5. Describe the morphology and give the cultural characteristics of *Histoplasma capsultatum.*
6. How would you establish a diagnosis of infestation with *Diphyllobothrium latum?* What is the source of this agent and how does man become infested?
7. What theories are advanced to explain the development of "drugfastness" by bacteria?

Group 6

Answer any five questions:

1. What bacteriological and serological observations would be of aid in the establishment of a diagnosis of (a) subacute bacterial endocarditis, (b) pulmonary tuberculosis, and (c) bacillary dysentery?
2. Name a bacterium, a spirochete, and a fungus—each of which grows best under anaerobic conditions. For any two of the above, describe the method you would

use in obtaining a pure culture, and list the identifying properties of the two organisms.

3. In the diagnosis of infectious disease, what is the significance of the following: (a) bile solubility, (b) agglutination of Proteus OXK, (c) the development of endosporulating spherules in tissues, (d) the presence of Halberstaedter-Prowazek inclusion bodies in the epithelial cells of the conjunctiva, (e) the agglutination of chicken red blood cells, (f) absence of hemolysis in the serum control tube of a Wassermann test?

4. How would you isolate the etiological agent of mumps?   What procedures, other than the direct isolation of the virus, are available for the demonstration of past or present infection with this agent?

5. Name two bacterial infections, two viral infections, and two helminthic infections in which skin tests may be of diagnostic value.   Explain the reaction in any two instances which involve different mechanisms.

6. What are the differences between the immunizing agents used for protection against: (a) yellow fever, (b) influenza, (c) tetanus, and (d) typhoid fever? How would you account for the differences in the efficacy and duration of immunity produced by these agents?

7. List four infections that man may acquire from household pets.   Describe two of the causative agents involved.

Group 7

Answer any five questions:

1. What are the animal reservoirs of tularemia in the United States?   How does the infectious agent reach man?   How may the dissemination of the disease be controlled?

2. Name three diseases for which contamination of food by microorganisms is commonly responsible.   Describe briefly the procedures necessary to prove that a suspected food sample is culpable.   Indicate the means which may be applied to make the food safe for use.

3. What is the evidence for and against the view that syphilis and yaws are cross-immunizing diseases?   If, in your opinion, the evidence is not conclusive, what additional information would you require?

4. Cite an instance in which two infective agents act in conjunction to produce disease, and another in which one interferes with the action of the other.   How are these results brought about?

5. For each of the following tests, name a disease in which the test is of outstanding value in the diagnosis: (a) nonspecific agglutination; (b) specific agglutination; (c) nonspecific complement fixation; (d) specific complement fixation; (e) skin test; (f) precipitation test.

6. Discuss atypical pneumonias from the standpoint of etiology and laboratory diagnosis.

7. What protozoal or helminthic infections are associated with the following: (a) protozoa that engulf red blood cells and have progressive motility by the rapid protrusion of pseudopodia; (b) lateral-spined ova in stools; (c) microfilaria in subcutaneous nodules; (d) encystment of parasites in muscles; (e) trypanosomes in lymph nodes?

Group 8

Answer any five questions:

1. Give the present status of immunization procedures for each of the following: measles, mumps, influenza, common cold, yellow fever. What are the difficulties and limitations involved?
2. Name one organism in each category that satisfies the characteristics listed below: (a) motile Gram-negative rod that liquefies gelatin and gives a positive urease test; (b) Gram-positive coccus, soluble in bile; (c) Gram-negative motile rod that produces acid without gas in glucose and mannitol; (d) Gram-positive aerobic, non-motile spore former; (e) Gram-negative non-motile rod that is indifferent toward carbohydrates and requires increased CO_2 tension for primary isolation; (f) a fungus that forms hyphae and tuberculated chlamydospores under some cultural conditions, and small intracellular budding yeast cells in tissues.
3. Describe the mechanisms by which a pneumococcus of one type may be transformed into a pneumococcus of another type. Comment on the possible significance of such transformation *in vivo*.
4. Name three helminths, pathogenic for men, that encyst in tissues. Show by means of a diagram the life cycle of any one of these.
5. Cite an example in which the reaction of one antigen with its antibody is interfered with by another antigen present in the same microorganism. Comment on the practical significance of this effect, and on methods of eliminating such interference.
6. On what basis may the specificity of a skin reaction to a bacterial product be established? Illustrate by a specific example.
7. Evaluate current concepts of the parts played in rheumatic fever by infection and by phenomena of hypersensitivity.

Group 9

Answer any five questions:

1. Aside from the use of serological tests, how would you distinguish between the following four groups of organisms: (a) the coliform group; (b) the Shigella group; (c) the Salmonella group; (d) the proteus group.
2. List the rickettsial infections of man known to occur in the continental United States. What methods are available to identify the etiological agent of any one of these diseases?
3. Name four diseases in which the spread is commonly through contact with carriers. For any one of these diseases, describe the methods that may be used to show that a given suspect is the possible source of infection.
4. State the distinguishing effects of the following toxins in the human host: (a) diphtheria toxin; (b) botulinum toxin; (c) tetanus toxin; (d) *Cl. welchii* toxin. What theories are advanced to explain the action of any one of these toxins?
5. Define: (a) transforming factor; (b) antiserum; (c) Weil-Felix reaction; (d) bacteriophage typing; (e) heterophile antibody; (f) infection immunity; (g) blastomycin.

6. Name two protozoal or helminthic infections of man in which the etiological agent is best demonstrated by each of the following methods: (a) blood smears; (b) stool examinations; (c) biopsy. Detail the identifying characteristics of any one of the parasites concerned with the infections mentioned above.
7. Discuss "B.C.G.," its nature, and the immunological basis for its use.

Group 10

Answer any five questions:

1. Name four species of pathogenic spore-forming bacteria. How may each be distinguished?
2. List five "milk-borne" infectious diseases. Select two diseases from the list, one caused by a coccus and one by a rod form, and tell how the source of infection could be determined by the use of serological and cultural methods.
3. Describe the methods of active immunization against (a) pneumococcal pneumonia, (b) gas gangrene due to *Cl. welchii*, (c) cholera. Discuss the advisability of each of these procedures.
4. What characteristics distinguish *Histoplasma capsulatum?* Comment on the significance of infection with this organism in reference to the diagnosis of pulmonary tuberculosis.
5. Give the method of identification of the causative agent of (a) vaccinia, (b) influenza, (c) endemic typhus fever.
6. State the material required, whence and when it should be obtained, and the bacteriological and serological laboratory procedures to be used in the diagnosis of (a) bacterial meningitis, (b) tularemia, (c) pertussis.
7. Discuss the function of antibodies in immunity.

Group 11

Answer any five questions:

1. (a) Describe three results that may follow the introduction of vaccinia virus into the skin of man. (b) State the significance of each.
2. If the acid-fast stain, prepared from a centrifuged specimen of urine from a suspected case of tuberculosis of the kidney, is negative, what further laboratory test would you do to demonstrate tubercle bacilli in this specimen?
3. While working in a barnyard, a patient stepped on a rusty nail that pierced his foot. A skin test indicated that he was sensitive to horse serum. What precaution would you take in giving tetanus antitoxin? A week later a staphylococcus infection developed in the wound. Would this indicate that another injection of tetanus antitoxin should be given? Give the reasons for your answer.
4. Discuss the advantages, in the sterilization of surgical instruments, that steam under-pressure has over (a) mercuric chloride, (b) 70 per cent alcohol, (c) boiling water.
5. (a) What are the ingredients used in the complement fixation test? (b) What part does each play? (c) What is the source of each ingredient?
6. What evidence must be obtained before a certain virus can be said to cause a given disease of man?
7. Discuss the significance of "sulfur granules" in the sputum of a patient ill with a chronic lung infection.

Group 12

Answer any five questions:

1. Define the term toxoid. Describe the prophylactic use of toxoids in two diseases of man.

2. In reference to the following diseases, give the source of material for cultivation of the etiological agent and the method of cultivation: (a) typhoid fever (fifth day of disease); (b) endocarditis (any one type); (c) meningitis.

3. How would you confirm a clinical diagnosis of Neisserian cervicitis by bacteriological methods?

4. How might the following pathogenic fungi be distinguished by microscopic examination of the exudate: (a) *Blastomyces dermatitidis;* (b) *Candida albicans;* (c) *Coccidioides immitis?*

5. Organism X is present in the feces of all cases of disease Y, but is found in only 10 per cent of normal people. Could this be considered conclusive evidence that organism X is the cause of disease Y? If not, what other evidence would you demand?

6. How would you isolate and identify *Hemophilus pertussis?*

7. A given infectious agent is said to "exist in five specific immunological types." What is meant by this expression?

STATE BOARD QUESTIONS FOR NURSES*

[*Note:* There are two groups of questions: (1) The essay type, answers to which are to be written out in full. (2) The new type, or "objective" type. Selected specimen questions of various kinds are given; the correct answers are filled in.]

ESSAY–TYPE QUESTIONS

Bacteria

1. What is a microscope and for what is it used?
2. Give several situations in which a knowledge of bacteriology is useful to nurses.
3. Define the meaning of the term bacteriology.
4. What are bacteria?
5. Where are bacteria found?
6. What conditions influence the growth of bacteria?
7. With what rapidity are bacteria produced?
8. How are bacteria treated so they may be seen under the microscope?
9. Classify bacteria according to: (a) Morphology; (b) Behavior to man; (c) Behavior to oxygen.
10. Name and describe the three important groups into which bacteria are divided according to shape, naming two belonging to each group.
11. How are bacteria distinguished from one another in the laboratory?
12. How may germs be cultivated outside the body?
13. Name a laboratory stain and give technic of staining a smear for the differential diagnosis of bacteria.
14. How do bacteria produce disease?
15. Name several avenues by which bacteria may gain entrance to the body.
16. What are the soluble poisonous products of bacterial growth called?
17. Which tissue of the body is directly affected by tetanus toxin?
18. How are disease organisms discharged from the body?
19. What do you mean by the term motile, as applied to bacteria?
20. Define spore.
21. What bearing have spores on sterilization?
22. What bacteria cause: Pneumonia? Tuberculosis? Acute Meningitis? Anthrax? Typhoid? Syphilis? Erysipelas? Tonsillitis? Gonorrhea?

---

* Reprinted by permission of the publishers from John A. Foote, *State Board Questions and Answers for Nurses* (Philadelphia: J. B. Lippincott Company, 1947). If the reader is in doubt about the answer to any question, he should use the Index and consult the relevant discussion in this Outline. Since no single source can supply data on every detail, it may be necessary to utilize in addition the standard textbooks listed on page ix. One of the most comprehensive sources of technical information is *Bergey's Manual of Determinative Bacteriology* (Seventh Edition. Baltimore: Williams & Wilkins Company, 1957).

23. Define pyogenic bacteria.
24. Which are more closely associated with pus formation, cocci, or bacilli?
25. What is meant by a "positive culture"?
26. Mention two important diseases where a bacterial diagnosis is considered essential.
27. What are some of the body defenses against bacterial invasion?
28. What good work is done by bacteria?
29. Name three types of bacteria which are useful to man.
30. What bacteria are found in the soil?   What is the most important one from a surgical standpoint?
31. Name two men who primarily gave to the world the idea of what is commonly called the germ theory of disease.
32. Name several bacteriologists and state one thing for which each is noted.
33. What is a mold or fungus?   What is a yeast?
34. Name two skin diseases that are caused by mold.
35. What are protozoa?
36. What is a parasite?
37. Name three animal parasites and the diseases produced by them.
38. What are Koch's rules or postulates?
39. What is meant by "filterable virus"?
40. Name five diseases caused by filterable viruses.
41. What is your idea of the difference between a saprophyte and a parasite?
42. What becomes of bacteria which cause disease after leaving the human body?
43. What is Vincent's angina and how is it diagnosed?
44. How do the causative organisms escape from the body in the following diseases: (a) scarlet fever; (b) measles; (c) cholera; (d) typhoid fever?
45. At what temperature do pathogenic bacteria grow best?
46. What is a culture medium?   Name one of the most common.
47. What membrane of the body is attacked by the Klebs-Löffler bacillus?
48. For what are the following commonly examined: (a) blood, (b) sputum, (c) vomitus, (d) urine, (e) feces, (f) spinal fluid, (g) vaginal smears, (h) throat cultures?
49. What is a blood culture?   How is it obtained?   What is its significance?
50. Translate "*Staphylococcus pyogenes aureus.*"   Where is it often found?
51. What is the difference between ordinary cleanliness and surgical cleanliness?
52. What is meant by droplet infection?
53. Name five diseases which may be carried by droplet infection.
54. What body fluids are examined to find the causative organisms in the following diseases: (a) cerebrospinal meningitis;   (b) diphtheria;   (c) tuberculosis; (d) gonorrhea?
55. How are disease germs carried from one person to another?
56. Name the point of entrance into the body of the germ in each of the following diseases: (a) tuberculosis, (b) syphilis, (c) diphtheria, (d) malaria, (e) local abscess.
57. What is meant by congenital infection?   Name one disease often transmitted this way.
58. Define the term carrier as applied to contagious and infectious diseases, and name two diseases where carriers play an important part.
59. What is meant by a focal infection?   Give an example.

60. How is the micro-organism of infantile paralysis transmitted?
61. Why is screening important in communities where malaria or yellow fever exists?
62. (a) What precautions are necessary to prevent the spread of diphtheria? (b) Where does the infection lurk?
63. What parts of the body are frequently affected by the tubercle bacilli?
64. What is the percentage of preventable constitutional diseases?
65. Why is it necessary for a nurse to know the symptoms of syphilis?
66. Show why the term "communicable" is preferable to the terms "infectious" or "contagious" when referring to diseases caused by germs.
67. What are the chief ways in which bacteria are carried around a hospital ward? Give three rules for nurses as a means of preventing the spread of disease.
68. What causes malaria fever?
69. Name a disease which may be produced by a filterable virus.
70. What diseases may be caused by flies and how may they be transmitted?
71. How is the bubonic plague spread from city to city and from country to country?
72. Why are mosquitoes a menace to health?

Tests

1. What do you mean by the Widal test?
2. When is the Widal test used?
3. What is the object of (a) a "Wassermann" test, (b) a "Widal"?
4. What is the Schick test for susceptibility to diphtheria?
5. What should be done when school children show a positive Schick test?
6. What is the Dick test?   How does it derive its name?
7. How may the pneumococcus pneumonia be diagnosed in the laboratory?

Immunity, Immunization, Serums, Vaccines

1. What is meant by immunity?
2. (a) Differentiate between natural, acquired, and artificial immunity.   (b) What is natural resistance?
3. Name three diseases to which the human body may be immunized.
4. Name three diseases which may be prevented by passive immunity.
5. What are serums?   Vaccines?
6. For what useful purpose may attenuated bacteria be used?
7. What is autogenous vaccine?
8. What are bacteriolysins?
9. What are antitoxins?
10. What are the natural protective agencies of the blood?
11. In what conditions does one find an increase in the leukocytes?   Give the functions of the leukocyte in an infectious disease.
12. What are diphtheria antitoxin, tetanus antitoxin, diphtheria toxin-antitoxin, diphtheria toxoid (anatoxin)?
13. How is diphtheria antitoxin obtained and why is it given?
14. What is prophylaxis?
15. What is meant by anaphylaxis?
16. What advice should a nurse give a consumptive relative to the cures and medicines so frequently advertised to cure his disease?
17. What prophylactic means are observed in scarlet fever and diphtheria?
18. What is hookworm disease?   How is it spread?

19. How is tetanus acquired?   What is its cause?   What is the prophylactic treatment?
20. Upon what theory is tetanus antitoxin administered?
21. What advantage does tetanus toxoid have over tetanus antitoxin as a method of securing immunity to the disease?
22. What is the prophylactic dose of antitoxin given to guard against diphtheria?
23. Explain the important difference between the way typhoid fever is spread by the house-fly and malaria by the mosquito.

Disinfection and Sterilization

1. What is the difference between an antiseptic and a disinfectant?
2. Name some points that should be considered in choosing a disinfectant.
3. Give exact method of disinfecting each of the following in the care of infectious diseases: (a) discharges and excreta; (b) linen; (c) utensils; (d) nurses' hands.
4. Name some of the most common disinfectants used.
5. What is the actual value of dishes of germicidal solution placed about a sick room?
6. How would you disinfect bowel discharges and urine from a typhoid patient? How care for dishes?
7. Name five disinfectant solutions, and give strength of each for some definite case.
8. What are good solutions for the disinfection of (a) clothing, (b) the hands?
9. What is a good disinfectant for sputum?
10. How would you care for mattress and pillows after infectious disease?
11. How would you disinfect yourself after nursing a contagious disease?
12. When and why is it necessary to disinfect urine?
13. How would you disinfect a room after a communicable disease?
14. What care would you give rubber goods?
15. Explain the difference between disinfection and sterilization.
16. What is an autoclave?
17. Name three valuable methods of sterilization.

SAMPLES OF NEW-TYPE QUESTIONS

Completion Type

In the following questions the correct answers should appear in the blank at the right.

1. The foul-smelling bacterium commonly found in a ruptured appendix is   *B. coli*   _____

2. An organic body upon which a parasite lives is called   A host   _____

3. The Widal test is a test for   Typhoid fever   _____

True-False Type

In questions of this type write True or False before each question; or if the letters T and F are printed with the question, the T or the F is circled, underlined, or otherwise marked as directed, indicating whether it be true or false.

1. The tetanus bacillus is the cause of lockjaw.   T.   F.
2. Colon bacilli in the intestinal tract are prone to produce abscesses in that region.   T.   F.
3. Tularemia, a disease of rodents, may be transmitted to man.   T.   F.

Selection Type (Single Choice)

Several separate statements, only one of which is correct, are submitted in this type of examination. The student is asked to select the correct answer and to indicate it in one of various ways: by using an X or a designated letter or figure; by underlining; or by striking out the incorrect words. An X is used here for correct answers.

1. A disease that confers immunity is:

    Scarlet fever.   X
    Erysipelas.   _____
    Pneumonia.   _____
    Colds.   _____

2. Negri Bodies are minute bodies found in persons or animals infected with:

    Typhoid fever.   _____
    Rabies.   X
    Tuberculosis.   _____

Place the number of the correct part on the following lines at the right.

1. Diphtheria is caused by (1) *B. subtilis;* (2) Streptococci; (3) Pneumococci; (4) Klebs-Löffler bacilli.   4
2. Agar is made from (1) Gelatin; (2) Peptone; (3) Seaweed.   3
3. An organism which harbors a parasite is called (1) Host; (2) Saprophyte.   1

Selection Type (Multiple Choice)

Check all the terms which help make a correct answer.

Diseases produced by fungous infections are:

    Athlete's foot.   X
    Impetigo.   _____
    Ringworm.   X
    Rheumatic fever.   _____
    Thrush.   X

Matching Terms Type

After each item in Column II, write the number of the item in Column I to which it refers.

Column I	Column II	
1. Gonococcus	Cause of syphilis	5
2. Virus	Cause of salpingitis	1
3. Koch's bacillus	Cause of whooping cough	4
4. *Bacillus pertussis*	Cause of tuberculosis	3
5. *Spirochaeta pallida*	Cause of rabies	2

In each group of four words, three are related to each other by a common factor. Underline the word which is not related to the other three and, in the space at the right, write the name of the factor which relates the three words.

1. Wassermann, Kahn, Dick, Kline   Tests for syphilis
2. Tuberculosis, mumps, smallpox, rabies   Virus diseases
3. Gonococcus, meningococcus, pneumococcus, staphylococcus   Diplococci

EXAMINATION QUESTIONS IN VETERINARY BACTERIOLOGY

1. Discuss briefly the relationship between the bovine and human form of tuberculosis.
2. What are the morphological and cultural characteristics of *Bacterium tularense* (*Pasteurella tularensis*)? How is diagnosis confirmed by laboratory methods?
3. What are the various methods of using tuberculin as a confirmatory test in the diagnosis of bovine tuberculosis?
4. (a) What are the morphological characteristics of *Brucella melitensis?*
   (b) What are the relationships between this organism and *Brucella abortus?*
   (c) What laboratory methods are used to confirm the diagnosis?
5. Name five bacterial diseases of animals transmissible to man and describe the pathogenesis of one.
6. How would you prepare an autogenous vaccine for the treatment of catarrhal mastitis in cows?
7. What is the cause of rabies and how is the prophylactic vaccine prepared?
8. Name and identify the organisms causing the following diseases:
   (a) Johne's disease.
   (b) Strangles.
   (c) Bacillary white diarrhea.
   (d) Bovine hemorrhagic septicemia.
   (e) Glanders.
   (f) Swine plague.
   (g) Cowpox.
   (h) Malignant edema.
   (i) Canine distemper.
9. Name five diseases caused by filterable viruses, and give the prophylactic measures for each one named.
10. Indicate the serological or bacteriological tests for diagnosis of the following diseases:
    (a) Bacillary white diarrhea.
    (b) Glanders.
    (c) Anthrax.
    (d) Nagana.
    (e) Contagious abortion.
11. Name five milk-borne diseases of animals transmissible to man.
12. Indicate some diagnostic procedures for catarrhal mastitis of dairy cattle.
13. Name the organisms associated with five protozoan diseases of veterinary significance.
14. Name five veterinary diseases for which a successful preventive vaccine has been prepared and indicate the content of each one.
15. Give the technic for making (a) Breed count, (b) Frost colony count, (c) Plate count, for numerical estimation of milk bacteria.

INDEX

Scientific names of microorganisms are shown in italics. Page numbers in italics indicate the principal reference(s) to a particular subject.

A

Abbé condenser, 46, 359
Abel, 206
Aberrant coliform, 249
Ablastin, *333*, 348
Abortin, 354
Abortion, contagious, in cattle, 12, 133, *177–179*, 265, 342
  in goats, 179
  in mares, 173
  in sheep, 173, 249, 265
Abortive poliomyelitis, 299. *See also* Poliomyelitis.
Abscess, 101, *189–191*, 192, 197, 206, 220, 359
Accelerated reaction to smallpox vaccine, 292
Acetobacter, *26, 150*, 151, 157, 359
  aceti, 151
  acetosum, 151
  industrium, 151
  pasteurianum, 151
  suboxydans, 151
  xylinum, 151
Acetone production, 156
Achorion, 101
  schoenleinii, 101
Achromatiaceae, 34, 114
Achromobacter, 29, 158
Achromobacteriaceae, 29
  Achromobacter, 29
  Alcaligenes, 29
  Flavobacterium, 29
Acid, normal solution, 65
  production in media, *37, 72*
Acid agglutination, 338
Acid dye, 52
Acid-fast bacteria, 221, 227
  stains for, 58
Acid-forming bacteria in milk, *131–132*
Acids, organic, 157
Acne, 220
Acquired immunity, *331–332*
Actinobacillus, 30
  mallei, 169. *See also Malleomyces mallei.*
Actinomyces, 32, 92, 114, 118, 152, *267–268,* 359
  antibiotics from, 93
  asteroides, 268
  bovis, 267
  farcinicus, 268
  hominis, 267
  madurae, 268
  waksmanii, 39
Actinomycetaceae, 32, 267
  Actinomyces, 32, 93, 114, 118, 152, *267–268,* 359
  Nocardia, 32, 268
Actinomycetales, 32, 34, 221, 267
  Actinomycetaceae, 32, 267
  Mycobacteriaceae, 32, 221
  Streptomycetaceae, 32
Actinomycetes, *267–268*
Actinomycetin, 92
Actinomycin, 92
Actinomycosis, 146, 267
Active immunity, 331
Adaptability of viruses, 288
Adaptation of bacteria, 23
Adaptive enzyme, 36

Adenitis, 359
Adenovirus, 290, *309*
Adsorption theory, 351
Aedes aegypti, 302, 303, 305
Aerobacter, 29, 139, 238
  aerogenes, 40, 41, 75, 76, 125, **132, 188,** 238, *240,* 247, 248
  cloacae, 146, 188, *240,* 247, 248
  liquefaciens, 240
Aerobe, 359
  facultative, 20, 365
  obligate, 20, 369
Aerogen, 36, 359
Aerosol, 120
African horse sickness, 306
African relapsing fever, 272
African sleeping sickness, *316–317*
Agalactia of sheep, 306
Agar, 62, 63, *68–79,* 188. *See also under names of individual types.*
  filtration of, 68
  for semisolid media, 64
  for solid media, 63
  plate count for milk, *134–135*
  plate method for disinfectants, **87**
Agar-agar, 359
Agglutinating titer, 338
Agglutination, 188, 203, *338–343,* 359
  Bacillus anthracis, 162
  Brucella abortus, 165, 166, 178, *179*
  Brucella melitensis, 165, 179, 180
  Clostridium chauvoei, 174, 176
  Clostridium septicum, 174, 176
  Diplococcus pneumoniae, 202
  Hemophilus pertussis, 236
  Klebsiella pneumoniae, 205
  Leptospira icterohaemorrhagiae, **277**
  Malleomyces mallei, 171
  Neisseria gonorrheae, 208
  Neisseria meningitidis, 210, 211, **212**
  Pasteurella avicida, *166–167*
  Pasteurella pestis, *166–167*
  Pasteurella pseudotuberculosis, **167**
  Pasteurella tularensis, 165
  Proteus vulgaris, 113
  Rickettsia prowazeki, 281
  Rickettsia rickettsi, 282
  Rickettsia tsutsugamushi, 282
  Salmonella choleraesuis, 171
  Shigella dysenteriae, 248
  Vibrio comma, 264
  in Weil-Felix reaction, 113, 281, **282, 343**
  in Widal test, 242
  tests, *339–343*
Agglutinin, *333, 338–343,* 349, 359
  absorption test, 342
Agglutinogen, 338, 359
Agglutinoid, 338, 359
Agitation, effect on bacteria, 85
Agramonte, Aristides (1869–1931), **11, 302**
Agranulocytosis, 359
Agrobacterium, 27
Air bacteriology, *118–120*
Air centrifuge, Well's, 119
Alastrim, 290, 291, *293,* 359
Albumin, 63, 67
Alcaligenes, 29
  abortus, 177. *See also Brucella abortus.*
  faecalis, 188, 247
  melitensis, 179. *See also Brucella melitensis.*
  vicosus, 133

403